NEW
ROME

NEW
ROME

THE EMPIRE IN THE EAST

PAUL STEPHENSON

The Belknap Press of Harvard University Press
Cambridge, Massachusetts

First published in Great Britain in 2021 as
New Rome: The Roman Empire in the East, AD 395–700
by Profile Books, Ltd
29 Cloth Fair
London EC1A 7JQ, UK

First Harvard University Press paperback edition, 2023
First Harvard University Press edition, 2022

Typeset in Garamond by MacGuru Ltd

Library of Congress Cataloging-in-Publication Data

Names: Stephenson, Paul, author.
Title: New Rome : the empire in the east / Paul Stephenson.
Description: First Harvard University Press edition. | Cambridge, MA : The Belknap
Press of Harvard University Press, 2022. | Includes bibliographical references
and index. | Identifiers: LCCN 2021024250 | ISBN 9780674659629 (cloth) |
ISBN 9780674294042 (pbk.)
Subjects: LCSH: Civilization, Greco-Roman. | Romans—Middle East. |
Byzantine Empire—History—To 527. | Byzantine Empire—History—
527-1081. | Islamic Empire—History. | Rome—History. | Middle
East—History—To 622. | Istanbul (Turkey)—History—To 1453.
Classification: LCC DF553 .S74 2021 | DDC 938--dc23
LC record available at https://lccn.loc.gov/2021024250

CONTENTS

✻

ACKNOWLEDGEMENTS

It has been my intention since this book was commissioned in 2011, by the late John Davey, to distinguish this work from many others in the field by integrating the latest research in a wide range of disciplines generally neglected by textual historians, notably that presenting and interpreting scientific data. To that end, the vast majority of notes that accompany the text relate to research published since 2011. That is also a reflection of the fact that I began work in earnest on the book rather later than planned, after the publication of two other works in 2016. I am grateful to Louisa Dunnigan, who stepped into John's shoes, to the copy editor Susanne Hillen, and to the teams at Profile and Harvard University Press, for their patience, for vastly improving the text, and for allowing me to keep the extensive notes, which were not part of the original remit for the book. The two anonymous readers commissioned by HUP were quick, generous with their attention to detail, and incredibly helpful in their commentary. I am also grateful to Michael Kulikowski for suggesting that I participate in this series, to which he has contributed two volumes in less time than I have taken to finish one. Michael also generously allowed me to have several of the maps he commissioned for *Imperial Tragedy* redrawn for this book.

In the time since the book was commissioned, I have benefited from the support of a number of institutions and the help and example of many colleagues. I single out only a few. I thank Olivier Hekster at Radboud University, Nijmegen, for his kindness and for teaching me much about Rome as we taught others together. I thank Anthony Kaldellis and Averil Cameron, who have not always agreed with each other in print, but who have both seen value in my contributions and have offered unwavering support. I thank Jonathan Shepard for introducing me to this world thirty years ago and for remaining a guide and friend. These scholars have written many letters on my behalf that have secured me support, access to libraries, and time to write.

As I completed the book, I also began new projects that have been generously supported by a Seeger fellowship at Princeton University, a Mellon fellowship at the Metropolitan Museum of Art, and the role of visiting scholar at Dumbarton Oaks. I thank Dimitri Gondicas, Director of the Seeger Center for Hellenic Studies at Princeton; Griffith Mann, Curator in Charge of the Department of Medieval Art and The Cloisters at the Met; and Tom Cummins, Director of Dumbarton Oaks. I am especially grateful to John Haldon at Princeton and Helen Evans at the Met, who offered support and guidance. I dedicate the book to my wife, who encouraged me to enjoy those fellowships, and my daughter and son, who tolerated my absences and have endured the process of watching me write when there are far better things to do outside.

ILLUSTRATIONS

Photographs are by the author except where stated otherwise.

Black and white illustrations

Colour plates

Plate 1. Mosaic of a solar chariot, Chapel of Sant'Aquilino, Milan.

Plate 2. Silver chalice, inscribed for Symeonius. (Walters Art Museum, Creative Commons License)

Plate 3. The stoa or *embolos* at Ephesus, once a grand colonnaded street, looking towards the theatre.

Plate 4. Mosaic showing Dionysus and Hermes, from a bath at Antioch. (Photo: Worcester Art Museum, excavation of Antioch and vicinity funded by the bequests of the Reverend Dr Austin S. Garver and Sarah C. Garver, with permission)

Plate 5. Fortunes (*tychai*) of Constantinople, Antioch, Rome and Alexandria. (From the Esquiline Treasure, British Museum, with permission)

Plate 6. Mosaic of Magerius, from Smirat, now in Sousse Archaeological Museum, Tunisia. (Photo: Wikimedia commons, public domain)

Plate 7. Mosaic of Dominus Julius, Carthage. (Photo: Wikimedia commons, public domain)

Plate 8. Mosaic at the episcopal basilica, Heraclea Lyncestis, Northern Macedonia.

Plate 9. Peutinger map, location and Fortune (*tyche*) of Constantinople. (Photo: Wikimedia commons, public domain)

Plate 10. Sea walls of Constantinople, looking towards the Theodosian harbour.

Plate 11. Basilica Cistern, today Yerebatan Sarnici, Istanbul.

Plate 12. Sarcophagi of porphyry (purple stone), Istanbul Archaeological Museums.

Plate 13. Sketch of the eastern face of the Column of Arcadius. (Freshfield Album, Trinity College, Cambridge, with permission)

Plate 14. Barberini ivory, Louvre Museum, Paris.

Plate 15. Ivory consular diptych of Flavius Petrus Sabbatius Justinianus. (Metropolitan Museum of Art, public domain)

Plate 16. Dome of Hagia Sophia.

Plate 17. Church of Saints Sergius and Bacchus, also known as Little Hagia Sophia, Istanbul.

Plate 18. The largest of nine David Plates, showing David slaying Goliath. (Metropolitan Museum of Art, public domain)

MAPS

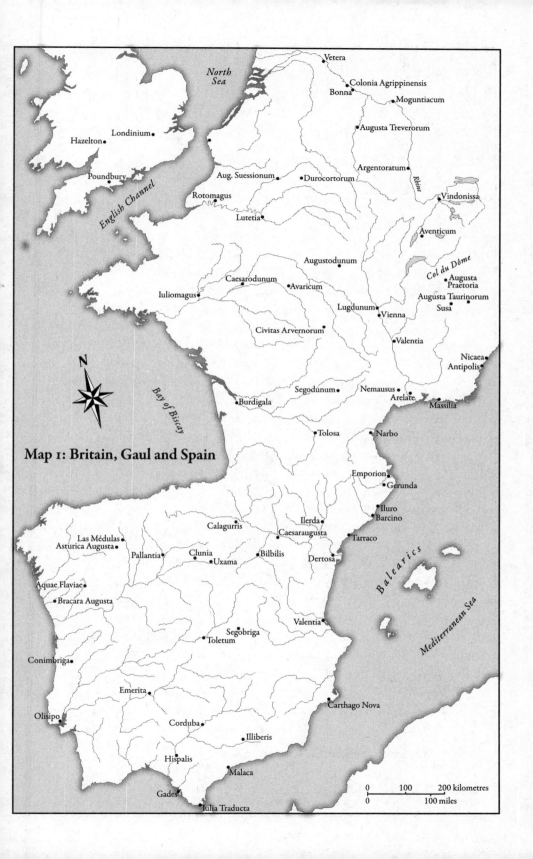

Map 1: Britain, Gaul and Spain

North
Sea

Hazelton
Londinium

English Channel

Poundbury

Aug. Suessionum
Rotomagus
Lutetia
Durocortorum

Vetera
Colonia Agrippinensis
Bonna
Moguntiacum
Augusta Treverorum
Argentoratum
Vindonissa
Aventicum

Augustodunum
Caesarodunum
Avaricum
Iuliomagus
Civitas Arvernorum
Lugdunum
Vienna
Valentia

Col du Dôme
Augusta
Praetoria
Augusta Taurinorum
Susa

N

Bay of Biscay

Segodunum
Burdigala
Nemausus
Arelate
Massilia

Nicaea
Antipolis

Tolosa
Narbo

Emporion
Gerunda

Iluro
Barcino

Calagurris
Ilerda
Caesaraugusta
Tarraco
Dertosa

Las Médulas
Asturica Augusta
Pallantia
Clunia
Uxama
Bilbilis

Rhine

Balearics

Mediterranean Sea

Aquae Flaviae
Bracara Augusta

Segobriga
Toletum

Valentia

Conimbriga

Emerita

Carthago Nova

Olisipo
Corduba
Illiberis

Hispalis
Malaca

Gades
Julia Traducta

0   100   200 kilometres
0   100 miles

Map 2: The Roman Empire, *c.* AD 400

Atlantic Ocean

North Sea

Baltic Sea

BRITANNIAE

BRITANNIA II

VALENTIA

FLAVIA CAESARIENSIS

MAXIMA CAESARIENSIS

BRITANNIA I

BELGICA II

GERMANIA II

BELGICA I

GALLIAE

Rhine

ITALIA

ILLYRICUM

LUGDUNENSIS II

LUGDUNENSIS SENONIA

Danube

RAETIA II

NORICUM RIPENSE

PANNONIA I

SEPTEM PROVINCIAE

LUGDUNENSIS III

LUGDUNENSIS I

MAXIMA SEQUANORUM

RAETIA I

NORICUM MEDITERRANEUM

VALERIA

DAC

AQUITANIA II

VIENNENSIS I

ALPES POENINAE

AEMILIA

VENETIA ET HISTRIA

SAVIA

PANNONIA II

AQUITANIA I

ALPES COTTIAE

Milan

LIGURIA

PANNONIA II

MOESIA I

GALLAECIA

TARRACONENSIS

NOVEMPOPULI

NARBONENSIS I

ALPES MARITIMAE

FLAMINIA ET PICENUM

DALMATIA

NARBONENSIS II

TUSCIA ET UMBRIA

PICENUM SUBURBI-CARIUM

PRAEVALITANA

DARD

LUSITANIA

H I S P A N I A E

CARTHAGINENSIS

CORSICA

Rome

SAMNIUM

CAMPANIA

APULIA ET CALABRIA

EPIRUS NOVA

BAETICA

SARDINIA

BALEARES

SUBURBICARIA

LUCANIA ET BRUTTIA

EPIRUS VETUS

Mediterranean Sea

SICILIA

TINGITANA

MAURETANIA CAESARIENSIS

MAURETANIA SITIFENSIS

NUMIDIA

AFRICA

Carthage

MACEDONI

A

F

R

I

BYZACENA

C

A

TRIPOLITANA

LI SUP

N

ASIA Dioceses according to the *Notitia Dignitatum*

| 0 | | 1000 | | 2000 kilometres |
| 0 | 500 | | 1000 miles | |

RIPENSIS
A
TERRANEA

SCYTHIA

THRACIA

Black Sea

OESIA II

HAEMIMONTUS

Constantinople

RODOPE

EUROPA

HELLESPONTUS

PAPHLAGONIA

HONORIAS

BITHYNIA

PONTICA

HELENO-
PONTUS

PONTUS
POLEMONIACUS

ARMENIA

LYDIA

GALATIA

GALATIA
SALUTARIS

PHRYGIA
SALUTARIS

PHRYGIA
PACATIANA

CAPPADOCIA I

CAPPADOCIA II

ARMENIA
II

ASIA

CARIA

ASIANA

PISIDIA

LYCAONIA

ISAURIA

CILICIA

CILICIA
II

EUPHRATENSIS

MESOPOTAMIA

OSRHOENE

Tigris

Euphrates

PAMPHYLIA

LYCIA

RETA

CRETA

CYPRUS

SYRIA

SYRIA SALUTARIS

Antioch

PHOENICE

PHOENICE
LIBANENSIS

ORIENS

PALAESTINA II

ARABIA

PALAESTINA I

Alexandria

LBYA INFERIOR

AEGYPTUS
AUGUST-
AMNICA

PALAESTINA
SALUTARIS

A E G Y P T U S

Nile

Red Sea

*Aral
Sea*

Caspian Sea

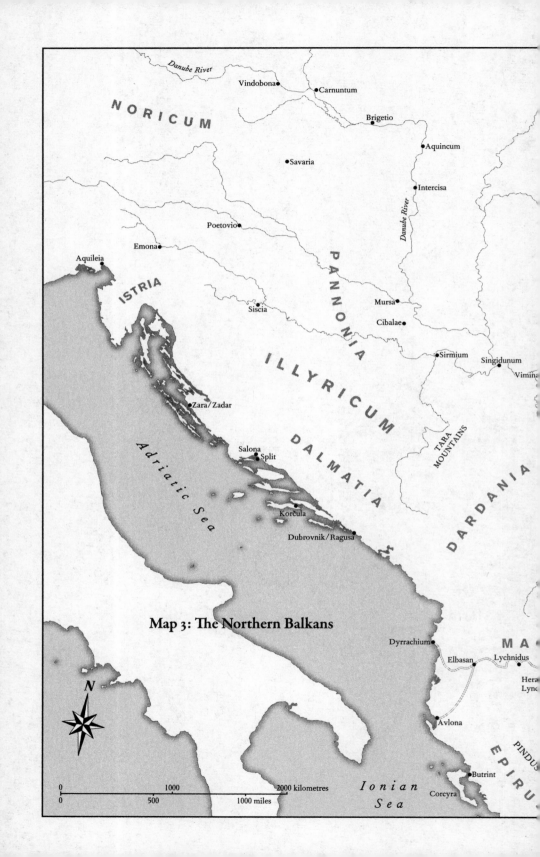

Map 3: The Northern Balkans

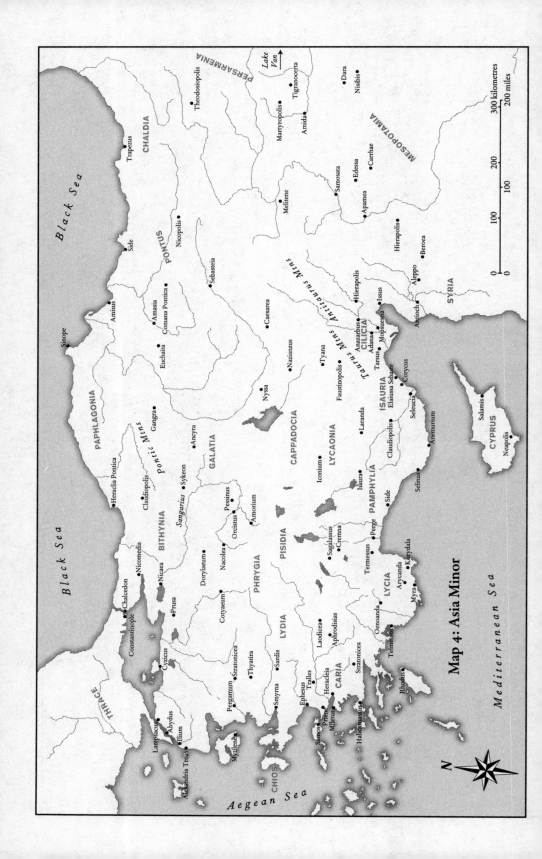

Map 4: Asia Minor

Black Sea

Black Sea

Aegean Sea

Mediterranean Sea

THRACE

CHALDIA

PONTUS

PERSARMENIA

Lake Van

MESOPOTAMIA

SYRIA

PAPHLAGONIA

BITHYNIA

GALATIA

CAPPADOCIA

LYCAONIA

ISAURIA

CILICIA

PAMPHYLIA

PISIDIA

PHRYGIA

LYDIA

CARIA

LYCIA

CYPRUS

CHIOS

Pontic Mtns

Pontic Mtns

Sangarius

Taurus Mtns

Antitaurus Mtns

300 kilometres
200 miles

300
200
100
100
0
0

N

Trapezus
Side
Sinope
Amisus
Heraclia Pontica
Nicomedia
Chalcedon
Constantinople
Abydus
Lampsacus
Ilium
Alexandria Troas
Mytilene
Nicaea
Prusa
Cyzicus
Claudiopolis
Gangra
Ancyra
Sykeon
Pessinus
Orcistus
Amorium
Dorylaeum
Cotyaeum
Nacolea
Pergamum
Stratonicea
Thyatira
Sardis
Smyrna
Ephesus
Tralles
Heraclea
Priene
Samos
Miletus
Halicarnassus
Rhodus
Telmessu
Stratonicea
Aphrodisias
Laodicea
Sagalassus
Cremna
Termessus
Oenoanda
Arycanda
Myra
Korydala
Perge
Side
Selinus
Anemurium
Claudiopolis
Isaura
Laranda
Iconium
Nyssa
Nazianzus
Caesarea
Tyana
Faustinopolis
Adana
Tarsus
Mopsuestia
Anazarbus
Hierapolis
Issus
Antioch
Aleppo
Beroea
Hierapolis
Apamea
Edessa
Carrhae
Samosata
Melitene
Amida
Marytopolis
Tigranocerta
Theodosiopolis
Nicopolis
Sebasteia
Comana Pontica
Amasia
Euchaita
Corycus
Elaiussa Sebaste
Seleucia
Salamis
Neapolis
Nisibis
Dara
PAMPHYLIA
PISIDIA

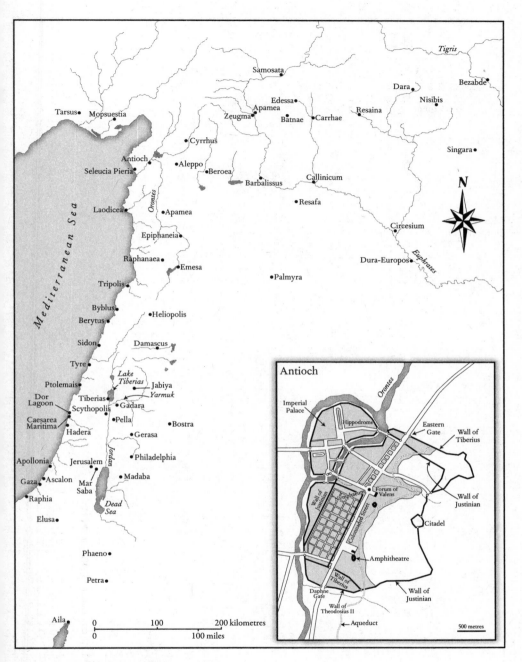

Map 5: Syria, with plan of Antioch

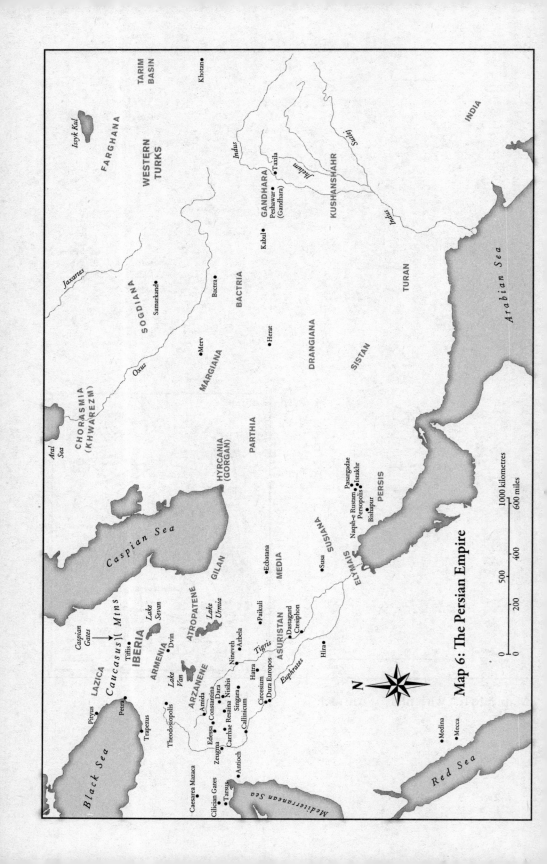

Map 6: The Persian Empire

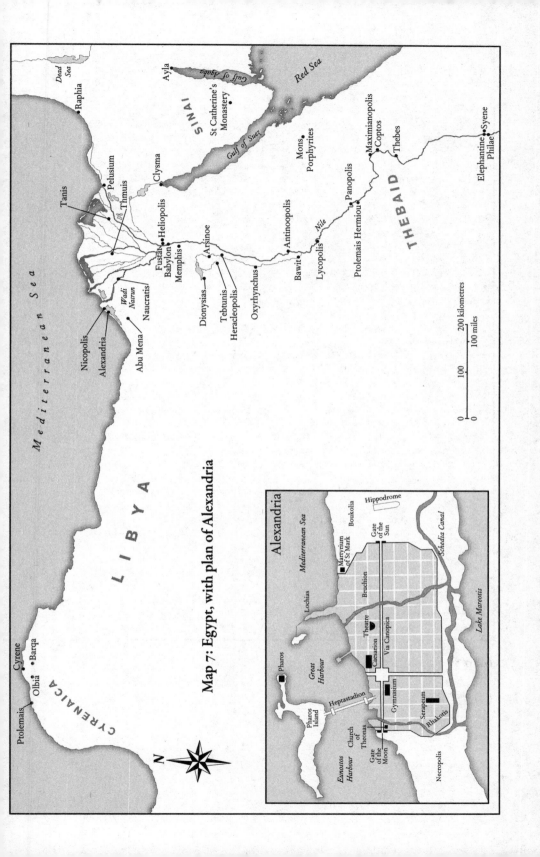

Map 7: Egypt, with plan of Alexandria

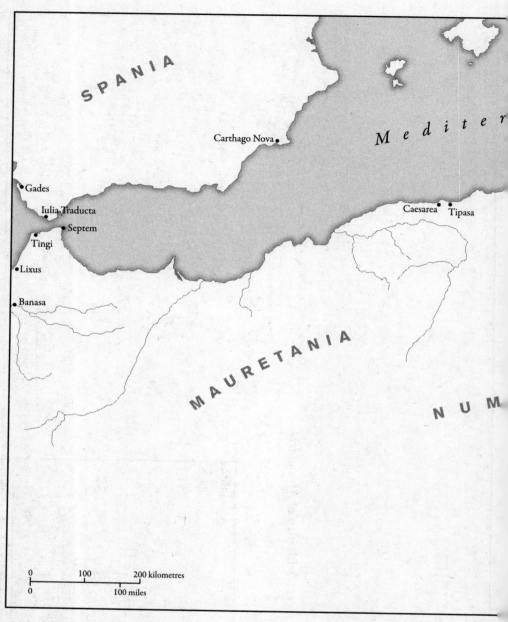

Map 8: North Africa

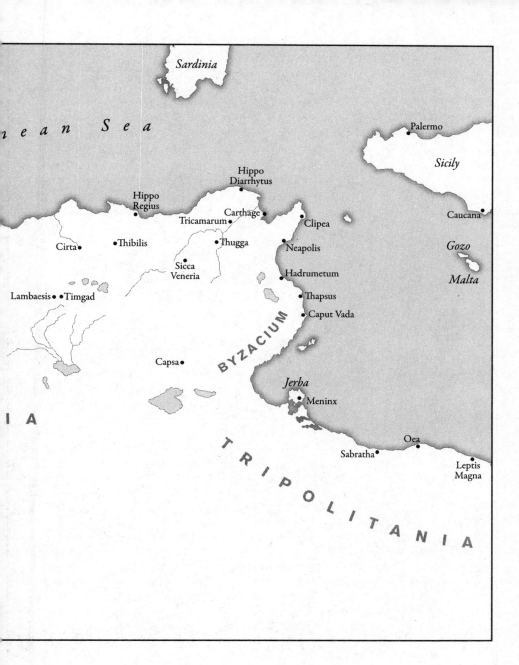

Sardinia

nean Sea

Palermo

Sicily

Hippo
Diarrhytus

Hippo
Regius

Caucana

Tricamarum •
• Carthage

Clipea

Gozo

Cirta•
•Thibilis
•Thugga

Neapolis

Malta

Sicca
Veneria

Hadrumetum

Lambaesis •
•Timgad

Thapsus

• Caput Vada

B Y Z A C I U M

Capsa•

Jerba

Meninx

I A

Oea

T R I P O L I T A N I A

Sabratha•

Leptis
Magna

200   400 kilometres
100   200 miles

Augusta Praetoria
Corium
Brixia
Mediolanum
Augusta Taurinorum
Ticinum
Placentia
Verona
VENETIA
Aquileia
Emona
ILLYRICUM
Veleia
Po
Mutina
Bononia
Ravenna
Pola
DALMATIA
Luca
Arno
Florentia
Ariminum
Ancona
Salona
Ligurian Sea
Osimo
Taginae
Adriatic Sea
CORSICA
Tiber
SARDINIA
Rome
Ostia
Alba
Interamna
Beneventum
Baiae
Neapolis
Pompeii
Brundisium
Otranto
Tyrrhenian Sea
Ionian Sea
Mediterranean Sea
Palermo
Messina
Rhegium
SICILY
Carthage
Syracuse
AFRICA
PROCONSULARIS
Caucana

N

Map 9: Italy

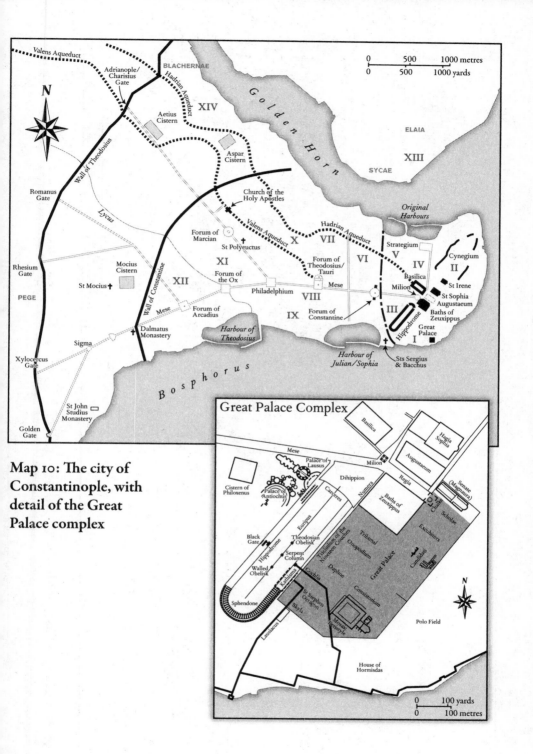

Map 10: The city of Constantinople, with detail of the Great Palace complex

Valens Aqueduct
BLACHERNAE
Adrianople/
Charisius
Gate
Hadrian Aqueduct
Aetius
Cistern
XIV
N
Golden Horn
ELAIA
XIII
SYCAE
Wall of Theodosius
Aspar
Cistern
Romanus
Gate
Lycus
Church of the
Holy Apostles
Original
Harbours
Valens Aqueduct
Hadrian Aqueduct
Forum of
Marcian
X
VII
Strategium
V
Cynegium
II
St Polyeuctus
Forum of
Theodosius/
Tauri
VI
IV
Rhesium
Gate
Mocius
Cistern
XI
Wall of Constantine
Mese
Basilica
Milion
St Irene
St Sophia
Augustaeum
Baths of
Zeuxippus
PEGE
St Mocius
XII
Forum of
the Ox
Philadelphium
VIII
III
Mese
Forum of
Arcadius
IX
Forum of
Constantine
Hippodrome
Great
Palace
I
Dalmatus
Monastery
Harbour of
Theodosius
Sigma
Xylocercus
Gate
Bosphorus
Harbour of
Julian/Sophia
Sts Sergius
& Bacchus
St John
Studius
Monastery
Golden
Gate

0   500   1000 metres
0   500   1000 yards

Great Palace Complex

Basilica
Hagia
Sophia
Mese
Palace of
Lausus
Milion
Augustaeum
Cistern of
Philoxenus
Palace of
Antiochus
Dihippion
Regia
Senate
(Magnaura)
Carceres
Numera
Baths of
Zeuxippus
Chalke
Scholae
Euripus
Excubitors
Black
Gate
Theodosian
Obelisk
Tribunal
Nineteen Couches
Onopodium
Candidati
Hippodrome
Serpent
Column
Daphne
Great Palace
Walled
Obelisk
Kathisma
Cochlia
Consistorium
Sphendone
St Stephen
Oratory
Skyla
Minor
Lausiacus
Polo Field
N
House of
Hormisdas

0   100 yards
0   100 metres

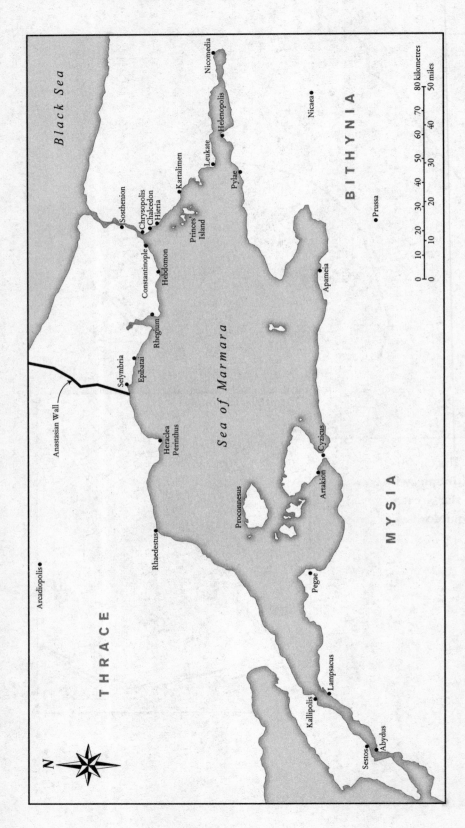

Map 11: Constantinople and its hinterland

Map 12: Greece and the Aegean

Atlantic Ocean

North Sea

Baltic Sea

REGNA FRANCORUM

Rhine

Danube

REGNUM SUEVORUM

VENETIA ET HISTRIA

AEMILIA

PANNON

ALPES COTTIAE

LIGURIA

FLAMINIA

DALMATIA

MOESIA I

REGNUM VISIGOTHORUM

TUSCIA ET UMBRIA

PICENUM

PRAEVALITANA

CORSICA

SAMNIUM

DAM

SPANIA

Rome

CAMPANIA

APULIA ET CALABRIA

EPIRUS NOVA

BALEARES

SARDINIA

LUCANIA ET BRUTTIA

EPIRUS VETUS

M e d i t e r r a n e a n   S e a

SICILIA

Carthage

MAURETANIA CAESARIENSIS

MAURETANIA SITIFENSIS

NUMIDIA

ZEUIGITANA

BYZACENA

N

Map 13: The Roman Empire, c. AD 550

TRIPOLITANIA

PEN

0		1000		2000 kilometres
0	500		1000 miles	

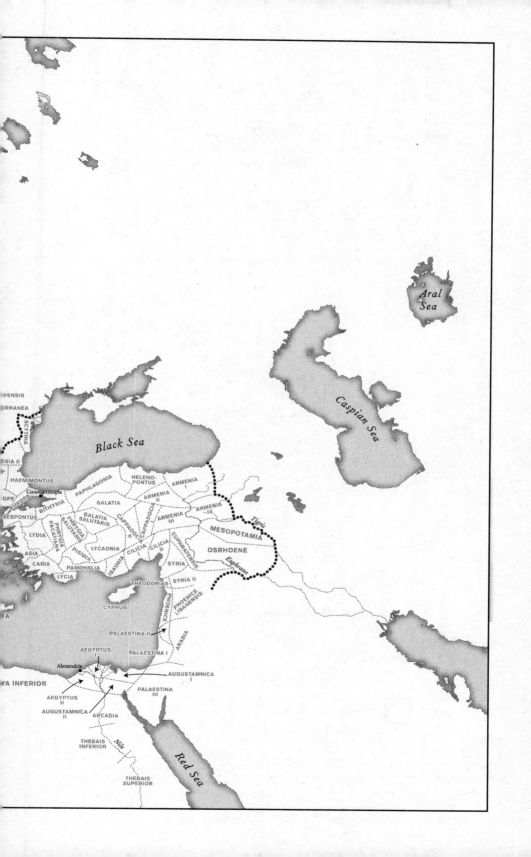

PENSIS
ERRANEA
SCYTHIA

Black Sea

ESIA II

HAEMIMONTUS
OPE
LESPONTUS
EUROPA
BITHYNIA

Constantinople

PAPHLAGONIA

HELENO-
PONTUS

ARMENIA

GALATIA

ARMENIA
II

LYDIA
PHRYGIA
PACATIANA
PHRYGIA
SALUTARIS
GALATIA
SALUTARIS
CAPPADOCIA
I
CAPPADOCIA II

ARMENIA
III

ARMENIA
IV

ASIA
CARIA
PISIDIA
LYCAONIA
ISAURIA
CILICIA
CILICIA
II

Tigris

MESOPOTAMIA

PAMPHYLIA
LYCIA

EUPHRATENSIS

OSRHOENE

Euphrates

THEODORIAS
SYRIA I

SYRIA II

CYPRUS
PHOENICE
PHOENICE
LIBANENSIS

Caspian Sea

Aral
Sea

TA

PALAESTINA II

ARABIA

AEGYPTUS

PALAESTINA I

Alexandria

AUGUSTAMNICA
I

YA INFERIOR

PALAESTINA
III

AEGYPTUS
II
AUGUSTAMNICA
II

ARCADIA

THEBAIS
INFERIOR

Nile

Red Sea

THEBAIS
SUPERIOR

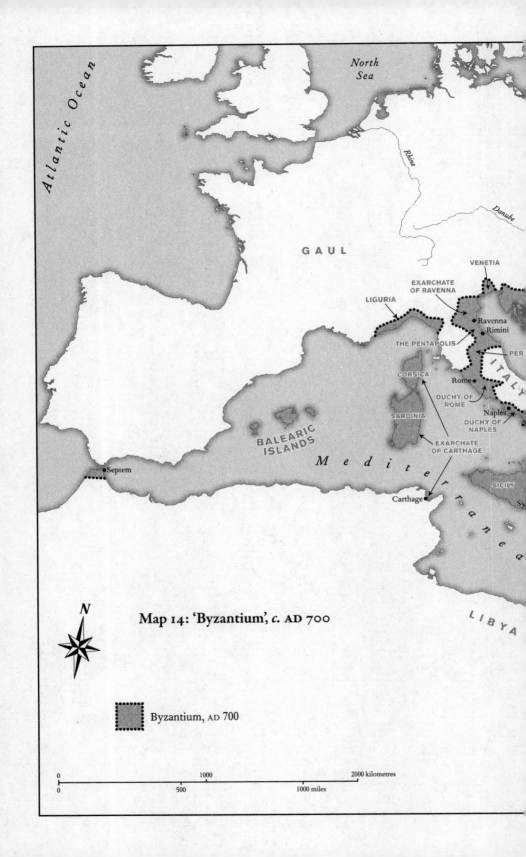

North
Sea

Atlantic Ocean

Rhine

Danube

GAUL

VENETIA

EXARCHATE
OF RAVENNA

LIGURIA

Ravenna
Rimini

THE PENTAPOLIS

PER

CORSICA

ITALY

Rome

DUCHY OF
ROME

SARDINIA

Naples

DUCHY OF
NAPLES

BALEARIC
ISLANDS

EXARCHATE
OF CARTHAGE

M e d i t e r r a n e a

Septem

SICILY

Sy

Carthage

LIBYA

N

Map 14: 'Byzantium', *c.* AD 700

Byzantium, AD 700

0		1000		2000 kilometres

0	500	1000 miles

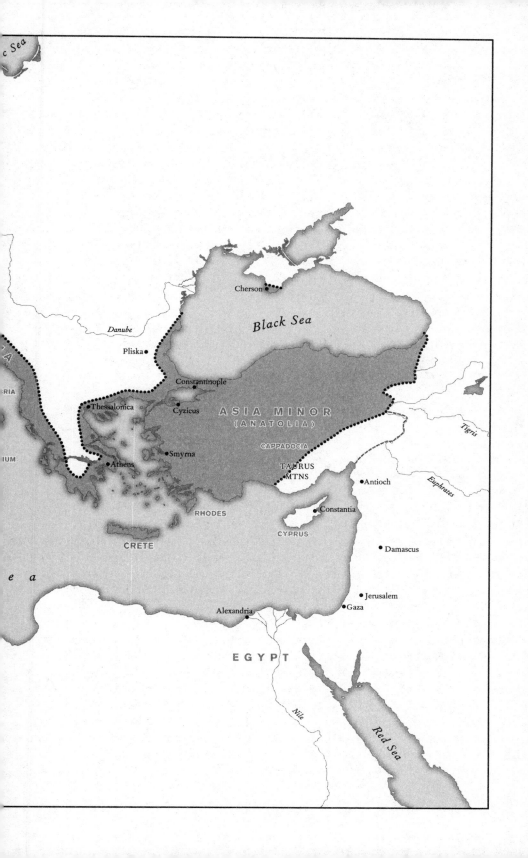

c Sea

Cherson

Black Sea

Danube

Pliska

RIA

Constantinople

Thessalonica

Cyzicus

ASIA MINOR
(ANATOLIA)

Tigris

IA

RIA

Smyrna

CAPPADOCIA

IUM

Athens

TAURUS
MTNS

Antioch

Euphrates

RHODES

Constantia

CYPRUS

CRETE

Damascus

e   a

Jerusalem

Alexandria

Gaza

E G Y P T

Nile

Red Sea

INTRODUCTION

Every schoolchild once learned that Rome fell to the Ostrogoths in AD 476, when Odoacer deposed the last legitimate western emperor, Augustulus, weedy 'Little Augustus'. On that occasion, an embassy was despatched by the senate from Rome to Constantinople, a city straddling East and West, Europe and Asia (Map 2). The city of Byzantium had been refounded by Constantine I and dedicated to him as 'the city of Constantine', Constantinople. Now it became the seat of the sovereign emperor of all Romans, for the ambassadors carried the imperial insignia and an indication that Odoacer would rule Italy on the emperor's behalf. In institutional terms, this is how the Roman empire came to be ruled from Constantinople. However, the empire in the fifth century was ruled not from any single city but by and from many cities, with greater authority – political and administrative, cultural and spiritual – concentrated in a few great *metropoleis*, 'mother cities', including Rome and Milan, Antioch and Alexandria, Carthage and Constantinople. This book sets out how this situation ended, and how many ancient cities fell or were transformed until the empire came to be ruled from a single great city, Constantinople, which was called New Rome.

A central theme of this book, that the end of the ancient city marked the end of antiquity, is hardly novel. There is no virtue in novelty that ignores decades of the finest scholarship. It is now clear that even where cities appear to have abrupt ends these were long in the making. In the view of Wolf Liebeschuetz, the 'disintegration of the ideal and reality of the classical city' was intimately linked to the transformative power of Christianity, which presented citizens with a compelling alternative world view to that which had united communities across the eastern Mediterranean. Christianity embodied a 'different set of values, and one not centring on the city and its political community'. If change began even before the fourth century AD, it accelerated rapidly through our period. After more than three centuries, the cumulative effects of foreign invasion and destruction, rapid climate cooling and famine, plague and depopulation, the loss of taxation revenues and urban institutions, the erosion of Hellenic culture and municipal autonomy, and the increased wealth and prominence of the church and the rapid growth of monasticism, affected all cities to some extent, and many ceased to be cities in any recognisable way.[1]

Life across the empire was sustained and characterised not only by what was produced or manufactured locally, but by what was shared internationally, including, as we shall see, glazed ceramics and terracotta lamps, natron glass and diverse marbles, fish products, olive oil and wine from preferred regions, and grain shipped in bulk. If comestibles have not survived, or have done so only in traces and fragments, many more of their containers and some of the coins used to purchase them have survived – evidence for a transformed, impoverished material landscape. The singular standout is Constantinople, which preserved the institutions of Roman civic life and as much of its material culture as could be contained within its walls. The empire of New Rome would become something quite different without Rome, and still more so without Antioch and Alexandria, its partners and competitors in the east. The insights of new scientific data throw aspects of this well-established thesis, the emergence of the empire we call Byzantium, into sharper relief and elucidate much that was hitherto quite opaque.

New scientific approaches are transforming our understanding of all past civilisations. It is impossible today to write a general history of any polity, society or culture without considering the natural world in which it emerged, developed and fell. To explain the end of the western Roman empire, among the thorniest of historical problems, historians have now turned to natural and environmental science, highlighting the roles played by climate and pandemic disease, supporting older hypotheses and undermining others. The consequences of climate change have been noticed and applied to the movement of 'barbarian' peoples from the increasingly arid lands of inner Eurasia, for example the Huns driving the Goths into the empire, and others following in their wake. Such considerations now precede explorations of what those barbarians destroyed and what they inherited and transformed of Roman civilisation. As people travelled, so they brought new pathogens and spread diseases unknown to the Romans, which devastated the citizens of Rome's myriad, densely populated and unsanitary cities. Any account of the decline of cities, the foundation for the Roman imperial system where Roman culture was generated and taxes were collected, must now take account of disease transmission, but also seismic and volcanic activity, climate forcing and pollution.

These compelling, urgent, novel concerns do not yet and likely never will offer scientific certainty and data in place of historiography and conjecture. Certainly, they help us answer older questions in different ways and force us to ask new questions. But 'scientism' cannot replace solid historical research, and to have lasting value scientific data must be integrated properly with more traditional approaches to the study of the past. Political, social and economic history, intellectual and religious history, and cultural and art history all still matter and deserve our careful attention, as do the insights offered by archaeology and the study of material culture. This book will attempt to integrate newer and older insights, to explore the environment of the eastern Roman empire and attend to the natural disasters that affected it, charting local, regional and cumulative impacts and human responses. Attention will be paid to metallurgy and health, magic and medicine, volcanoes and climate forcing, plague and earthquakes, historical climatology, geophysics and epidemiology. However, the book will also offer a political and cultural narrative based principally on written sources and works of art, integrating human responses to a range of phenomena, including warfare and religious disputes, developments in engineering and architecture, the rise of apocalypticism in literature and art, and developments in the imperial image.

Weaving these threads together into a fabric that has coherence and integrity presents a major challenge. The loom, being the parameters of this study, is AD 395 to c.700, and encompasses the last tranche of Roman antiquity, from an apparent division of empire at the death of Theodosius I between eastern and western emperors in their respective capitals of Milan and Constantinople, to the conquest of the Near East and Mediterranean by the forces of Islam. The warp of the fabric, the first yarns stretched out across the loom, are the themes and focus of the book: environmental and material concerns, informed by archaeology and the new science of Roman history. Through these subjects, other threads, thicker and thinner, lighter and darker, will be woven, introducing far greater detail and higher resolution in some parts than others, reflecting the nature of surviving sources. By design, a tapestry will emerge that has style and form, presenting imagery, history, in three registers or bands. Inevitably, this composition has many flaws and gaps, holes into which fingers will be jabbed. Like almost all ancient textiles, its colours will fade quickly, its edges will fray, its connecting threads will prove weak and will snap. It will disappear sooner than its maker imagined possible during the decade it took to complete as scholarship moves on and new interpretations are offered.

Works of art in three registers were popular in the Roman world and many have survived. If tapestries are rarely preserved, things fashioned of more enduring materials remain, including the Barberini ivory and the largest of nine silver David Plates, both of which will be discussed in the book. If the central register, the second part of the book, is a political narrative, full of detail and intrigue, then the upper and lower registers present the wider contexts in which this story might be understood, broadly conceived as the before and after. Urban and imperial themes, material and ideological, dominate the first part of the book. However, the focus is on people, the Romans who lived and died, suffered and worked, traded and were traded. The earliest chapters sketch out life at the end of the 'Lead Age', the natural world and environment, and faith and family. A commonwealth of cities is described, their streets teeming with life – human and faunal, floral and microbial – defining and delimiting the civilised world, linked by effective communications, dominating productive hinterlands, and ruled by wealthy families known as decurions. These families owed, and ostentatiously professed, allegiance to, and collected taxes for, administrators located in the greatest cities of Rome, Antioch, Alexandria, Carthage, and eventually above all others Constantinople, the New Rome. The last chapters depict the rise of Islam and its aftermath. Cities still dominate the landscape in this period, although there are fewer and many are transformed, while others are ruined or displaced. An inventory informed by archaeology demonstrates that the wide world of linked cities had become something else, something narrower and less urban, a new world founded on new ideas about government and God, art and war, and much else besides. The final chapter recaps and evaluates the emperors at New Rome, seeing them as they wished to be seen through three centuries, but also as they did not, their power constrained, their sovereignty contingent, their lives ended often violently, their reputations destroyed.

PART 1

LIFE IN THE LATER ROMAN WORLD

1

LIFE AT THE END OF THE 'LEAD AGE'

In the last decades of the fourth century, an octagonal imperial mauso-leum was constructed at Mediolanum (Milan), which had supplanted Rome as the western imperial capital. Today known as the Chapel of Sant'Aquilino, the mausoleum housed the remains of Gratian and Valentinian II, sons of Valentinian I. The mausoleum abutted the Church of St Lawrence, today San Lorenzo Maggiore, a mighty edifice built using materials from Mediolanum's abandoned amphitheatre. Ambrose, bishop of Milan from AD 374 to 397, presided over both of their Christian burials. Above him, in the cupola of the mausoleum, was a figure in a chariot streaking across the sky in the style of a solar god (Plate 1). This was Elijah, the Old Testament prophet borne to heaven on a chariot of fire (2 Kings 2). It was an appropriate image for a mausoleum, alluding to the apotheosis of a Roman emperor, traditionally shown ascending to the heavens on a *quadriga*, a chariot pulled by four horses. However, now it announced the triumph of Christianity and the certainty of resurrection for the faithful. According to Malachi (4:1–5), the last book of the Christian Old Testament, for Christians like Gratian and Valentinian, 'the sun of righteousness will rise with healing in its rays', raising them to heaven, even as it burned others to ashes. 'See, I will send the prophet Elijah to you before that great and dreadful day of the Lord comes.'[1]

Ambrose was a sophisticated commentator on scripture, and he took care to draw out all meanings from a sacred text or image. Having gazed upon Elijah's fiery chariot, he offered a further layer of interpretation, a gentler solar metaphor. Before his prophesied return, Christ would be the true sun, a constant source of grace, whose radiance cascaded from heaven in sanctifying beams upon his followers. Ambrose composed a hymn of enduring beauty now known to English speakers as *Splendour of God's Glory Bright*.

> O splendour of God's glory bright,
> O Thou that bringest light from light,
> O Light of light, light's living spring,
> O day, all days illumining.
> O Thou True Sun, on us Thy glance,
> Let fall in royal radiance,
> The spirit's sanctifying beam,
> Upon our earthly senses stream.

Our physical sun, in contrast, is an erratic provider of energy. Variations in the earth's tilt, the way it wobbles on its own axis, and the path of its orbit around the sun lead to more or less solar radiation reaching the earth's surface. This has resulted in very long periods of far higher or lower temperatures, including the last glacial maximum that lasted for eighteen millennia, between *c.*33,000 and 15,000 years ago. The sun also releases larger and smaller amounts of energy over shorter periods of hundreds of years. Greater solar emissions, which directly heat the earth's atmosphere, are indicated by an increase in the number of solar flares and the duration of sunspots. Periods of very intense solar activity are known as solar maxima and those of very little activity are called solar minima. The last solar minimum, the Maunder Minimum, took place between AD 1645 and 1715, coinciding with the 'Little Ice Age'. Finally, the sun has a natural short-term cycle, with eleven years between small peaks and troughs in the amount of energy released. The sum of the energy emitted by the sun that reaches earth, solar irradiation, can be measured through proxies known as cosmogenic radionuclides, which are radioactive isotopes. These are formed in the upper atmosphere when stable isotopes collide with cosmic rays and fall to earth to be dispersed by aerosol deposition in snow and rain or by photosynthesis. Counter-intuitively, greater solar activity results in the production of fewer radionuclides like carbon-14 (radiocarbon), which is preserved in organic matter including the wood of ancient trees, and beryllium-10, which is trapped in glacial ice. Measurements of deposited radionuclides demonstrate that from *c.* AD 360 until *c.*690 the earth experienced a long period of declining sunlight, culminating in a solar minimum of a magnitude that had not been seen for more than a millennium, and would not be seen again until the 'Little Ice Age'.[2]

This centuries-long period of declining solar irradiation corresponds to the period of our study. It followed a sustained period of far more sunlight initiated by a solar maximum in *c.*270 BC, the so-called 'Roman Warm Period'. Also known as the 'Roman Climate Optimum', it coincided with a protracted phase of Roman imperial expansion and general economic prosperity. In contrast, our period included what has been called the 'Late Antique Little Ice Age' and culminated in the low point of Roman power and prosperity. Quite what impact increased or decreased solar activity had on the emergence or decline of the Roman empire, a complex civilisation that encompassed the whole Mediterranean and much of Europe, cannot fully be ascertained. In the simplest terms, more sunlight and higher temperatures allow more land to be brought under cultivation. The zone in which vines and olives can be grown extends further north as the sun shines, and retreats south as sunlight dwindles. Lowland trees like the beech might grow higher up slopes, and then retreat back down them. Alpine glaciers will move in the opposite directions. By one estimate, in Roman Italy alone an increase in temperature of a single degree Celsius would have brought five million additional hectares of land under cultivation, feeding more than three million more people. Consequently, for each degree that temperatures declined, towards the solar minimum that marked the end of our study, it might be expected that as much cultivable land was lost.

To those living across the Mediterranean world in late antiquity, whether it was slightly hotter or colder was ultimately less important than whether it was significantly wetter or drier. Without adequate precipitation lands became not additional fields feeding millions but parched wastelands. Levels of historical precipitation can be measured approximately in tree rings and cave mineral deposits like stalagtites, which show, broadly speaking, that the 'Roman Warm Period' delivered rain sufficient to support a significant expansion in agriculture across the empire, whereas from around the middle of the fourth century in the eastern Mediterranean, two centuries before the posited onset of the 'Late Antique

Little Ice Age', there was generally less rain. Since there was great regional variability in rainfall across the Roman world, scholars have yet to reach agreement on the data and how or whether this should be aggregated to produce a global picture.[3]

At the regional and local levels, much evidence is clear and compelling. For example, a wetter and warmer period allowed for greater cultivation of large parts of Anatolia, in modern Turkey, in the fourth and fifth centuries AD. The plain around Iconium (modern Konya), in south central Anatolia, was far more heavily settled at that time than before or after, and olive trees were planted and cultivated in the Anatolian interior. As more arid, cooler conditions arrived in the sixth to seventh centuries, olive cultivation stopped. Similarly, in Syria and Palestine, climatic conditions in the period before the sixth century allowed the cultivation of lands that had hitherto been marginal and uneconomic, for example the Judean hills and slopes above the Dead Sea, where olives and cereals were grown. This ended in the seventh century, after which the climate became much drier. Olive production, like all forms of arboriculture, requires a heavy initial investment and patience as trees mature, taking decades to produce an economic yield. It implies confidence that rural communities will endure, that there will be sustained demand supported by political and economic stability, and that efficient trade networks will deliver products to local markets or further afield. This confidence was lost by the end of our period.[4]

Metallurgy and the early Anthropocene

If the impact that a constantly changing natural environment had on the fate of the Roman empire is now under careful scrutiny, the impact Rome had on its environment and nature has barely been considered. However, this can be measured with remarkable precision. As our study begins, an age of industry and pollution, of mining and smelting, and of long-distance shipping of metals, minting of millions of coins and large-scale building of infrastructure was ending. Roman metallurgy has left signals across northern Europe and the northern Atlantic world in the form of anthropogenic heavy metal contamination of soil, sediment and ice. Contamination is so substantial and significant that it has been identified as the start of the Anthropocene, the period through which we are living, a discrete chapter of the Holocene, our current geological epoch. Lake beds, peat bogs, salt marshes and ice fields produce very reliable pollution records. Cores extracted from Irish and Swedish lake beds, an Icelandic salt marsh, Faroese peat bogs and Arctic glaciers all show the same sudden and dramatic rise in the deposition of atmospheric lead pollutants between *c.*100 BC and AD 100. Lead is released by the smelting of a range of metallic ores, including those mined for copper and gold, tin, zinc and silver, and from lead itself. In each location the levels of lead pollutants fall away rapidly towards AD 400, only beginning to rise again after 800, and not reaching Roman levels until *c.*1700. In none of these locations is there any evidence for contemporary mining and smelting of metallic ores, which would have produced the contamination.

Roman-age pollution in the north Atlantic world is the direct result of fluctuations in the intensity of smelting that took place thousands of kilometres to the south, releasing into the atmosphere lead aerosol particles that were conveyed great distances within the northern hemisphere's atmospheric transport system and deposited by precipitation. The origin of the lead in Greenlandic ice has been confirmed by geochemistry (isotope analysis). Spain was the source of up to 70 per cent of the heavy metal pollution at its peak in the first century AD (see Map 1).[5] Contamination is far greater the closer one gets to its source. In an ice core taken from the Col du Dôme glacier in the French Alps, the magnitude of lead contamination is one hundred times greater than that recorded in Greenland in the first century BC, reaching a lower peak in *c.* AD 100, before falling steadily and dramatically to its lowest point in the sixth century.[6]

The rapid rise in atmospheric lead pollution mirrored the rise of the Roman silver denarius, the coin minted in the greatest numbers and at the peak of its fineness (at almost 98 per cent silver, considered pure by the mints) from the time of Augustus in the first century AD. Careful analysis of Roman silver coinage supports the notion that far less silver was smelted after AD 100. From the reign of Nero onwards, Roman silver coinage was increasingly debased. The ratio of silver to copper alloy decreased, and the amount of recycled silver used, obtained by melting down older coins, increased. The Roman silver denarius had a fixed exchange rate with the aureus, a gold coin. During the Republican period this was largely notional, but from the time of Julius Caesar gold coins were produced in greater numbers. This too appears to be reflected in the environmental record.[7]

Gold production produced copious contamination. The largest known Roman gold-mining operation was located at Las Médulas in north-western Spain (Map 1). A sediment core extracted from a small glacial lake around thirty-five kilometres from the mines shows evidence for the first gold metallurgy at Las Médulas in around 300 BC, with a rapid increase in lead contamination from around 100 BC, a peak in *c.*15 BC, and a decline to background (pre-300 BC) levels by *c.* AD 120. At the same time there were dramatic increases in both antimony and arsenic. The peak of pollution corresponds with that identified in the Greenland ice core, but the concentration of lead, being so close to the smelters, is far greater. Around 150 kilometres to the north of Las Médulas, a peat bog saw an increase in lead pollution that was thirty times greater than its local background level in *c.* AD 100, which had fallen back to the baseline by *c.* AD 500.[8]

Silver and gold were noble and rare metals, whereas lead was a dull, base metal, the plastic of its age, and was employed in quantities and for purposes unknown before and since. It has been suggested that the Roman period should be called the 'Lead Age', an archaeological successor to the 'Iron Age'. Lead was used extensively in Roman construction, because it is malleable and resists corrosion when in contact with air and water. Molten lead was poured around iron clamps to join column drums together, and to secure marble facades to blockwork. Lead sheets and solder were used to form and seal waterproof joints. Most famously, lead was used in Roman waterworks: to form pipes that transport water at pressure, to plumb fountains and baths, for rain gutters and roofs, and as tanks to store water, including potable water, for various purposes. It has been determined that the piped water of the city of Rome may have contained up to forty times the lead of natural spring water before AD 250, falling to a multiple of fourteen by the year 500, as pipes became choked with scale, cracked and failed, and the broader water system fell into disrepair.[9]

In addition to its uses in construction and for waterworks, lead was used as a bulking agent in copper alloys, appearing in volume in later Roman copper coins. Lead tokens, and occasionally coins, were struck in great numbers. Because it was heavy, lead was used to form weights and also to give heft to bronze steelyard weights. Because it was easy to shape, lead sealings were used to secure and verify pouches, bags and letters. Lead bungs stoppered liquids and lead labels were attached to sacks, with letters scratched into the soft metal. Its low melting point (327°C) and physical properties meant that lead was used in the extraction and refining of silver

Figure 1. The lead coffin of a Roman infant, buried in Syria

and gold, and to give fluidity to bronze and copper for the manufacture of sculpture, vessels and jewellery. The Roman army used lead in abundance, including lead shot. Lead compounds – 'white lead' (*cerussa*) and 'red lead' (*minium*) – were used as pigments in paints and cosmetics. As a dark and dull base metal drawn from the same ore, galena, as noble shimmering silver, lead was considered chthonic and inauspicious. Thin sheets of lead formed into tablets were used for inscribing curses, which were folded and buried. Stone sarcophagi might contain a lead liner. Those who could not afford stone sarcophagi might choose to be interred in decorated lead coffins, or lead-lined wooden coffins, their lids soldered shut and protective designs imprinted upon the surfaces to preserve and protect interred bodies. Lead coffins have been found in small numbers at various places across the late Roman world, and in far greater numbers in Britain, where lead was naturally abundant, and the Levant, where it was not (Figure 1).[10]

The ubiquity of lead meant that Roman soil was polluted by it even where there is no evidence for mining or smelting. Analysis of sediments conducted at excavated sites in the modern state of Israel, where there are no known lead ore deposits, shows that the Roman occupation levels are replete with lead, unknown in earlier times and for a millennium afterwards. Sediment cores taken from the harbours at Alexandria, Egypt (Map 7), indicate that very little lead was present in the local environment before the foundation of the city (2 to 6 parts per million [ppm]), followed by some use of lead in the Ptolemaic city (200 to 300 ppm), spiking to around 600 ppm in AD 100, followed by a steady decline to pre-Roman levels by AD 500. Roman levels of lead pollution are higher than those recorded in Alexandria's modern harbour. This is not evidence for lead aerosols from smelting facilities deposited by precipitation, nor was Alexandria plumbed so extensively as Rome with lead pipes. Rather, lead was present in abundance in the harbour itself. Lead sheathing was applied to the hulls of ships, perhaps two-thirds of all Roman ships, as protection against shipworms (*teredo navalis*). Some ships carried spare rolls of lead to patch and repair, while others had composite anchors and oars of wood and lead, lead sounding weights (for measuring water depth), bilge pumps with lead parts, sinkers for fishing nets and lines, and even ingenious lead braziers for safer cooking on board. Some of these fell to the sea floor, as did a great deal of lead scrapings from regular cleaning of barnacle-encrusted lead-sheathed hulls, contaminating the sediment.[11]

Lead exposure posed a significant health risk to sailors, but that paled in comparison with the risk of living near sites where metallic ores were mined and smelted. Both Pliny the Elder and Vitruvius observed how unhealthy lead-workers were. At the height of activity, before AD 100, many thousands of those workers were slaves, who suffered not only from lead poisoning but also from the inhalation in high concentrations of sulphur dioxide, which produces irritation and inflammation of the respiratory system and affects long-term lung function. For those who lived far from smelting facilities, lead poisoning was less common, but still possible. Lead is equally toxic however it enters the body. Between 30 per cent and 50 per cent of inhaled lead particles remain in the body, compared with 5 to 10 per cent of lead particles that are ingested. Contrary to a popular theory, it is unlikely that many Romans ingested toxic levels of lead from their water pipes. Although lead is soluble in water, calcium carbonate deposited by hard water provided a barrier between water and lead. Moreover, calcium prevents the gut from absorbing lead. Drinking hard water transported in lead pipes did not present a major health risk to Romans, although in soft-water areas the risks were higher, and lead carbonate might form a less protective scale inside pipes. If not from their water, however, Romans contrived many additional ways to ingest and absorb lead. It was used for medicinal purposes, in cooking and for mixing sauces, and for preserving and sweetening wine. Roman saucepans manufactured from a mixture of lead and tin were used to produce reductions of must (unfermented grape juice) called, according to its concentration, *sapa, defrutum* or *caroenum*, all full of lead. Salt was produced in lead brine pans, heated to evaporate water, before the salt was chipped and scraped away.[12]

Flora and fauna are excellent indicators of environmental pollution, including lead contamination. Modern surveys have shown that honey bees absorb twice as much lead as drones, effectively removing particulates trapped by plants and mixed with pollen. Unfortunately, the lead also ends up in honey, which Romans ate in abundance and which was a staple for weaning children. Bees and other insects will not live where the atmosphere is too toxic, as Pliny the Elder was aware, and animals were not pastured close to smelters. Even if the most contaminated sites were not used for arable or pastoral farming, substantially increased levels of lead aerosols deposited across a broad region by precipitation will still have affected bees and birds, fish and shellfish in polluted rivers and bays, and cattle and sheep eating grass growing in contaminated soil. Humans who consumed these creatures or their products, including milk and cheese, absorbed lead and stored it in their bones and teeth. Lactating mothers transferred a good deal of this trapped lead, which was released from their bones into their breast milk and hence to their infants.

The lead that permeated the bones and teeth of Romans allows us to distinguish them from those who lived before and after them in the same settlements. Analysis of skeletal remains from Poundbury in Dorset (see Map 1), in south-western Britain, some distance from the nearest smelting facilities of Somerset, showed greatly elevated bone lead concentrations, more than twice those of pre-Roman skeletons from the same location. At third- to fifth-century cemeteries at Poundbury and at West Tenter Street in London, lead concentrations were in a range from 100 to 250 µg/g, compared with c.14 µg/g measured in the ribs of thirty-four Neolithic farmers buried in a long barrow at Hazelton, Gloucestershire. Natives of Roman Britain were also far shorter than their Neolithic predecessors and, just as strikingly, shorter than the population that followed them. Women were on average only 152 cm (5' 0") tall and men 164 cm (5' 5"). The average length of a Roman's thigh bone was between two and three centimetres (about an inch) shorter than that of an Anglo-Saxon. While changes in diet and disease burden after the Roman conquest were consequential, data suggest that Roman-age Britons were on average eating more proteins, including a range of seawater fish and molluscs, than their Iron Age ancestors, which should have led to an increase in stature. One must wonder, therefore, at the impact of extremely elevated lead concentrations, since any level of lead contamination is known to stunt growth in children.[13]

New Roman metallurgy

For centuries after the mines of northern and western Europe had been lost to the Roman empire, silver and lead, copper, tin and iron, and small amounts of gold were extracted from mining operations in inhospitable locations across the eastern empire, notably in the Balkans, the Taurus Mountains, the mountains of Armenia, in Egypt's eastern desert, and in the desert of what is today southern Jordan. A peat bog in the Tara mountains, today in western Serbia (see Map 3), bears witness to a rapid increase in smelting in the Balkans during the later Roman period. Concentrations of lead pollutants fall from a peak in the middle of the third century, but rise again steadily from around AD 400 to reach a higher peak in the first half of the seventh century. In the Taurus Mountains, in what is today known as the Bolkardağ mining district (Turkey, Map 4), a fifteen-kilo-metre-long valley was lined by mines at an altitude of 2,000 to 3,000 m. Radiocarbon dating of fragments of a wooden ladder found in one mine show that it was still in use at the end of the eighth century. Pottery and coin finds provide evidence for far later occupation. Basil of Caesarea had expressed concern for the ironworkers of the Taurus range and the levels of taxation levied on their production in the 370s. Dense pine forests supplied fuel for the smelting works at Bolkardağ, and the smelters appear to have alternated between sides of the valley to allow trees to regrow.[14]

Mining communities lived hard, unhealthy, often unfree lives. Miners might still be slaves, perhaps criminals or prisoners of war condemned to their profession, but increasingly they were people who happened simply to be born to miners, and therefore were bound to the mines. A section of the *Theodosian Code*, a compilation of laws produced in the 430s, entitled *De metallis et metallariis*, addresses the regulation of mines and miners, metals and metalworkers, and stone quarries. A law of 424 ruled that 'if any person should be born from a miner and from any other stock, he shall necessarily follow the ignoble birth status of a miner'. Similarly, if miners should 'seek to migrate to foreign parts, they shall be recalled to the family stock and the household of their birth status'. There was by then an established body of law relating to the flight of indentured miners, starting with a ruling of 30 April 369, which instructed provincial judges to provide support in identifying and apprehending vagrants.[15]

The law codes are equally clear that private enterprise in mining was encouraged, as per *CTh*. 10.19.3, dated 365, which specified that such gold might preferably be sold to the treasury. The cost of doing business was fixed at eight *scrupula*, 'scruples of gold dust', per man employed. The *scrupulum* is the smallest unit in the system of weights and measures, and in this case refers to 1/24 of an ounce. This levy was later reduced to seven *scrupula*, a rather more difficult calculation than one-third of an ounce. Additionally, gold production was taxed at a fixed rate per pound. Extensive, frequently successful prospecting for gold took place in the mountains of Armenia, between Roman and Persian lands, which changed hands several times in our period. John Malalas, a sixth-century chronicler, informs us that rich 'gold-bearing mountains lie on the border between Roman Armenia and Persarmenia' and that a large new seam had been discovered in the reign of Anastasius (491–518), which was at that time under Roman jurisdiction but was formerly part of Persia (see Map 6). 'Previously certain people leased these mountains from the Romans and Persians for 200 *litrai* of gold', but now only the Romans derived revenue from them.

The need for new sources of gold in the fifth and sixth centuries also saw the opening of new mines in Egypt's eastern desert, to which hundreds of workers were transported, and sustained in newly built villages. They were well fed. Abundant bones of sheep and goats have been found, as have those of cattle, which are almost never found where there is no fodder. Grain was imported, fresh water was available, and cheese was made locally. The workers' task was to hack mineral-rich quartz veins from granite using iron picks, to pound the quartz with slabs to identify ore, then to grind the ore to powder, wash it and bag it for transportation to a location where sufficient fuel was available for smelting.[16]

Metallurgy required lavish amounts of fuel and prudent management of supplies. The Romans were careful husbands of woodlands, practising coppicing and pollarding. Tall trees were saved for construction of important buildings and ships – Hadrian raised inscribed warnings around the tall cedars of Lebanon to prevent theft and misuse – but these were increasingly rare in the later empire. The best woods for specific purposes and the calorific value of fuel woods were well known. A great deal of wood was converted to charcoal to sustain high temperatures required for smelting and other industrial processes, which reduced the volume of fuel by about four-fifths. Charcoal was also preferred for domestic purposes by those who could afford it, since it produced far less smoke. By one estimate, the annual iron production across the Roman world would have required 38,400 km² of forest, after wood had been converted to charcoal. Silver-lead production would have required just as much fuel, meaning metallurgy commanded the resources of around 2 per cent of the total landed area of the empire at the start of our period (3,800,000 km²). This was but a small fraction of domestic fuel requirements.

Demand for firewood per head of population, based on an estimate of total population of around fifty million, would mean that between 10 and 12 per cent of the empire's lands in any given year would have to be given over to wood production. Total demand was huge but manageable. Far less fuel was used for cremation after the second century AD, when a preference for inhumation was established across the empire. However, abundant fuel was needed for the firing of bricks and for ceramic kilns, for the manufacture of glass and lime mortar, and for the heating of water for public baths. In sum, between 80 and 90 per cent of wood used annually was firewood, with between 10 and 20 per cent of harvested wood used as timber. Although smaller settlements were generally self-sufficient, many large cities, such as Alexandria, could not be supplied with enough wood locally or even regionally. Consequently, wood was exported and traded extensively, and deforestation, and consequent soil erosion, became a problem in some regions. Alternatives to firewood were used where available, for example coal in northern regions where it was mined, or peat and animal dung for heating at lower temperatures. Lands not considered cultivable might produce a good deal of fuel locally. Peat and reeds were harvested from bogs and marshes, for example. There were also innovative fuels, including 'pomace', the detritus from olive oil pressing, which was used in North Africa for the manufacture of amphorae, large ceramic vessels, in which the oil was transported.[17]

The mining district in what is today southern Jordan, a rich source of copper used in the production of late Roman coins, will have relied on imported fuel. Roman Phaeno (see Map 5) was a site of extensive and intensive mining and smelting, with more than 250 separate mines. During the fifth century AD, slag discarded by earlier miners was re-smelted, and older mines were reopened and deeper shafts dug, with access routes up to 1.5 metres high to allow animal access for the removal of mined ore. The intensification of the smelting of ore and slag has left clear traces in the environmental record. Water was piped to the mines and stored in a reservoir, now dried up. A sedimentary core extracted from this shows that heavy metal pollution reached levels that rival those of nineteenth-century industrial smelting plants. Concentrations of lead and copper in sand and silt beneath slag deposits were dangerously high, with readings for lead as high as 45,000 parts per million (ppm) and for copper as high as 17,000 ppm. The quality of life for miners and their families was poor and health was severely compromised. Analysis of bones and teeth has revealed toxic to lethal concentrations of lead (as high as 63 μg per gram of bone) and copper (as high as 117 μg). It has also revealed that these were not captives or convicts transported to the mines, but locals who were born, lived and died at the site. Contamination is so severe that even today it affects the health of Bedouin.[18]

Health and diet

In contrast to the peak of heavy metal pollution reached locally at Faynan, the fifth century was a low point for global atmospheric lead pollution as measured in the northern Atlantic. Regardless of their level of exposure to lead, subjects of the eastern Roman emperor could not expect to live long and healthy lives. Life expectancy at birth was always below thirty years, and could be as low as twenty-three depending on when and where one was born. This was due to the high rate of infant and childhood mortality, but also to spikes in mortality for young adults. Bioarchaeology, the scientific study of human remains, provides reliable insights into life course and life expectancy. The rate of neonatal mortality was high, but not extraordinarily so in historical terms. Rates of mortality among young children were higher than among infants. Weaning was tough on Roman children, leading to a spike in mortality when this began between the ages of four and nine. Poor dietary choices will have played a role, for example the use in weaning of honey, now known to be linked to infant botulism, and of various unpasteurised dairy products that may harbour a variety of pathogens (*E. coli*, salmonella, listeria, campylobacter), as well as lead contamination. The transition to solid foods also led to parasitic and other infections associated with poor hygiene and sanitation in the preparation and storage of food. Surely the greatest threats were dirty water, unwashed hands and flies, which transmitted faecal particles pulsing with bacteria and other pathogens that resulted in gastroenteritis and diarrhoea, still among the most frequent killers of children under five worldwide. In total, up to 40 per cent of all those born would have died before the age of eighteen.

Children died at greater rates for reasons that will not have been known either to their mothers or to doctors in antiquity. Lead is more toxic to children, since more enters their soft tissue. Adults store up to 94 per cent of absorbed lead in their bones and teeth, whereas in children the figure is only 70 per cent. Poor maternal nutrition leads to lower neonatal birth weight and, modern research has demonstrated, greater susceptibility to childhood diseases and longer-term conditions. Vitamin and mineral deficiencies such as anaemia and scurvy, related to vitamin A and vitamin C deficiencies respectively, affected many children and made them far more susceptible to other diseases. Far more children than adults would have died from pneumonia and bronchitis, and to waves of infectious diseases such as measles, smallpox or mumps, which led to meningitis. Malaria was endemic to many regions in the Mediterranean, affecting most of all those who had not developed tolerance.[19]

Nevertheless, in the absence of effective birth control, there were always considerably more infants and children than adults in every village, town and city. The mean age of death for those who made it to adulthood was around forty for men and thirty-five for women. Very few lived past fifty, 'senility'. Women died earlier than men due to complications associated with frequent pregnancies and childbirth, which led to both immediate and chronic infections, compromising immune systems already sorely tested by the ubiquity of excrement and bad water. Breastfeeding continued late, evidently until the age of four, placing immense strain on mothers, who may have nursed several children at any one time and repeatedly, and may themselves have been malnourished. Men suffered greater levels of injury and conditions such as degenerative joint disease and osteoarthritis. Deterioration of knees is very common in both sexes, whereas men suffered more frequently in their shoulders, and especially the right shoulder, suggesting repeated stress from physical labour, notably lifting. Evidence for trauma, including blows and stab wounds, and broken bones, is somewhat more common among men than women. Dental diseases were common in both sexes and brushing teeth was not widely practised. Malaria and malnutrition may both manifest in spongy or porous bone tissue (porotic hyperostosis), found in bones of people of all ages. Bone lesions indicate diseases (for example tuberculosis, pneumonia and bronchitis are all associated with rib lesions), which manifest as porous 'woven bone'. Most diseases, however, do not affect the bones, so we have no indicators of cause for the vast majority of deaths. Written records, notably epigraphy, indicate a clear seasonality to death rates, with a distinct spike in late summer and autumn for all age groups, although lower for older people, who had developed immunity to certain diseases. There was, however, a further spike in mortality for the old in winter.[20]

Richer members of society, who had access to sufficient food at all times, who did not undertake hard physical labour, who could afford to employ others to undertake tasks for them, from working their fields to breastfeeding their infants, were more likely to live longer and to have healthier children. However, emperors succumbed to infectious diseases and health conditions before they reached senility, empresses died during or as the result of childbirth, and princes and princesses died in infancy and childhood for reasons that are rarely recorded. Indeed, scholars have had great difficulty identifying quite how many imperial children were born. The Theodosian imperial dynasty is an excellent case study. Galla, the second wife of Theodosius I, died before she was twenty-five while giving birth to her third child, John, who also died. Gratian, her first son, also died as a child. Only her daughter, Galla Placidia, lived to become an adult. Theodosius's first wife, Flacilla, had died aged thirty, having given birth to three children. Her daughter died as an infant, and her two sons, Arcadius and Honorius, survived to adulthood. Arcadius married Eudoxia, who gave birth to five children between 397 and 403 and died the following year during childbirth, aged under thirty. Her first-born daughter died in infancy, but daughters Pulcheria, Arcadia and Marina, and a son, Theodosius II, lived to adulthood. Theodosius II married Athenaïs Eudocia, who gave birth to Licinia Eudoxia in 422, a second daughter Flacilla, who died in infancy in 431, and a son, Arcadius, who died as a child. Theodosius I died aged forty-eight, Arcadius aged thirty-three, and Theodosius II died aged fifty-one.[21]

Wealth and status, therefore, did not guarantee a long and healthy life. As in all eras, however, life was shorter and harder for the poor. Histria on the Black Sea coast, close to Rome's lower Danubian frontier, offers insights and illustrations. Skeletons exhumed from the city's late Roman graveyard displayed signs of hard, repetitive labour, poor diet and personal hygiene, infection, trauma and chronic pain. A young woman, perhaps as young as twenty-four, was buried with an infant who died at around three months. Despite her youth, the woman displayed signs of severe joint degeneration, and appears to have died with, and perhaps as the result of, a wound to her right leg. She had earlier suffered a wound to her pelvis and a blow to the head with a blunt object, both healed without signs of infection. Another woman, buried in an adjacent grave, died between the ages of thirty-three and forty-six. At the time she would have been suffering from both acute and chronic pain. She had lost a tooth, had dental caries in two more, and had four abscesses. Her remaining teeth – molars, canines and incisors – were all heavily or abnormally worn. Her back and limbs showed signs of osteoarthritis and degeneration, and her right hand would have appeared deformed from damage and wear. She would have walked hunched over, although once stood 1.59 m (5' 2") tall. The man buried next to her was around the same age at death and 1.71 m (5' 7") tall. He had lost ten teeth and had caries in four more. He suffered from severe osteoarthritis and eleven of his vertebrae were fused.[22]

As we have seen in Roman Britain, average height is a proxy for health and nutrition across broader populations. Measuring long bones can provide relatively reliable estimates of average height. Across the eastern Mediterranean in our period, men and women were on average far shorter than today, but also shorter than in both earlier and later centuries. Mean heights for men have been estimated at between 1.64 m (5' 4") and 1.70 m (5' 6"), and women between 1.52 m (5' 0") and 1.60 m (5' 3") at sites in Italy, Greece and modern Turkey. Bioarchaeology also provides evidence of what children and adults consumed, and how this and the broader environment affected their health. Bones of growing children turn over far more rapidly, leaving quite a clear and recent signal of what they were eating or being fed at the time of death, which can be determined by measuring collagen and nitrogen isotopes. Adult bones change slowly and often retain signals of far earlier diets. Unsurprisingly, adult diet does not change much over time and analysis reveals a standard Mediterranean diet of grains (bread), olives and wine, supplemented by some animal proteins, including a good amount of dairy, fish and seafood. It is not clear how many chickens were eaten, although this would only have been once they stopped laying eggs. Chicken bones are fragile and less likely to have survived in abundance in archaeological contexts.

Written sources also indicate that the year-round diet of all citizens comprised bread and vegetables, supplemented by fish and occasionally meat. Herds and flocks were driven seasonally into larger cities for sale and for slaughter. Most people ate two meals a day, in late morning and early evening, with the latter frequently being a hot meal if sufficient fuel was available to cook it and one could afford it. Firewood, as we have seen, may have been less abundant and too expensive, and dried dung, reeds or pomace might be burned at lower temperatures to heat food. Higher temperatures were required for baking bread, but ovens used for that task early each morning might be used later as they cooled for other forms of cooking. Bread was not baked at home, but hearths could be used for stews and gruels or for reheating food. Bakers sold bread each morning and grocers offered preserved and processed goods, including cheese and butter made from sheep and goat milk. There were several grades of bread based on the type and quality of grain used, ranging from the 'pure bread' (*artos katharos*) made with unadulterated, finely sieved flour, through 'middle bread' (wheat flour cut with bran or barley) to the coarser grained barley (*krithinos*) and bran (*rypartos*) breads. Barley bread was more commonly consumed by the urban poor, since barley was only two-thirds the price of wheat. Millet and rye were less expensive still, and the latter was a staple ingredient in gruel.[23]

Indications of diet and parasitic infection can also be found in human faecal waste excavated at public latrines and private toilets. A toilet soil pipe might discharge into a sewer, but more frequently waste was washed into a nearby septic tank, allowing for modern investigation of Roman intestinal diseases and parasites. At Ephesus (western Turkey, Map 4), samples from a private toilet in a mansion ('Hanghaus 2') showed evidence for roundworm and whipworm, and those from the Varius Baths, a large public baths with a latrine that seated thirty-seven at once, also indicated whipworm, with eggs discovered in every sample studied. Both roundworm and whipworm are soil-transmitted helminthes; their presence demonstrates the dominance of soil-grown, farmed foods, cereals and vegetables in the Ephesian diet, and also highlights the perils of recycling human waste for use as fertiliser. Another common parasite, tapeworm, is transmitted by the consumption of undercooked meat and fish containing its larvae. There was no evidence at Ephesus for tapeworm, suggesting that transmission of parasitic infections was because of poor personal hygiene – ineffective washing of hands and cooking utensils – rather than via poorly cooked food. Results at Sagalassos, the only other eastern Mediterranean site to date to have been investigated in the same manner, are similar. However, these findings cannot be projected across the entire Roman world, since ten intestinal parasites, including tapeworm, have been identified at Roman sites in Britain, Austria, France, Germany and the Netherlands.[24]

A range of vegetables was grown close to many large cities, since perishables could not be transported far. We have good evidence from Constantinople that vegetables were grown both within the walls and immediately outside them, notably lettuces and leaves that were sown in spring and would not travel well (see Maps 10 and 11). The *Geoponica*, a tenth-century compilation based on late antique sources, lists these alongside beetroots, cabbages, carrots, cress, dill, fenugreek, leeks, mustard, parsnips, swede (rutabaga), and much else. Many staples do not feature in this text, including grapes and olives, although we know from other sources that these were both grown close to the city and transported there in great volume. According to the miracles of Daniel the Stylite, the saint's column stood on vineyards owned by a wealthy imperial steward at Anaplus (modern Rumelihisarı), a suburb of Constantinople. Fertile gardens and lands for hunting also sat just outside the walls. Broad beans, pulses and chickpeas were dried and consumed in winter, as were pickled and preserved fruits and vegetables. Some fruits and vegetables did not grow well in Constantinople's climate, which experiences at least a week of snow each year and temperatures regularly below 5°C in January and February, and for much of our period it was rather colder for longer periods.

Markets across the Mediterranean were rich in the same types of produce, harvested from market gardens and orchards or driven from fields and pasture. To take one last example of domestic consumption, at Caucana (Punta Secca) in south-eastern Sicily (Map 9), the port (as we shall see) from which Belisarius sailed to reconquer North Africa in 533, an archaeobotanical study of food remains left at the site of a tomb suggest that those feasting, perhaps annually or biannually, with their dead relatives ate a range of local cereals, legumes, fruits and nuts, including wheat, barley, chickpeas and lentils, almonds, peaches and figs. Herbs and greens consumed or added to recipes included garlic, fenugreek, hawthorn and cornflower. A table and bench allowed for only a few diners, and a hearth allowed food to be cooked, or perhaps reheated rather than roasted on site, as the bones of various animals are not scorched. These were principally the bones of sheep, goats and pigs (about 83 per cent), with far fewer chickens (14 per cent), and a little beef, horse and tortoise. There were also remains of limpets, cuttlefish and snails. Charcoal from the hearth was of cypress. The tomb was covered by a heavy slab pierced to allow the pouring of libations, and fragments of wine amphorae of Sicilian, Egyptian and

North African types were found nearby, as were diverse pieces of African red-slip cooking and table ware, and three African lamps, one decorated with a unique design that seems to depict a backgammon board. The tomb held the body of a women, aged between twenty and twenty-five; three bones of an unborn baby, indicating that the woman was about six months pregnant at the time of her death; and the bones of a female child aged about four, buried later than the others. DNA analysis confirmed that this was the woman's daughter. She died when her mother's body had lost its flesh and the pelvis could be detached and moved aside to make room for her small daughter's flexed body. They were buried at home, in a house built around AD 580–600 and abandoned by 650, overwhelmed by aeolian sand.

One of the slabs that sealed the tomb at Caucana was inscribed with a curving swastika-like design within which, in the middle, a square contained the Greek word HAGIOS, 'holy', scratched three times in capitals, but backwards to increase its protective power. Alas, the magical inscription did not prevent tomb robbers from plundering the tomb. Excavation has revealed how carefully those who remembered the departed had restored it to good order, following the robbery and before it was covered by shifting sands. In a small community in southern Sicily, one might imagine that the family knew the robbers.[25]

Magic and medicine

As well as seeking to protect their dead, people sought tangible things to protect them in their daily lives, including objects made from copper alloys, clay and stone. Pectoral crosses, rings and amulets were considered effective media for conducting divine energy for the good of health. Many things absorb the sun's energy, but some store and conduct it better than others. So it is with divine energy – in the words of St Ambrose with which we began the chapter, 'the spirit's sanctifying beam' – which some things by their nature, by contact or proximity, or by how they are fashioned, are able to store and conduct better than others. Pilgrim tokens were felt to be especially effective at storing and radiating divine energy, for they may be formed from the soil on which a holy athlete once engaged in spiritual combat with the Devil and his demons. *Ampullae*, small pilgrim flasks, improved upon tokens by containing and preserving holy fluid, for

Figure 2. Pilgrim flasks (ampullae) from Abu Mena, Egypt, and Resafa, Syria, sixth to seventh centuries

example sweat or ooze from a tomb, or water blessed by contact with the saint's tomb, or oil from a jug near a tomb (Figure 2). Water and oil were carried away by pilgrims who visited the tombs of saints, such as those of St Menas (near Alexandria, modern Abu Mena in Egypt, Map 7), or of St Sergius at Sergiopolis (Resafa in Syria, Map 5), or from Constantinople's healing shrines at Pege and Blachernae (Map 10). If pilgrims travelled home by sea, they might share the protection afforded by the actions of shipowners and sailors. Every Epiphany the Christian shipowners of Alexandria took water collected from the river Jordan, not the Nile, and had it blessed by a priest before sprinkling it on the prows of their ships to protect them and their sailors for the coming year. Anchors discovered in shipwrecks bear protective devices, an example being a sixth- or seventh-century stone anchor found off the coast of Ascalon (Ashkelon, Israel, Map 5) that is inscribed with a cruciform monogram incorporating the letters AΩ, symbolising Christ.[26]

Prayer was the most common means employed to preserve health. At the time of his marriage, Synesius of Cyrene, one of the wealthiest men in the North African province of Cyrenaica (Maps 8 and 12), whom we shall meet several times in the following pages, wrote a prayer. In it, he thanked the Lord for preserving the life of his beloved brother, and he asked for the safekeeping of the household of his father, Hesychius, including his two sisters, and the children he would have with his new wife. 'Beneath your hand protect also the partner of my marriage bed, free of illness and harm, faithful, of one mind with me.' Synesius's prayer did not save his family: three of his children, all sons including a pair of twins, died in a single year, probably of infectious disease.[27]

Prayer was also the most common and immediate response to illness. The ancient tradition of incubation – sleeping at temples – continued in Christian shrines and churches. Many reported visits from the Mother of God (*Theotokos*) or other healing saints while dreaming in their shrines. Divine protection and the summoning of holy healing powers through magic were considered necessary and effectual because demons caused illness and disease. The *Testament of Solomon*, a popular work, detailed the thirty-six demons relating to particular illnesses and conditions as well as prescriptions for repelling them, including magical incantations, spells, and the use of seals, rings and amulets with magical inscriptions and designs. Demons often took the form of wild animals or, very commonly, snakes and reptiles. Amulets and rings, therefore, might feature a serpent, because like cures like. Other armbands, jugs and flasks featured holy riders, who would wield a lance to pierce a serpent demon in the manner of earlier, and later, dragon-slayers. In a now lost fresco at the Monastery of St Apollo at Bawit in Egypt, and dated to the sixth or seventh century, St Sisinnius was shown slaying the demoness Alabasdria, while her snake-tailed daughter hovered above. Alabasdria, a cognate of Lilith, is identified in the *Testament of Solomon* as the demon responsible for the deaths of children during childbirth.[28]

For three decades at the end of the fifth century, Daniel the Stylite received visitors at his double column in the northern hinterland of Constantinople, including parents who sought his powerful prayers to drive demons from their sick children. Among those to visit Daniel was the poet and praetorian prefect Cyrus of Panopolis – we shall meet him again at the court of Theodosius II – who sought help twice for his daughter, who was twice cured. A boy aged seven who could only crawl was able to leap to his feet to embrace the saint's column. Others sought remedies at a distance. One father was 'so rich in the poverty of Christ' that if anyone in his household fell ill, 'be it son or daughter of man-servant or maid' he wrote asking for the saint's prayers. On receiving the holy man's written reply he would lay the letter 'as if it were the miracle-working hand of

Jesus on the sufferer and immediately he received the fruits of his faith'.
During the usurpation of Basiliscus (AD 475–6), Daniel descended from
his column and processed through the city to Hagia Sophia, stopping en
route at Constantinople's most famous monastery. Founded in 450 by a
certain Studius, and by 460 occupied by a group of sleepless monks who
prayed around the clock, it was dedicated to St John the Baptist, having
as its official name the Monastery of the Prodromus ('forerunner') of
Studius.[29] When the monks offered to ferry Daniel to Hagia Sophia by
boat from their pier, an angry crowd gathered outside shouting 'Bring the
just man here if you love Orthodoxy; do not begrudge healing to the sick'.
And so he proceeded on foot, healing a girl, bedridden for three years, at
her mother's request, before he had reached the Forum of the Ox.[30]

The *Miracles of St Artemius* report a range of ailments relating to men's
health that the saintly martyr cured when visited at his shrine in Constantinople. Unlike those performed by the living Daniel in the fifth century,
these were posthumous miracles performed in the seventh century, some
two centuries after Artemius's death. In the third miracle we learn of a
man who had a boil on his testicles. 'Although he visited many doctors', his
cure came in the form of a poultice formed from holy wax and the appearance of the saint himself, who lanced the boil with a scalpel. Another
man, Isidore, 'a certain sailor for many years had problems with his testicles' caused by 'an evil spirit'. Isidore visited the Church of St John the
Baptist, where Artemius's body had been interred and where, 'in the aisle
to the left, the saint is accustomed to make his rounds as if he were chief
physician in charge of a hospital'. There, 'one night the saint in full view
approached the man, while many of those awaiting a cure looked on' and
witnessed what happened, as Artemius drew out the demon, who took
the form of a black crow. A silver dealer, Acacius, and his wife brought
their seven-year-old son and several mattresses to the church one Easter,
and were on the point of departing without a cure when Artemius heard
the boy himself, despite his 'lisping, inarticulate speech', asking 'St Artemius, take away my hernia'. He did.[31]

Finally, there were doctors. Two works by Nicander of Colophon,
written in the second century BC, were widely read during our period.
Nicander's *Alexipharmaca* offers advice on the treatment of ailments with
a full range of plants, fruits, honey and seafood. His advice for the alle-
viation of symptoms of lead poisoning included a fresh infusion of sage,
or roasted pepper and rue grated into wine. His *Theriaca* is replete with

Greek lore relating to reptiles and serpents that was copied repeatedly thereafter. Readers will have learned from it which herbs and ointments would repel snakes and how to create salves and herbal remedies if bitten. Medical treatises updated ancient knowledge in the field of pharmacology, with prescriptions available for a range of ailments from dermatitis to bubonic plague. Alexander of Tralles – brother to Anthemius, whom we shall meet as an architect of the church Hagia Sophia – wrote a twelve-book medical encyclopaedia in the sixth century that is practical rather than theoretical, dealing with common ailments to the eyes, chest, lungs, heart and digestive system. He made careful observations on fevers and intestinal worms, but also advocated the effective use of magical amulets. Paul of Aegina produced an extensive medical compilation and commentary in the seventh century, for example citing those who advised that a 'plate of lead, applied to the loins, will prevent libidinous dreams'. Burned lead or lead oxides appear in a number of potions and salves in his work, such as for the treatment of ulcers. Other concoctions might be pounded and mixed or stored in a leaden mortar or pot, all highly toxic. Doctors will have used these works in private practice, offering services for a fee.[32]

When the imperial mausoleum at Mediolanum was constructed, Roman authority was still exercised in northern Britain and in eastern Syria. Roman metals were shipped across the empire, and the pollution produced by their manufacture was captured in ice and sediment from the Arctic to Alexandria. Life at the end of the 'Lead Age' was fragile, even for the wealthy and powerful. The imperial occupants of the Milanese mausoleum were both killed in their early twenties. Synesius of Cyrene lost children, despite his wealth and prayers. The vast majority of Romans were poor and suffered chronic hardship and poor health. Wherever a child was born, however hard her parents prayed or whatever magic they deployed, her chance of living a long and healthy life was tragically small. Yet as we shall see in the following chapter, if the inconstant sun faltered or crops failed, families persevered and faith endured, sustaining communities across the Mediterranean world in the face of invasions and natural disasters. These were communities bound also by their adherence to Roman law.[33]

2

FAMILY AND FAITH

❦

John, a worried father in sixth-century Egypt, recognised that he had made a mistake in betrothing his daughter Euphemia to a local boy, Phoibammon, who was engaged in 'lawless deeds, which are pleasing neither to God nor to man and are not fit to put into writing'. Therefore, John paid Anastasius, 'the most illustrious advocate of the City of Oxyrhynchus', to write a deed of separation and dissolution, which John signed in his own sloping handwriting and sent to Phoibammon. We do not know if Euphemia wished, as her father did for her, 'to lead a peaceful and quiet life', nor do we know the nature of Phoibammon's unmentionable crimes. Euphemia and Phoibammon were probably both children at the time of their betrothal, which was allowed from the age of seven and would endure until sexual maturity was achieved or considerably later. In the previous chapter, which considered health and physical development through the study of scientific data, childhood was held to end at the age of eighteen. In Roman legal terms, however, childhood ended rather earlier, at twelve for girls and at fourteen for boys. After that, and until the age of twenty-five, both young men and young women were referred to as youths or adolescents (*epheboi*), or legally as minors (*aphilikes*). Youths, when sexual maturity was reached, were permitted to marry, although it seems to have been unusual to marry so young.[1]

Family

Children and youths were members of their parents' family until they married, at which point they formed their own family, that is a husband and wife and their children by birth or adoption. A household might also include the parents of the husband, less commonly those of the wife, and others of various ages, including domestic servants, slaves and their children. The family was the foundational social structure and was greatly strengthened by legislation introduced between the fourth and eighth centuries. The fifth-century Roman father no longer had the authority of the first-century paterfamilias, but he was head of household and represented and protected the rights of others. The equivalent of common-law marriages existed: if a man introduced a woman into his house, if they had sexual relations together and he entrusted her with the management of his household, the couple were understood to have entered into a contract of marriage. From the sixth century this was often accompanied by a solemn religious ceremony known literally as 'crowning' (*stephanoma*), after the marriage crowns placed on the heads of both partners. However, the legal status of marriage was unaffected by whether a crowning had taken place. Marriage impediments were established, with prohibitions on marriage within certain degrees of kinship recognised but not always effectively policed.

A daughter was to be provided with a dowry upon marrying and while this became part of the family's common property it remained hers if her marriage ended. Divorce by mutual consent was allowed under Roman law, although during Justinian's reign it was made available only if both partners took monastic vows. Separation was permitted in circumstances that favoured the husband, for example if the wife was found to have committed adultery or had otherwise behaved badly outside the household. Examples of such bad behaviour included dining or bathing with other men, or hunting wild animals. Synesius of Cyrene prayed that the Lord keep his wife 'in ignorance of clandestine associations. May she keep my bed holy, unsullied, pious, inaccessible to unlawful desires'. A husband's impotence for three years might be grounds for a wife to end her marriage. When a husband died it was usual for the wife to become head of the household, with protected rights over property and care for her children, including planning for their futures and the provision of dowries. Where a child could not be cared for, adoption was possible and absolute, as Aurelia Herais, a mother from Oxyrhynchus in Egypt, recognised in 554, a year after the death of her husband. In handing over legal custody of her nine-year-old daughter to a couple better able to meet her needs, she agreed 'I have no legal power henceforth to take her away from you'. Property and inheritance rights were transmitted to both daughters and sons. Married daughters would receive an inheritance through their husbands, as head of family. The principle of partible, not patrilineal, inheritance distinguished east Roman society.[2]

The adoption letter for Aurelia Herais's daughter is an example of the many precious papyri excavated at various sites in Egypt, where the dry sands have offered excellent conditions for their preservation. Papyri allow us to access ordinary lives and investigate social relations in a manner that is impossible elsewhere in the empire at this time. A dossier of papyri from Syene (now Aswan), in the far south of Egypt, illuminates many aspects of family life, retirement and partible inheritance. In this dossier is a will, dated 12 March 584, drawn up for a husband and wife, Aurelius Patermouthis and Aurelia Kako, which specifies that when husband or wife dies, 'the survivor shall receive and keep, of all the possessions left by the deceased',

> those which were acquired by inheritance from parents and by right of purchase and by sweat and labour, comprising houses and objects of gold and silver and copper and mountain copper and clothes and woven materials and minor objects of all kinds, large or little, and shall be owner and master of these and administer and manage and keep them in repair and out of them shall feed and clothe himself and bury the deceased and execute his bequests, none of our children being able now or in future to proceed against the survivor or bring a claim or call him to account or obstruct him until his death, and that after his death all the possessions left by us shall go to our common children in equal portions, no difference being made between one child and another.[3]

On the same day, Patermouthis and Kako entered into a legal contract with Jacob, Kako's father, to look after him in the family home for the rest of his days. In exchange, the couple received all his possessions, effectively disinheriting all other relatives. This led, twelve years later, to legal proceedings and arbitration between Patermouthis and John, Kako's brother, recorded in another papyrus. Jacob died within a year of the legal agreement, and in March 585 the couple were back with their lawyer, so that Aurelia Tapia, Kako's mother and Jacob's widow, could sell to them a number of properties 'by the law of purchase and for eternal possession' for the sum of ten solidi. The first named of these was a 'half-share of the living room belonging to me in the house of my mother, the other half of which belongs to Menas and Tselet, my siblings, facing north towards the stair, on the second floor; and also my share of the terrace on the fourth floor above the bedroom of Talephantis'. A fourth sibling, George, had earlier sold his quarter share of the half living room to Tapia. Patermouthis is said to have been a boatman, presumably of goods and passengers on the Nile. He appears in other documents as a soldier. In both jobs, he would have been away from home a great deal, leaving Kako to care for her elderly relatives as well as her own children, while she also oversaw the day-to-day running of their property portfolio. The surprising verticality of village life also emerges from these and many papyri, which detail multi-storeyed structures with up to five floors.[4]

Less surprising is the presence of slaves, although identifying them in the written record can be challenging. The Roman legal distinction between freeborn and slave persisted and Justinian's lawyers recapitulated and simplified existing laws on slavery, for example the law that a sick slave abandoned by a master must be regarded as free in the event that he or she recovered. In the same vein, when Justinian learned that families in Thessalonica were reclaiming their own children, whom they had entrusted to the church to raise, and that they were treating them as slaves, he ruled that the children be freed and the parents punished. There were to be no limits on the freedom of freedmen and none on who could manumit a slave. A former slave-owner remained patron to his freedman, but must allow him sufficient time and allowances to support himself and his family.[5]

Outside the law, references to slaves are often opaque, thereby revealing how embedded and normalised was the institution of slavery in Christian Roman society. Fundamental criticism of slavery was expressed rarely and it remained without any general consequences in practice. However, some discomfiture may be reflected in the terminology for slaves, which is diverse and often ambivalent: apart from *doulos* and *doule*, the male and female versions of the standard word for slave, we come across many terms that might be translated as servant, attendant, houseman, housemaid, boy and girl, in contexts that imply we are reading about slaves. Diminutives are common not only because they diminish the status of the slave, but also because the largest number of slaves was always below the age of eighteen. These slave children suffered similarly to freeborn children but might also endure workplace injuries and the earlier onset of chronic conditions brought on by manual labour.[6]

Social relations and the land

By the end of the fourth century, the Roman empire had ceased to be a 'slave society', meaning that its economy was no longer dependent upon slave labour. However, slaves were employed on some large landed estates in parts of the eastern empire. Their condition is referred to in some Egyptian papyri as *paidaria*, where we find freeborn villagers of modest means inheriting and selling their co-workers. The vast majority of workers of the land, however, were paid agricultural workers and tenant farmers. A preserved letter of petition from one such worker, Anoup, is another example of the wealth of information preserved in collections of papryri. Anoup wrote to Flavius Apion in the following terms:

> To my good lord, the Christ-loving lover of the poor, the most praiseworthy and esteemed patrician and duke of the lands of the Thebaid, Apion, from Anoup your wretched slave from the parcel of the estate known as Phakra. No injustice or wickedness has ever attached to the glorious household of my kind lord, but it is ever full of mercy and overflowing to supply the needs of others. On account of this, I the wretched slave of my good lord, wish to bring it to your lordship's knowledge by this present entreaty for mercy that I serve my kind lord as my fathers and forefathers did before me and pay the taxes every year. And by the will of God in the past eleventh and tenth fiscal indictions my cattle died and I borrowed the not inconsiderable sum of fifteen gold coins to be able to buy the same number of cattle. Yet when I approached my kind lord and asked for pity in my straits, those responsible to my lord refused to do my lord's bidding. For unless your pity extends to me, my lord, I cannot stay on my allotment and fulfill my services with regard to the properties of the estate. But I beseech and urge your lordship to command that mercy be shown to me because of the disaster that has overtaken me.[7]

Unpacking this petition offers us countless insights into life on the land in sixth-century Egypt. Anoup twice presents himself as the wretched slave (*eleeinos doulos*) of his master, which is a nicety that does not reflect true servile status, but does serve to highlight the complications of identifying slavery in the written record. In fact, this is a standard rhetorical formulation written by the scribe employed to draft the petition, and it reveals nothing of how Anoup conceived of his relationship to Apion. Next, Anoup reveals that he is only the most recent of his family to have worked the land owned by the Apions and to have paid the necessary taxes on that land. Anoup refers to taxes (*ta demosia*) not rents, which reveals much about how he understood his relationship to the land and to the landowner. He recognised that the Apions stood in place of the state, collecting rents to pay the total tax bill for all the lands they held, but also collecting the annual poll tax paid by all adult men (but not women and children) who lived and worked on their lands. The total working population on the Apion estates has been estimated at more than 10,000 men, women and children. Anoup was not just a paid labourer on an Apion estate, although he was certainly that, but he was also a tenant farmer who rented an allotment and kept his own cattle on pasture associated with his settlement. He was a registered agricultural worker (*enapographos georgos*), bound to the land he worked by private contract with the Apions, but also by a separate legal agreement that obliged the landowner to pay taxes on his behalf. Since he had taken on an enormous debt to replace lost cattle, Anoup was unable to fulfil obligations under his contract with the Apions. He faced eviction from his rented plot, perhaps too the seizure of his belongings. However, he would not have been released from his obligation to work as an agricultural labourer on the lands to which he had tied himself, perhaps in a different condition, that of *paidarion*. Ejected from his allotment, Anoup and his family might have joined other *paidaria* living and working on the portion of the estate farmed and managed for the Apion household, the lands kept 'in-hand' (*autourgia*).

As Anoup reveals, his initial plea for mercy had been refused. This would have been the decision of a revenue collector (*pronoetes*) employed by the Apions, whose job was to ensure that the landowner met his own financial obligations to the state and turned a profit. Between sixteen and twenty revenue collectors handled the Apion estates, carefully recording how much and from whom they collected. In 542, according to another precious papyrus record (*P. Oxy.* XVI. 1918), the Apion estates generated

20,010 solidi, of which 6,917 solidi were sent as a tax payment to Alexandria.[8] We do not know whether Anoup's petition for mercy to Flavius Apion was successful, but others were. We have a letter signed 'From the landowner', which orders an overseer to 'have the goodness to grant a respite to the people of Pempo', a village estate owned by the Apions. In 550 (*P. Oxy.* I. 133), the same Flavius Apion lent 200 *artabas* of seed corn to the village of Tacona, specifying only that they would pay it back the following year. It is clear that the large landed estates of the wealthiest families grew, in part, because they could and did protect families like Anoup's, or the villagers of Pempo and Tacona, from the annual vagaries of agricultural life such as the loss of livestock or crop failure. This was especially true in the troubled years of the mid-sixth century, when (as we shall see in later chapters) climate events appear to have affected harvests in many places, and when epidemic disease may have diminished the producing population. It is equally clear that the state came to recognise this as necessary to protect its various taxation revenues, including the *embole*, being the levies of grain that supported the state grain dole, and the *annona*, levies of provisions and goods that supported the army.[9] Efficient accounting systems were sustained even in turbulent times.[10]

Justinian issued an edict on governing Egypt in 539, which observed that any threat to the revenues generated from that province threatened 'the very cohesion of our state'. Where revenues were underwritten by the landowner, who was required to use past profits to make good any annual shortfall, the state was guaranteed necessary income. Any act of generosity by Apion, such as his allowances to the villagers of Pempo or Tacona, was evidence that the system was functioning as intended. The relationship between state and elite, therefore, was not antagonistic but cooperative. It placed substantial burdens on the landowners, which they could not refuse, but from which they benefited economically and politically even as it cemented their social status.

As we shall see in the following chapter, those of a certain level of landed wealth were obliged to become decurions, members of their local city councils, and commit substantial sums to maintaining local infrastructure and services. In the case of Flavius Apion, that city was Oxyrhynchus, one of the larger cities of the Nile Valley south of Alexandria, the same city where Euphemia had been betrothed to Phoibammon. One of the great expenses at Oxyrhynchus, as elsewhere, was maintaining and supplying fuel to the public baths, and we read in one papyrus of AD 566–7 (*P.*

Oxy. XVI. 2040) that the Apions paid a quarter of the twenty-seven solidi required, with three-quarters shared between other decurion families and the church. In this way, and countless others, the surplus generated from the land supported the urban population of the Nile Valley, although 90 per cent of those who lived in those cities did not own land, and a good number of them paid rents on hundreds of urban properties, including houses, apartment buildings and workshops, owned by the Apions. The quotidian, intimate links between people in the cities and those working the lands around them is well illustrated in another document (*P. Oxy.* I. 135) signed by Aurelius Pamouthius, a lead-worker (*molybourgos*) in Oxyrhynchus, who pledged to pay eight solidi to the Apions if Aurelius Abraham did not remain on the land to which he was bound as an *ena-graphos georgos*, along with his wife and herds. The relationship between the two men is not otherwise explained.[11]

Wealth, faith and community

Flavius Samuel was a soldier who held lands at the village of Aphrodite in the Nile Valley, some distance to the south of Oxyrhynchus, in the first decades of the sixth century.[12] In 526 (*P. Michael* 43), Samuel leased a twenty-eight *aroura* farm – one *aroura* is around two-thirds of an acre, so the farm was a little under 19 acres or 7.5 hectares – for a period of eight years to a fellow villager, Aurelius Phoibammon, son of Triadelphus. If we assume that Samuel was still a soldier and absentee landlord, the lease may have resulted from the vacation of the land by a third villager, John, son of Phrer, named in the papyrus as Samuel's former tenant. Phoibammon agreed to pay a rent of 140 *artabas* of grain per year, more wheat than barley, plus other comestibles, including fifty cheeses, some wild mustard, and half the fruit produced by trees on the leased land. An *artaba*, the standard dry measure of grain in Egypt, equated to 4.5 *modioi* elsewhere in the empire, hence a little over thirty kilos. In other words, Phoibam-mon was to pay 4,200 kilos in grain. Although this may appear high, the terms of the lease were quite standard, granting about half the annual crop as rent to Samuel and his daughters. At the same time, however, Samuel received a loan from Phoibammon of eighteen solidi and fifty-eight *artabas* (*c.*1,750 kilos) of wheat that modified the terms of the lease immediately – Phoibammon received a rent rebate of fifteen *artabas* per year – and additionally the rent was to be reduced by a further ten *artabas* per year were Samuel unable to repay his debt. Very quickly, Phoibammon paid far less than market rent for the farm. Moreover, over the following year, on two occasions Samuel borrowed even more from Phoibammon, a further thirty *artabas* (*c.*900 kilos) then eighteen *artabas* (*c.*550 kilos) more of wheat. Samuel had quickly become heavily indebted to Phoibammon, who now fully possessed the farm as surety for his loans.

We hear no more about Samuel, but Phoibammon continued to grow in wealth and influence in Aphrodite, where he served on the village council. He leased land from those who did not wish to or could not farm it themselves, including absentee soldiers and functionaries, and the local church, and he both rented and purchased numerous plots. In 572, in the last dated papyrus to mention him, now clearly an old man, Phoibammon purchased pasturage with his nephew Victor, son of his brother Colluthus. In an undated papyrus Phoibammon acquired with Colluthus a share of a large house in the village, which had a good amount of storage space, including a special storage room for bread. In such a space he could store his surplus and seed corn from year to year, and when necessary withdraw large amounts to lend or sell to others. His reserves were sufficient that he could pay his taxes in advance of the harvest. Phoibammon had a second brother, Aurelius Paulus, and a wife, Anastasia, whom he called 'the dove', and with whom he purchased a vineyard. Anastasia's maternal uncle was Apollus, first man in the village, whose family is the subject of another, still larger cache of preserved papyri. Apollus's son Dioscorus was the village of Aphrodite's great scholar and poet who composed a petition that secured imperial favour for the villagers. Both father and son worked for periods for Count Flavius Ammonius, the region's equivalent to Flavius Apion. When Apollus died he left two-thirds of his estate to Dioscorus and one-third to Phoibammon.[13]

The story of Phoibammon is that of a man of the middling sort told through his appearances in the papyri of Aphrodite. These precious documents were once retained in a family archive and are now dispersed to libraries and museums as far from Aphrodite as Florence and Vatican City, Cairo and London, Michigan and New Jersey. His story, like those of Anoup and Samuel, Patermouthis, Kako and all their litigious relatives, would have been familiar in villages and towns across the empire. However, we can recover them only very rarely outside Egypt. For insights into life in the villages and towns elsewhere in the empire we must rely on archaeology, hagiography (the lives of saints), and references in literary sources. At the top of this list might be placed Libanius's observations on the Orontes valley and the hinterland of his beloved Antioch, where the land is 'deep and rich and soft, yielding easily to the plow, wonderfully exceeding the expectation of its farmers ... providing tall trees in all their beauty and sheaves of grain taller than trees ... much wine goes hence to our neighbours, and even more oil is carried in freight ships all over the world'.[14]

Libanius elaborates on how these lands were farmed and goods were exchanged at large villages or small towns (*komai, metrokomai*), 'with populations greater than not a few cities',

> Using the services of artisans even as in towns and sharing their goods with one another by means of the festivals to which they invite others and are invited, each one in turn summoning the others to it. They enjoy and take pleasure in the same things, and profit by sharing with each other their surplus and gaining in return what they need, selling some things and buying others, more fortunate by far than those who trade by sea, since they earn their wealth with laughter and bustle instead of among the waves and breakers, needing few of the city's products because of the commerce among themselves.[15]

Elsewhere, Libanius observed that the fertile lands of these 'large villages each belonged to many masters', just like Aphrodite in Egypt. The remains of more than seven hundred villages have been identified on the lands of the northern Syrian limestone massif, where olives and vines were grown. Many of the villages are well preserved because they were abandoned early, and because they were built in stone, a sign of prosperity. The abundance of limestone has been advantageous for the archaeologist, as stone structures survive far more readily than those built from organic matter (e.g. wood, mud, wattle and thatch). Prosperity is also evident from the small stone churches that stood in each village, some serving only four or five structures, and in the silver that those churches acquired and buried.[16]

Among the artisans in the larger villages and towns were metalworkers, blacksmiths, carpenters, textile weavers and stonemasons, whose crafts and wares would have supported life at home and on the farm, making tools and sacks, repairing carts and horse tack, building storehouses and

fashioning quern stones. Traces of all these have appeared in the archaeological records of sites across the eastern Mediterranean and North Africa.
There were not in these *komai* the grand colonnaded streets, the monumental fountains or baths, or the municipal and educational institutions
that together characterised city life. We shall turn to these in the following chapter. But a church there would be, and in it the donated wealth of
locals who in their daily lives and travels would encounter another funda-
ment of late antique life, the holy man, who would intervene in social
relations and arbitrate disputes, whose prayers healed and helped, even
solving crimes and apprehending criminals.

The Life of St Theodore of Sykeon is instructive in this regard, both as a
repository of fascinating anecdotes about an early seventh-century rural
community in Galatia, Anatolia, and as a example of how a saint's life
shapes and refines information to serve its principal purpose, to propagate
its protagonist's story and promote sanctity above other considerations.
Theodore, 'God's gift', is persuaded at a very young age by visions of St
George to emulate Christ by taking to the wilderness, but his desert is the
hinterland of his own village, where he takes up residence in an abandoned
chapel. Digging a hole under the altar, he fasts and lives there from Christmas to Palm Sunday. He then ventures outside, to a local hillside, where
he lives in another hole for two years, known only to one man. Returning
emaciated, with worms in his hair and pus leaking from his head, he is
identified as holy by his local bishop, Theodosius. Theodore's holiness had
been clear to a host of local holy men consulted by his mother, Mary, even
before his birth.[17]

Theodore's is a small world. The village of Sykeon is located precisely
at the very start of the *Life* 'in the country of Galatia' twelve miles distant
from and 'under the jurisdiction of the town of Anastasiopolis, which
belongs to the province of Galatia Prima, namely that of Ancyra'. Apart
from a pilgrimage he makes to Jerusalem, covered in a single short chapter,
Theodore's miracles are performed for locals and visitors to Sykeon from
villages six, eight or ten miles away. His powers could reach further, as was
made clear on the occasion when his prayers found a man near Nicaea and
stopped him from leaving so that he could be apprehended and returned
to Sykeon with the money he had stolen. At the height of his powers, Theodore had a cage built from tools donated by local farmers, which he hung
above his cave, open to the elements. Within the cage, he donned iron
garments to weigh him down. His feet were frozen to the floorboards,

his skin torn from them when he moved. He fainted from sunstroke in summer then admonished his servant for shading him. It is striking that the young man undergoing his self-imposed discipline (*askesis*) retained a personal attendant, perhaps a slave, but equally it is clear that Theodore came from a family of innkeepers wealthy enough to buy the young man a golden belt and expensive clothes, which he had cast aside even as he kept his devoted servant. He, like Phoibammon, was of the middling sort, wealthy enough for his decision to set aside worldly things to matter.

In the absence of a village doctor, Theodore provided healing services, driving away the demons that caused diseases of the body and mind. He took miraculous counsel and medical advice from the doctor saints Cosmas and Damian, who were noteworthy in their time for taking no money for their medical services. On one occasion Theodore drove a demon from a woman whose husband paid her no attention. To the hagiographer she was damaged, tainted by sin. Most characters serve only as props for Theodore's miracles, and women are presented as virgins or whores. Theodore's sister, Blatta, is a blessed virgin, whose youthful death he celebrates, and his grandmother is too solicitous about Theodore's health until she becomes a nun and leads a community, later in life, following his example. Her late-won holiness is demonstrated by her ability to predict the moment of her own death. The worst treatment in the *Life* is reserved for Theodore's mother Mary, who conceived him out of wedlock with a rich traveller. Mary is quite unlike her namesake. Her failure to recognise her son's calling, her frequent maternal interventions to prevent his latest folly, and her decision ultimately to leave the village and remarry in Ancyra, are all roundly condemned. When news of his mother's death is brought to the saint, he refuses to accept it but fasts for a week so that God might forgive her sins. Theodore's saintly life began only when he turned away from his family, but his *Life* adds to our understanding of the bonds of family and the centrality of faith to village life.[18]

As Theodore's fame and standing grew, he established a monastery befitting his status and despatched a deacon to Constantinople to buy liturgical silver. Finding a chalice and paten of high quality, 'the quality proved by the fivefold stamp upon it', the deacon was pleased with his purchase, but the saint was not and 'condemned them as being useless and defiled'. When first used for the liturgy, the truth was revealed: the silver turned black.

The brothers, seeing this, glorified God who made invisible things visible at the hands of His servant. When the archdeacon took them and locked them up they appeared once more as pure silver; then he returned to Constantinople and gave them back to the dealer in silver and told him the reason. The latter made inquiries of his manager and his silversmith who fashioned the vessels, and found out that they came from the chamber-pot of a prostitute.[19]

John Chrysostom had already complained of such vulgarity, the ostentatious display of wealth to no useful end: 'Do you pay such honour to your excrements as to receive them in silver?'[20]

The wages of sin would buy silver vessels as readily as donations, votives for prayers answered and memorials for the departed. Monastery churches like Theodore's certainly acquired riches, an example being the Monastery of Holy Sion, near Myra in Lycia, a mountainous coastal region in southern Asia Minor. At least seventy-five silver vessels and church furnishings were discovered at Korydala in Lycia, modern Kumluca, around forty kilometres from the monastery. A connection with the monastery church, however, is demonstrated in eight inscriptions that refer to Holy Sion. Thirty silver objects have the fivefold silver stamps in a manner mentioned in *The Life of Theodore of Sykeon*, a practice that was introduced under Anastasius in *c.*500 and continued until 670. The stamps prove that the items were all made between 550 and 565. All the stamped silver, and three unstamped objects, a total of thirty-three pieces, were given by one man, Bishop Eutychianus. The largest and most expensive of these are large patens, dishes for delivering the Eucharistic bread. One of these stamped items, now in the Antalya Museum, has gilded repoussé medallions with acanthus leaves around the rim and a large gilded chi-rho at its centre. Two further patens bearing Eutychianus's name but without silver stamps imitate this dish, although one is larger (now in the George Ortiz Collection) and one smaller (now at Dumbarton Oaks). The relationship between the three patens is intriguing, perhaps intractable, but at least one may have been a local reproduction of the stamped piece, which was surely made in Constantinople, and all three may have been finished locally. Two further stamped patens are a pair, clearly manufactured together in the same metropolitan workshop, with rims formed from alternating gilded and silver lobes and with large gilded crosses at their centre. These both bear weight marks, but not full hallmarks. The indicated weight is still accurate to within two ounces, which might well have been lost during the original working of the silver. Both are inscribed, one to the memory of Rufinus and another to the memory of John and his daughter Procle.[21]

At Kaper Koraon, an unremarkable village in northern Syria, one of the seven hundred such villages identified on the limestone massif, up to fifty-six separate silver liturgical objects once formed the treasury of its church dedicated to St Sergius. They bear the names of at least fifty named individuals. The objects – chalices, patens, ewers, fans, spoons and flasks – are now spread through several museums, including the British Museum, which purchased a chalice in 1914 at the recommendation of T. E. Lawrence, 'Lawrence of Arabia', to whom it had been offered for sale in his hotel room in Antioch a year earlier. The inscription around the chalice rim reads: 'In fulfillment of a vow of Sergius and John', meaning that the two dedicatees had prayed for a favourable outcome, and upon achieving good fortune had fulfilled their vow by giving a silver chalice to the local church. Several more silver items are now in the Walters Art Museum in Baltimore, almost all of them bearing the names of local donors. Around the rim of a silver chalice an inscription bears witness to the fulfillment of a vow by Symeonius, a *magistrianos* or middle-ranking bureaucrat who may have acted as a postal or port inspector (Plate 2). St Sergius is named on a number of vessels and his image appears on others. On a ewer in the Walters, the inscription reads 'Ewer of Saint Sergius, prayer of Daniel, Sergius, Symeonius, Bacchus, and prayer of Thomas [from the] village of Kaper Koraon'. On a flask, St Sergius is shown as one of four figures in repoussé decoration with an inscription that reads 'In fulfillment of a vow and for the salvation of Megale and her children and nephews, and for the repose of the souls of Heliodorus and Acacius'. It is supposed that Megale is Heliodorus's widow, and Acacius his brother.

These fifty-six silver objects, and perhaps more not recovered or no longer known, were made for and donated by members of just five families between 540 and 640. Seventeen of the items, for example the chalice of Symeonius, bear an imperial fivefold silver stamp. It is likely that those objects were made in a state-run workshop, perhaps to be finished and inscribed by a local silversmith. Most of the items, however, bear no stamps. The richest donor appears to have been Megas, whose biography is uncertain but is rich with possibilities: perhaps he was Megale's brother, a local boy who became a very important person at court in Constantinople and a high official at Antioch. Megas's total donation to the Church of

St Sergius, in terms of the bullion value of the silver, was 75 solidi, a substantial gift to such a modest foundation. The silver vessels are generally lightweight and metallurgical analysis has shown very high percentages of copper used in their manufacture, evidence that the five families were not extraordinarily wealthy, but well-off and devout. Theirs is the empire, the culture and civilisation of new Romans, with which we are concerned. Their world view was forged not by developments in Rome and its western provinces, of which Britain was the furthest and darkest, but in the east, at Antioch and Alexandria, and only lately Constantinople.[22]

Monks and eunuchs

When Theodore of Sykeon had successfully severed ties with his family he founded a monastery. By the beginning of the seventh century, monks were a familiar presence in communities across the eastern empire, although their movement had begun quite differently, as a means to live apart from society. Monasticism derives from *monazein*, 'to live alone', and early monks emulated Antony, whose celebrity had been established during his lifetime, before the end of the third century, but was spread far further posthumously by the account of his life attributed to Athanasius of Alexandria. Written around 360, the life emphasised Antony's struggles with demons and his performance of miracles, establishing a new genre of literature: hagiography. Many other stories about Antony by other writers circulated, and these placed emphasis on his role as mentor or *abba* ('father') to younger monks. Antony had withdrawn from society and family to the Egyptian desert where he lived alone until his reputation attracted followers. Quickly, a means of ordering communal (cenobitic) monastic life was written, by Pachomius, a contemporary of Antony. Cenobitic monasticism stressed obedience to a superior, regular attendance at religious services, and manual labour, as well as poverty and celibacy. Asceticism, on the other hand, was based on the competitive spiritual athleticism of individuals. It was sustained by holy men, who typically evinced abhorrence for all matters carnal, fearing the bodies of women and mortifying their own flesh. But moderation was generally more appealing to those who wished to devote themselves to prayer, and by the efforts of Basil of Caesarea, cenobitic monasticism became the dominant form by the end of the fourth century. Basil composed a set of rules for nuns and monks that emphasised moderation and self-sufficiency, and he supported the establishment of urban monasteries. Quite early on, therefore, the desert came to the city, or at least to the suburbs, with both holy men and communities of monks established just outside the walls of Constantinople by the end of the fourth century, and at Antioch, Alexandria and Jerusalem rather earlier.

For many in the fourth century conversion to Christianity coincided with the adoption of a monastic lifestyle. For some this came at the end of life, but for others devotion was pledged in youth, so allowing them to avoid military service, paying taxes or fulfilling curial duties. A law of 373, issued by the emperor Valens, observes that 'certain devotees of idleness, having deserted the compulsory services of the municipalities, have betaken themselves to solitary and secret places, and under the pretext of religion have joined with bands of monks'. They were to be apprehended and recalled to service. Valens also issued laws mandating military service for monks and conscripted a number of his opponents, even sending some to the mines to counter manpower shortages. Among those conscripted to fight was a young monk named Isaac, who was beaten with a cudgel, a common punishment for military deserters. After Valens's death at the disastrous battle of Adrianople, which Isaac is reported to have foretold, Isaac became a commanding figure in Constantinople, a frequent visitor to court, who founded his own monastery on an estate outside the Constantinian walls.

Isaac found an opponent in John Chrysostom, and just as surely John found one in Isaac, who represented the type of monastic he railed against: worldly, politically engaged and very well connected. The clash appears to have caused some embarrassment to Isaac's hagiographer, who placed his death in 383, although it seems certain that he lived until at least 406 and witnessed Chrysostom's fall from power and exile. We shall return to these matters in a later chapter. Isaac's designated successor as abbot was Dalmatus, who gave his name to the monastery, now enclosed within the city's Theodosian walls. Dalmatus was also a former soldier, who had served as a member of the imperial guard under Theodosius I. Chrysostom had managed to bring another powerful monastic, the deaconess Olympias, over to his cause. Olympias had founded a nunnery right in the heart of the city, adjacent to the great church, to which she dedicated fifty of her own servants (and, one presumes, manumitted slaves). It swiftly grew to 250 women. Olympias's nunnery endured into the seventh century, when a narration by Sergia, then the abbess, reports the recovery and translation of the relics of the founder and saint to her foundation. The Dalmatus Monastery would welcome the usurper and emperor Leontius following his removal by Apsimar at the end of the seventh century.

The best evidence for early monasticism in and around Constantinople comes from the *Life of Hypatius*, founder and abbot of the so-called Rufinian (*Rouphinianai*) Monastery at Chalcedon, over which he presided for the first half of the fifth century. The foundation was in part an occupation of an abandoned palace, built by the praetorian prefect Rufinus between 392 and 395 as a suburban retreat. It was there that the Council of the Oak, which condemned John Chrysostom to exile, was held in 403. Hypatius advocated an engaged communal form of monasticism, encouraging gardening and basket-weaving, as well as obedience, in his followers. He condemned extreme forms of asceticism in others, although engaged personally in some fierce struggles with demons. Hypatius's monastery thrived, even welcoming Theodosius II as a visitor. Less successful initially were the 'Sleepless Ones' (*akoimetoi*), whose devotion to singing psalms throughout the night was the foundation of a reputation for rigorous asceticism. This group appealed to the urban poor, and the number of the sleepless grew quickly. However, their leader, Alexander, not only failed to find a powerful patron, but was openly critical of the wealthy, and alienated other abbots by allowing their monks to join his community. Their monastery was dissolved and the community survived only at the intervention of Hypatius, who secured the sleepless ones a parcel of land some distance from the city. Soon, however, they would return to the city to take possession of the Monastery of the Prodromos *tou stoudiou*, the Studius Monastery.[23]

Where the elderly could not be cared for by their relatives or married children, monastic life was an alternative. Widows and widowers took monastic vows, but couples who no longer cared for children might also choose to join communities and this was one circumstance in which divorce was always legally permitted. The wealthiest might found and endow family monasteries. Monasticism may also have been a lifelong vocation chosen by a child's parents. It is a trope of hagiography that wilful, holy children, like Theodore of Sykeon, knew their own minds, choosing the life of the ascetic against parental wishes. But it is also the case that parents frequently committed children to monasteries to afford them an education and better prospects, as well as regular food and shelter.

The rise of monastic education mirrors the demise of traditional schools, established in the empire's cities, which we shall trace in later chapters. This had a profound impact upon how and where learning happened, and indeed upon what it meant to be educated. Another hagiographical trope tells of a child begging to be allowed to embrace the monastic life before it was legally permitted. From the end of the seventh century, children were permitted to enter monasteries only on reaching the age of ten. This was raised from seven to protect monks from temptation and children from predation. One must imagine that this continued, however. Pederasty was considered morally repugnant and punished severely under the law, as was homosexuality, with no clear distinction drawn between two consenting male adults engaging in sexual relations and a man sexually abusing a male child.[24]

Genital mutilation was considered an appropriate punishment for sexual deviance or criminality. Towards the end of our period, many other forms of mutilation became common, as a means of punishing criminals, and in order to distinguish and exclude rivals and political opponents from high office. The removal of hands, tongue cutting and nose slitting were all deployed to exclude men from imperial power, and were considered acts of kindness, alternatives to killing. Whereas the loss of a writing hand or a tongue could end a career, castration afforded the opportunity to rise to high office, for a division of ranks and titles at the imperial court were reserved entirely for eunuchs. The majority of eunuchs in the eastern empire were boys who had been castrated prior to puberty. It was only these individuals who would have the classic physical markers of the eunuch: a high-pitched voice and beardlessness, but also increased proportions, as limbs were elongated since the ends of long bones did not close. Those castrated after puberty could grow beards and did not have high-pitched voices, but did lack genitals. They might thereby be considered neither men nor eunuchs. It was the assignment of eunuchs to the position of chamberlain (*cubicularius*) in the later Roman empire that had cemented their position in the imperial system. For this reason, eunuchs are often thought of in the specific role of body servant to rulers (and other members of the elite) but they could fulfil a range of functions at court and beyond. Eunuchs were entrusted with special missions, operating both within the confines of the court and beyond the empire's frontiers. In fact, eunuchs could fill almost all the other posts available to the bearded, with some notable exceptions. Eunuchs could not become *eparchos* (urban prefect, effectively the mayor of Constantinople), *quaestor* (judge and legislator), or various types of *domestikos* (senior military officer). Still, eunuchs could serve as generals. The most famous eunuch general is Narses, who was less of a threat to Justinian than Belisarius because an emperor had to be physically undamaged: a whole man.

Opinions on eunuchs varied. On the one hand they were compared to angels, since these were incorporeal, sexless beings. Just as Michael, an archangel, might lead the Lord's host, a eunuch like Narses might lead the emperor's armies, or, like Archangel Gabriel, a eunuch might serve as an emperor's chief of staff and go-between, ferrying those who are required into the imperial presence. On the other hand, eunuchs faced prejudice, and were commonly equated with women or branded as homosexual. Monastic texts may warn of the danger of the presence in monasteries of eunuchs who have faces like women, and who should be segregated from the other monks to prevent temptation.

The number and distribution of eunuchs across the empire was always relatively small, but their continued presence through our period entailed specific legislation and practices. Eunuchs were not allowed to marry and laws on the matter assert that this was because the purpose of marriage was to have children. Eunuchs could not father children, but they did use fictive kinship such as godparenthood and fostering to form social bonds. Or they might become brothers with other eunuchs or non-eunuchs through a 'brother-making' ritual (*adelphopoiesis*). Such bonds were also available to monks and to high clergy. Excluding eunuchs from marriage did not make them less like men but rather highlighted their distinction from husbands. It was possible for men to remain unmarried, and therefore not to produce legitimate children. Unmarried men might produce children and married men also produced children outside their marriages. Illegitimate children, bastards (*nothoi*), could not inherit from their fathers, but could do so from their mothers in the same manner as other children.

If monks and eunuchs stood apart from the family, that foundational unit of late Roman society, they created new kinship groups and communities governed by ritual and under the law. All Roman Christians were expected to contribute to their common wealth. In 517, Patriarch Severus of Antioch had demanded of his congregation that even the poorest should donate a pound of silver, worth about four solidi, for the revetments that adorned his cathedral. Likewise, precious hoards of silver

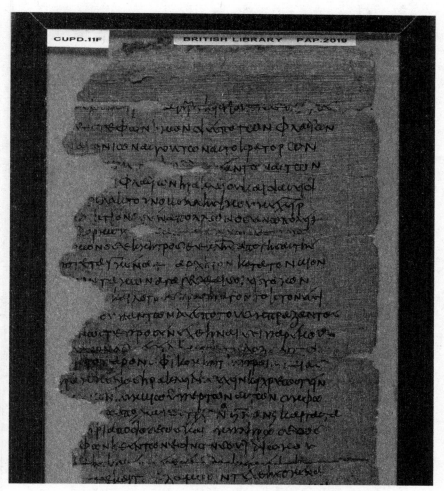

*Figure 3. Papyrus from Apollonopolis Magna, Egypt, dated to the
last year of Heraclius's reign (AD 641); a marriage contract*

chalices and ewers preserve evidence for the faith and hopes of families
of modest wealth in otherwise unremarkable villages. If donating silver
was well beyond the means of the poor, still the Egyptian papyri, better
than any other extant source, reveal how monetised and liquid was the
agricultural market and how widely available was credit, with people of all
levels of wealth lending to and borrowing from each other as well as from
commercial lenders. They inform us that Anoup, a smallholder and agricultural worker, was able to borrow such a large sum as fifteen solidi, and
that Phoibammon was able to acquire a portfolio of land and property in

Aphrodite, becoming a man of influence. They tell us that Aurelia Herais gave up her daughter for adoption after her husband's death, but that John would not give his daughter Euphemia in marriage to another Phoibammon because of unmentionable crimes. Access to legal advice in family matters was available across the empire to those of modest means as well as to the wealthy and remained so to the end of our period, even as threats rose and cities fell. In the thirty-first year of the reign of Heraclius, early in the year 641, as an invading Arab army prepared to besiege Alexandria, far to the south in the Nile city of Apollonopolis Magna (modern Edfu), Philemon and Thecla gave their daughter in marriage to Heraclides, a local man. The marriage contract has been preserved, another precious papyrus document from a small family archive created at the end of antiquity (Figure 3).[25]

3

AN EMPIRE OF CITIES

T he Roman empire was a network of greater and lesser cities that churned with life, human and animal. The porticoed streets and baths were thronged with citizens, most of them with chronic ailments and bad breath. Dogs wandered freely, scavenging in packs after dark. Youths spilled out of classes, libraries and gymnasia, while their wealthy fathers milled around in public buildings and the baths. Lawyers and doctors, orators and administrators, professionals and philosophers conducted business or conversed with men of leisure, too rich to need a profession, but often burdened with duties. There were priests and magistrates who tended the imperial cults and implemented imperial laws, local worthies who governed for Rome and shared its wealth. The streets were filled also with men for hire, as well as slaves, tinkers and beggars, money-changers and traders. There were artisans who threw pots and blew glass, manufacturers of metal goods, master builders and manual labourers, whose arthritic bodies were interred prematurely in suburban and, latterly, urban graveyards. The larger cities enjoyed the presence of deviant thinkers and their dangerous thoughts, theurgists and astrologers, preachers and heretics. Above all else, there were children and infants, boys and girls, at home with siblings or playing in the streets, making noise and mischief, falling down and falling ill, dying young.

The empire's cities were nodes in a vast communications nexus, linked by roads and sea routes to each other and to regional or provincial capitals, and to the great cities that encircled the Mediterranean. Abundant goods produced at distant locations were shipped in bulk at low cost across its waters, then hauled at greater cost along its roads, so that the empire's cities shared a common material culture and diet. Each city's kitchens were stocked with local coarse wares and similar fine ceramics, glazed plates and bowls from Asia Minor or Cyprus. Houses and shared rooms were lit by terracotta lamps from North Africa and windows were glazed with Levantine glass, while bakeries used flour ground from cereals harvested thousands of miles away in Egypt or Sicily.

Defining the city

What did subjects of later Roman emperors imagine a city to be, and how was it distinct from other settlements and communities? Across the eastern empire, where Greek was the spoken language or lingua franca, the word *polis* was used to identify a city. It was applied to communities that were in origin autonomous and responsible to the state only in respect of the supervision and collection of taxation revenues from the defined territory within which its jurisdiction operated. The legal status of the city was recognised by emperors, and could therefore be rescinded or extended. Legally recognised cities, like foreign powers, had the right to send ambassadors to the emperor. Each city had its own history, identity and citizenship. But *'polis'* also embraced a whole complex of cultural meanings related to tradition and memory, cult and ritual, education and entertainment, defence and the arrangement and adornment of physical space. Pausanias, a Greek geographer of the second century AD, recognised that Panopeus in Phocis, Greece, legally was a city, but drew attention disparagingly to the fact that it did not look like one (Pausanias 10.4.1). It had 'no building for magistrates, no gymnasium, no theatre, no market-place (*agora*), no water collected in a fountain' and its people lived in hovels best described as 'mountain huts'.[1]

Those who lived centuries after Pausanias also knew what a city should look like. It should have a grand public building for magistrates, a gymnasium and bath complex, and a public fountain (*nymphaeum*) serviced by a well-maintained water supply, which might require aqueducts above and below ground. Its citizens should not live in hovels, and the richest should build urban palaces and villas, their floors covered in mosaics and marble, their walls in murals, their windows glazed, and with their own private water supply. A colonnaded street, or stoa, was essential. It was the perfect place to stop and chat, its roof keeping citizens shaded from the sun and dry when it rained. A colonnade connected buildings and citizens. As Libanius observed towards the end of the fourth century on his native city of Antioch, 'truly it is good to speak, and to hear is better and converse best'. He pitied those who did not benefit from a colonnade, as 'although they can be said to live in a city, they are actually separated from each other no less than those who live in different cities, and they learn news of those who dwell near them as they would of those who are living abroad'.[2]

The number of its inhabitants was not definitive in distinguishing between a *polis* and another type of settlement, although most cities had populations in the thousands, some in the tens of thousands, and very few in the hundreds of thousands. The whole empire at the start of our period had, perhaps, five cities with populations above 100,000, another fifteen above 30,000, and thirty more of above 10,000, for a total of around fifty large cities. Thirty-five of these cities were in the empire's eastern provinces including Egypt, and another five in North Africa, most significantly Carthage, with a population of up to 50,000 in the fifth century. Only around ten large cities were situated in the western provinces, most in Italy and Spain. Rome was by far the largest, but it was much diminished, with a population of around 100,000 in the later fifth century, perhaps one tenth of what it was at its height. According to some estimates, by the seventh century that number would shrink again to around 25,000. City populations changed over time, often dramatically and for reasons we shall explore, including combinations of fire and earthquake, epidemic disease and invasion, famine and drought, the silting of harbours, rivers and canals, and much else besides.[3]

The density of urban settlement in the east was far greater than in the west, but even there, sitting adjacent to regions that were densely urbanised, we find areas where cities failed to thrive. Isauria was a highland region in south-eastern Asia Minor, adjacent to Cilicia and opposite Cyprus. Encompassing principally the Taurus mountain range, which divided Anatolia from Syria, Isauria was only gradually and never fully incorporated into the Roman state. It retained its own culture and language, Luwian, a dialect of Hittite. For centuries it had supported transhumant pastoralism and banditry, and had been a regular source of manpower for the Roman army. The measure of an Isaurian warlord's power was not his status in a city but the impregnability of his fortress and the men and weapons he could gather within it. The neighbouring mountainous region, 'Rough' Cilicia, was also a source of bandits and soldiers, but was known too for the quality of its wine and olive oil, produced in rock-cut presses along its rocky littoral and shipped in amphorae manufactured in a few flourishing coastal cities, such as Elaiussa Sebaste, Corycus and Seleucia. Sitting near the ends of river valleys, these cities also serviced the surplus pastoral products of the interior and acted as entrepôts and way stations in active commerce east and west along the coast, and southwards across the straits to Cyprus. The Taurus Mountains, as we have seen, were also a source of metals, notably iron, but also tin, copper, silver and lead.[4]

Ephesus, one of the great cities of antiquity, looked the part, with its grand colonnaded street, the *embolos* (Plate 3). Situated on the Aegean coast it also boasted a substantial harbour, a huge stadium and theatre, libraries and schools, several substantial public baths, and two marketplaces between which the *embolos* ran. A large basilica church, the Cathedral of the Virgin, hosted two great Church councils in 431 and 449. The late antique city of Ephesus occupied a vast area that preserved and added greatly to its classical plan. Its walls enclosed about 415 hectares, which was a very large area, around half of which was inhabited space. It was considerably smaller than the area contained by Rome's Aurelian walls (1,373 hectares), but larger than the enclosed areas of Thessalonica (385 hectares), Carthage (321 hectares), and Milan (250 hectares). We cannot know how many people lived within the Ephesian walls, but estimates have ranged from as few as 30,000 to more than 200,000.

The 'Hanghäuser', or terrace houses, at Ephesus were luxurious mansions on a steep terraced bank overlooking the city's main colonnaded street. They are interesting in having several private toilets that remained in use for centuries, including a remarkable five separate toilets in 'Hanghaus 2'. Even a single private toilet, rather than use of a commode or chamber pot, indicated significant wealth since it required access to a substantial dedicated water supply. The size of a grant of water to a private individual was in accordance with his status, the size of his property, and the number of baths and latrines. The grant was calculated and recorded as the specific diameter of a water conduit to be connected to the private residence or baths. Water was distributed principally using terracotta pipes, and these would have been connected to lead pipes with a bronze stopcock, to allow the flow of water to be stopped, to change its direction, or to create small

Figure 4. Public latrines near the lower agora at Ephesus

spouts at a fountain. Lead pipes would be stamped to indicate by whose authority and to whom the water was granted. This relied on the employment by each city of a dedicated staff of water inspectors to oversee and maintain the whole water delivery system and to identify, correct and punish any abuses.[5]

A total of eighteen private and public toilets have been investigated at Ephesus. Public toilet facilities, latrines, were generally constructed in association with public baths, but they also appear at other locations where the municipality could control and guarantee the flow of water. At Ephesus, five latrines were part of public baths, another was adjacent to a bath house, and two more were free-standing, including a fifth- to sixth-century construction at the lower agora that could accommodate twenty people at once (Figure 4). The Vedius Baths had two separate latrines that could each seat sixty patrons, the Harbour Baths had space for forty, and the Varius Baths had thirty-seven seats. Latrines were public places where wall paintings, decoration, inscriptions and graffiti attest to a vibrant toilet culture. Jokes and poems feature and there are also signs indicating that certain seats were reserved for members of named professional guilds operating nearby. Privacy was not expected and there were no partitions, but the standard toilet seat was 57 cm to 60 cm (22 to 24 inches) wide, edge to edge, allowing each person sufficient space. The Roman preference for sitting over squatting allowed those who wished to take their time or to strike up a conversation.[6]

Antioch

Few cities surpassed Ephesus, but two cities of the east were greater and had long rivalled Rome: Antioch and Alexandria. From its foundation in *c.*300 BC by Seleucus I Nicator, formerly a general of Alexander the Great, Antioch was represented by a female personification, a *tyche* or Fortune. Seleucus had commissioned Eutychides of Sicyon, a pupil of Lysippus, to cast her in bronze, wearing a mural or walled crown, holding in her right hand fruits of the earth, fully covered in a flowing robe, and seated on a rock with her legs crossed above a second figure, a young man with flowing hair swimming, a personification of the River Orontes (Figure 5). This type spread rapidly through the Hellenistic world until every city had its own Fortune, each slightly different, although this is not always evident from the generic *tychai* that featured on countless coins. Antioch on the Orontes once had a population approaching 800,000, but today it is a small town in the far south-east of Turkey, which until the outbreak of the Second World War was in Syria. It has the richest of early Christian histories. The term Christians was first used of followers of Christ in Antioch (Acts 11:26), and the city was an important early base for Christ's apostles. Both Peter and Paul, the latter from nearby Tarsus, were resident in Antioch, and it was there that they clashed over Gentile converts. The Gospel of Matthew was written in Antioch in *c.* AD 80–90, and its focus on ecclesiastical organisation reflects the development of the early church in the city. Indeed, Matthew, writing in and of Antioch, was the first to use the Greek term *ecclesia*, meaning 'assembly' or 'gathering' for the Christian congregations, which shortly came to mean 'church'.[7]

A necessary emphasis on Christianity must not blind us to the fact that Antioch was a city whose inhabitants professed a profusion of beliefs, and who lived in a world rich in mythology and imagery, where Christians and pagans were trained in the same intellectual traditions by the same

Figure 5. Tyche of Antioch, sitting above Orontes, a young man with flowing hair. Bronze copy of a statue by Eutychides

scholars. By the later fourth century AD, Antioch boasted a hippodrome, a palace, and all the other features of a Roman capital, including at least eight imperial baths, and many other public and private establishments making full use of its abundant water supply. While the Orontes would have supplied water for bathing, drinking water was piped in from the suburb called Daphne, whose famous springs included those called Castalia and Pallas. Issuing from limestone rocks a hundred feet above the city, the sweet water flowed at around 1,500 litres (400 gallons) per second and could be delivered by gravity to every part of the city by aqueducts and pipes built by Caligula, Trajan and Hadrian. Libanius, the great orator and a native of Antioch, praised its climate, its fertile land and the mountains, its city streets, the old city and its new island city, and its suburban gardens and villas, but he knew that water defined his city, 'Indeed the thing by which we are especially supreme is that our city has water flowing all through it.'

We surpass the beautiful waters of other cities by the abundance of ours, and the abundant waters of other cities by the beauty of ours ... Each of the public baths pours forth a stream as large as a river; some of the private baths have as great a stream as these, and the others are not far behind them. Whoever has the means to erect a bath on the site of earlier ones does so the more confidently because of these streams ... Wherefore all the tribes of the city pride themselves on the particular adornments of their baths more than on their very names.[8]

Adornments included statuary and abundant mosaics, which attest to the vibrant mixed culture of the city and its inhabitants. Six baths have been excavated, five in a relatively small area on the Orontes Island, where they sit beside the imperial palace, the circus and the stadium. From one of these baths comes a fourth-century mosaic, which shows Dionysus, a rival to Christ for the affections of locals, being carried off as an infant by Hermes (Plate 4). Hermes is cloaked and wears a winged headband but is otherwise nude; Dionysus is also nude, but his divinity is signalled by a radiant nimbus, or halo, around his head. Hermes is looking back, anxious to deliver his package to the water nymphs of Nysa as instructed by the baby's father, Zeus. 'Thus Hermes carried upon his arm the little brother who had passed through one birth without a bath ... He gave him in charge of the daughters of Lamos, river nymphs – the son of Zeus the vineplanter.' These lines are just two of more than 20,000 in the forty-eight books of the *Dionysiaca* by Nonnus of Panopolis, the longest extant poem of antiquity, a compendium of Dionysiac myths in dactylic hexameter that is perhaps a century later than the mosaic here described. Nonnus, a name popular among Christians, means 'pure' or 'holy'. He was certainly a Christian when he wrote his 'semi-pornographic classicizing mythological epic'.[9]

John Malalas, writing in the sixth century, attributed much of the construction on the Orontes Island to Diocletian at the end of the third century, including a large palace and 'a public bath, which he called the Diocletianum, in the level part of the city near the old hippodrome'. These two structures were part of what Libanius called the 'New City', the result of a building boom that Diocletian financed at Antioch, including a new Olympic stadium in the suburb of Daphne, within which were temples to Olympian Zeus and Nemesis. A suburban palace was placed nearby. The Temple of Apollo was restored, and a new subterranean temple to Hekate excavated, with 365 steps. Malalas concludes his list with three arms factories 'for the manufacture of weapons for the army', a mint 'so that coins might be struck there ... a bath, which he called the Senatorial, and three more baths'.[10]

Thanks to Diocletian, therefore, Antioch might well have remained superior to Constantinople as the dominant imperial city of the east, and therefore the newest of the New Romes. Libanius implied as much when he wrote of the new palace at Antioch that 'if this palace stood by itself in some insignificant city, such as are numerous in Thrace, where a few huts form the cities, it would give the one that possessed it good reason to claim a proud position in the catalogue of cities'. He was well aware that in his lifetime one of those insignificant Thracian cities had been adorned with such a palace and that a succession of emperors had made it their home in preference to Antioch. Therefore the city on the Orontes became instead the spiritual capital of the Monophysite East, the seat of 'heretical' bishops throughout the Christological controversies of our period. Sitting between Constantinople and Persian Ctesiphon, it was the headquarters of commanders charged with the defence of the East. It was also the home of the most famous holy man of late antiquity, Symeon the Stylite, who for forty years until his death in AD 459 lived atop a pillar in the desert, fasting and praying, but also receiving pilgrims, arbitrating disputes, and performing miracles.[11]

Alexandria

In the later fourth century AD, the Fortunes (*tychai*) of Constantinople, Antioch and Rome were fashioned in gilded silver as furniture fittings, joined by a fourth, the Fortune of Alexandria. Rome and Constantinople both wear plumed helmets, perhaps indicative of political power and victories in war (Plate 5). Their arms are exposed, emerging ungilded from their tunics, and there is nothing beneath their feet. The *tyche* of Alexandria most closely resembles that of Antioch. She wears a mural crown and holds evidence of the earth's bounty, sheaths of wheat that each year were shipped across the Mediterranean from her port. Her foot is planted firmly on the prow of a ship.[12]

Alexandria was founded for Alexander the Great and swiftly became a megalopolis with the most culturally rich and diverse population in the

Mediterranean world. The city had a population estimated at between 300,000 and 500,000, while perhaps another five million lived in the Nile Valley. A port city, it welcomed large numbers of migrant workers and had an ethnically diverse immigrant population. Every Epiphany, the Christian shipowners of Alexandria took water from the River Jordan, blessed by a priest, to sprinkle on the prows of their ships to protect them and their sailors for the coming year. The church at Alexandria claims foundation by St Mark, although this is first recorded by Eusebius of Caesarea at a time when Alexandria's bishops were seeking to legitimate an accumulation of authority in their own hands in the face of resistance by diverse groups. Around 320, a dispute arose between the bishop of Alexandria, Alexander, and one of his priests, Arius, who was a popular preacher. Arius earned the censure of his bishop for challenging his authority and therefore sought and received support from another Eusebius, who had recently become bishop of Nicomedia in Bithynia. This is the origin of Arianism, which began as a dispute over authority. However, the reason the issue resonated beyond Alexandria was that it challenged the preferred Christology of other powerful figures. Arius insisted that God the Father preceded the Son and had created the Son from matter. It was a position that many found compelling, but others found horrifying.[13]

Numerous councils were convened and sides were taken before the emperor Constantine, in 324, despatched his personal pastor, Ossius of Cordoba, to mediate between the parties. Ossius bore a letter that set out the emperor's thoughts in all their simplicity. The letter has been preserved, revealing that the emperor considered the matter 'extremely trivial and quite unworthy of such controversy'. Those on both sides of the argument felt otherwise, since for them the nature of Christ affected the salvation of all for whom he died. To those opposed to him, Arius was a heretic who claimed that the Son was less than fully God, denied the divinity of his sacrifice, and hence barred the possibility of *theosis*, the deification of mankind. To Arius, those who maintained that the Father and the Son were equally eternal (coeternal) and of the same substance (consubstantial) were misleading their congregations into an earlier heresy, that of Sabellianism, which treated the Godhead as unitary and the Son and Holy Spirit as aspects of divinity. Salvation could only be guaranteed by correct belief, in Greek *orthodoxia*. Orthodoxy and the issue of salvation were not, to either party, trivial matters.[14]

Eventually, a council was convened in Nicaea in 325, for which

Constantine paid. A solution was advanced, upon which the emperor insisted agreement be reached, based on the term 'consubstantial' (*homoousios*). By using the very term that Arius had rejected as the root of the heresy of Sabellianism, the fate of the preacher was sealed. He was condemned and exiled with two bishops who refused with him to agree to the statement of faith, which came to be known as the Nicene Creed. The only surviving eyewitness account of the Council of Nicaea, which was later called the first ecumenical council of the church, was written around a decade afterwards by Eusebius of Caesarea, who was at that time composing a biography of Constantine. Eusebius had good reasons to remain opaque on what took place at Nicaea, for his own role in the proceedings was hardly less controversial that that of his namesake, the bishop of Nicomedia. Constantine's biographer had been named in letters that circulated prior to Nicaea as one sympathetic to the teachings of Arius, and had been provisionally excommunicated. In fact, rather than resolving the issue of Christ's nature by imperial decree, as Constantine had hoped and insisted, Nicaea opened fissures between Christian communities in the empire's great cities, which grew deeper and wider throughout our period.

The greatest challenge the Arians of Alexandria faced was a young and brilliant new bishop named Athanasius. Athanasius was ordered to restore Arius and his followers to full communion with the Alexandrian church. Athanasius refused, instead fleeing from Alexandria into the desert to live among the monks and thus avoid the officers sent to enforce the emperor's will. Athanasius is alleged to have organised raids from there to sack Arian churches, upending altars and destroying liturgical vessels. He continued to resist after Arius's death. Athanasius drew the bishops of Rome into the dispute, initially on his side, until the emperor faced them down. Athanasius returned to Alexandria from his monastic refuge on the death of Constantius II, only to be exiled again by Julian, recalled by Jovian, and finally both banished then swiftly recalled to Alexandria by Valens. There he died in 373. Within a decade, Arianism had been driven from Constantinople and the Nicene Creed affirmed at the Council of Constantinople, the second ecumenical council of 381. By then, however, it was too late: Arianism had become the dominant faith among the empire's 'barbarian' neighbours, whose armies would establish successor kingdoms in the west.

Theodosius I legislated against heretics and pagans. According to Sozomen, Theodosius 'forbade the heterodox to meet in churches and teach about the faith and appoint bishops', even as he legislated, for the first time, to ban pagans from entering temples. The emperor found a zealous supporter in Theophilus, bishop of Alexandria, who provoked a stand-off with a prominent group of the city's philosophers, those devoted to the Iamblichan tradition of Neoplatonism for whom the temples and rituals of their religion remained essential. A riot ensued, led by the philosopher Olympius and the grammarians Helladius and Ammonius, all distinguished teachers. When the rioters took refuge in Alexandria's oldest temple, the Serapeum, Theophilus secured an order from Theodosius, which led to their capitulation. Helladius and Ammonius moved to Constantinople, Olympius was never heard of again, and the Serapeum was razed to the ground. This did not, however, signal the end of philosophical education in Alexandria, and others, for example followers of Plotinus, pursued a purely contemplative path to union with the divine that required no temple rituals. Among them was Hypatia.[15]

Hypatia of Alexandria (c.370–415) is a figure compelling to history who has a clutch of modern biographers. We do not know what she looked like, but typically the beauty of her mind inspired others to describe her as possessing great physical beauty. In a man, to be attractive was advantageous and revealed a blessed soul, whereas for a women who encountered young men, taught them and inspired them, beauty was a danger and a temptation. It provoked those who engaged in serious conversation with her to 'wish to do evil and shameful things with her', as Evagrius Ponticus (d. 399) wrote at the time. Hypatia was a gifted mathematician, the daughter and student of Theon, with whom she worked on an edition of and commentary on Ptolemy's monumental mathematical and astronomical treatise, the *Almagest*. To her devoted Christian pupil and correspondent Synesius of Cyrene, Hypatia's paganism was of no significance, whereas her mastery of Neoplatonic philosophy was of profound importance. In his letters she was simply 'the Philosopher', including in a letter of 413, when he informed her about the death of his son, and asked for her help.[16]

I account you as the only good thing that remains inviolate, along with virtue. You always have power, and long may you have it and make good use of that power. I recommend to your care Nicaeus and Philolaus, two excellent young men united by the bond of relationship. In order that they may come again into possession of their own property, try to get support for them from all your friends, whether private individuals or magistrates.[17]

Synesius knew that Hypatia was well connected, as would be any teacher of the young and rich, but she was clearly far more than that. Hypatia was influential in her own right, visited by 'leading men' (*archontes*), imperial officials and other powerful visitors to Alexandria, and she was a confidante of Orestes, the Augustal prefect. Such erudition and influence possessed by a woman threatened many in Alexandria, not least its bishop Cyril, who had recently been elected to the see his uncle Theophilus had held for almost three decades. In March 415, a rumour circulated that Hypatia was using 'magic, astrolabes, and musical instruments to beguile many people', including Orestes. A monk named Peter gathered a mob that seized Hypatia as she returned from giving a public lecture. 'Dragging her from her carriage, they took her to the church called Caesareum, where they completely stripped her, and then murdered her with roof tiles', a weapon often used by street fighters. Outrage at her brutal murder was widespread and the shock of it appears to have calmed the city. There were no further recorded riots before Orestes ended his term as governor, nor during Cyril's long tenure as bishop, which ended in 444. Cyril redeemed his reputation by leading the opposition to Nestorius, bishop of Constantinople, who had sparked a further Christological crisis, the 'Two Christs' controversy, which was addressed at the Council of Ephesus in 431.[18]

The murder of Hypatia, two decades after the destruction of the Serapeum, would appear to signal the triumph of a more militant Christianity over Alexandria's long philosophical tradition. However, the confrontation was as much about rich and poor as it was about pagan versus Christian. The rich embodied culture and education (*paideusis*, to which we shall turn in the following chapter), appreciated art and literature, clung to myths, understood allegory, and craved philosophy; the poor saw demons, employed household magic, and understood the power of witchcraft. The poor of Alexandria vastly outnumbered the rich, but the rich ran the city, province and empire. The privileges of wealth were evident in Alexandria, a city of traders and of fortunes made and lost.

City and empire

The political will of every Roman city, including Antioch and Alexandria, was subordinated to that of a greater power, the empire. The expansion to the east of Roman power, of the *imperium Romanum*, entailed the subjection of those great *poleis* and others such as Athens and Corinth, Ephesus and Pergamum that had been established centuries earlier as self-governing city-states. Roman civilisation, however, advanced not by destruction, but by the appropriation of the culture of those *poleis*, its translation into something meaningfully 'Roman', and its diffusion far and wide. The brilliance of the Roman imperial system was not just its ability to harness overwhelming violence to further its ends, although that was vital to expanding and sustaining its dominance, but also that it tolerated and integrated alien ideas and practices, binding them into a larger but looser web.[19]

Rome ruled by the co-optation of local elites, who were in frequent contact with the centre. These elites represented their own interests, and by extension those of their fellow citizens, by their petitions to the emperor and by implementing his responses. The dispatch of messages and messengers to the emperor, who in our period resided at Constantinople, required an effective and relatively efficient communications system, such as that described in the following chapter. Most commonly, a city's petitions concerned retaining or reinstating privileges, often those relating to taxation or regional status. On particular occasions cities might request imperial liberality, for example in the aftermath of a fire or an earthquake. A very lucky city might also gain the privilege of changing its name to echo that of its patron and his family members, such as Salamis on Cyprus, which became Constantia after earthquake damage was repaired at imperial expense and it was granted a four-year remission of taxes. It could be lengthy and expensive to secure a benefit, requiring lobbying of various members of the administration or having the ear of a powerful individual at the right moment.

Synesius of Cyrene spent three years at the court of Arcadius, from 397 to 400, before he secured the tax breaks for which he was sent to petition. First, he presented golden crowns to the emperor on behalf of his city, an expected gift from a petitioner, given nominally in honour of Arcadius's fifteenth year of rule. Shortly afterwards, he courted the powerful commander Paeonius, presenting him with a silver astrolabe of Synesius's own devising, which had been manufactured for him in Alexandria. This was accompanied by an essay entitled *On the Gift*, which praises its addressee as a rare combination of philosopher and general, wisdom and strength. When Paeonius failed to deliver, Synesius found a still more powerful patron, Aurelian, for whom he wrote two treatises, *On Providence* and *On Kingship*, both flattering Aurelian and critical of the current regime. Aurelian would become praetorian prefect of the east and would grant Synesius his tax concessions at the end of his three-year stay at court. Synesius also supplied letters of recommendation to Aurelian on behalf of others, for example his relative Herodes, a man of means seeking personal tax relief. Synesius himself was relieved of local duties and their financial burdens following his successful embassy. However, he looked back on his time at court with regret. 'My life has been books and hunting, except when I was an ambassador. Would that I had not seen three years wasted from my life!'[20]

Competition for imperial favour within and between cities was serious business, providing not only benefits but also freedom from subordination to neighbouring cities or more distant metropolitans. Those who earned imperial disfavour suffered accordingly. It was known for a city's walls to be destroyed, its city status rescinded, or for a neighbouring city to be given superintendent status over it. This might mean meeting the tax burden of another *polis* as its subordinate. The city of Byzantium, for example, before it became Constantinople, suffered for backing Pescenius Niger against Septimius Severus in a civil war. Byzantium lost its civic status and its walls were razed and remained ruined for half a century. It was subordinated for a period to neighbouring Perinthus. Such punishments do not, however, reveal the strength of the central administration so much as its essential weakness: enforcement was carried out locally or regionally, by rival *poleis*, under the supervision of a provincial governor and the threat of overwhelming violence, a trump card Roman emperors held but less frequently played.

Cities also petitioned to regain status, as did the city of Orcistus between 324 and 326. The petition was successful and was reaffirmed in 331, apparently against the encroachments of the neighbouring city of Nacolea, which had been accustomed to the subordinate Orcistus paying towards the upkeep of its cults. Orcistus's request to be afforded once again the name and privileges of a *polis* (in fact, in the Latin text *civitas*) was granted for a number of reasons that are highly revealing of the perception of cities and their essential function in the fourth and fifth centuries. In his own words, Constantine, the emperor who received it, 'found their petition thoroughly welcome' because Orcistus was 'adorned with magistrates in annual offices and abounded with *curiales* and people'.

Moreover, by its own description it sounded like a city, with 'an abundant supply of water and baths, both public and private, a forum adorned with statues of former emperors' and an inn (*mansio*) that served both private and public needs. This was especially useful, as Orcistus sat at a crossroads. In their petition, the citizens of Orcistus described their city as 'situated on the border between Phrygia and Galatia'.[21]

Cities were essential to the governance of empire at the start of our period, but sustaining the interest of local elites in their cities was ever more challenging, as more and greater opportunities emerged in a rapidly expanding imperial bureaucracy. The qualifications required for local office were considerably lower than in earlier centuries, and emperors legislated to ensure that even those of relatively modest wealth represented the interests of their cities and contributed to the maintenance of urban infrastructure. A law issued in AD 439 required that those who owned land worth 300 solidi become members of their local city council (*boule*), holding the status of decurions (in Greek *bouleutai*, in Latin *curiales* or *decuriones*). How much land 300 solidi would buy varied by quality and location, but if we imagine that it amounted to about 40 hectares (100 or more acres) then it comprised a substantial holding but not a great estate. The importance of the city councils, and the reluctance of wealthier citizens to serve, is evident from the vast amount of legislation devoted to the subject, much of it surveyed by A. H. M. Jones, who concluded that 'From this vast and tangled mass of legislation two points clearly emerge',

that the imperial government considered the maintenance of the city councils essential to the well-being of empire, and that many members of the city councils strongly disliked their position. To the emperors the decurions were, as Majorian put it: 'the sinews of the commonwealth and the hearts of the cities'. In the former capacity they collected and underwrote the imperial levies and taxes, repaired the roads, administered the public post, conscripted recruits for the army, managed the mines ... As the 'hearts of the cities' they maintained those amenities of urban life, in particular the baths and the games, which were in Roman eyes essentials of civilized life.[22]

A portrait bust of one such man, named Eutropius, has survived. A console – a stone base that was inserted into a wall facing the street – has also survived, on which the inscription, beginning with a cross, lauds his paving. 'On account of your sleepless toil, Eutropius, progeny of sacred Ephesus, you have obtained this modest thanks because you adorned your fatherland with twisting marble streets.'

Curial wealth and display

Decurions, therefore, held public office as a function of private wealth, and both wealth and office were hereditary. Decurions were entitled to call upon imperial resources, for example state troops, to aid in the collection of taxes, although the wealthiest will have supported private retinues. As landowners, they served as living links between cities and the land, but they might also have substantial property portfolios in their home city. The rich invested in their home cities because they were required to do so, as decurions, but also because they wished to do so, purchasing estates and villages from which they might benefit upon retirement and acquiring the means to entrench their families at the top of the social scale 'back home'. The Apion family of Oxyrhynchus may, once again, serve as an example, as their affairs are remarkably well documented.

The Apions flourished from the middle of the fifth century into the seventh, when the papyrological record ends and they disappear from history. They played substantive roles in the events we shall explore in later chapters. The first documented Apion, Strategius, was a local decurion who served as curator of imperial estates near Oxyrhynchus, in particular those owned by the wife of Theodosius II, the empress Eudocia. During the reign of Anastasius, 'Apion the Egyptian', a descendant of Strategius and already a patrician, served as quartermaster-general of the army sent against the Persians in 503. After the failure to relieve Amida, he was summoned from Alexandria to Constantinople to explain himself and later was exiled. He was among those leading men recalled from exile at the start of his reign by Justin and entrusted with high command, becoming briefly praetorian prefect of the east. Apion's son, Strategius, was Augustal prefect of Egypt in 523, and is named elsewhere as *magister militum* and patrician before he was appointed by Justinian as count of the sacred treasuries, head of one of the state's major financial bureaux. Among his duties was financing the construction of Hagia Sophia in Constantinople. Strategius's son, Flavius Apion, was raised in Constantinople, where the Apions had built an urban palace, and where he was named consul in 539.

The leaves of his consular diptych, preserved at Oviedo in Spain, have the largest surface area of any surviving diptych, using the natural curve of the elephant's tusk from which they were carved. According to a petition that we have already explored in some detail, Flavius Apion became a patrician and Duke of the Thebaid, a huge region comprising the thirteen southernmost administrative districts (nomes) of Egypt. His status was more than sufficient to confer upon him the rank of senator at Constantinople. Perhaps he moved to Egypt from Constantinople only in order to take up his role as Duke of the Thebaid. He certainly maintained the family's palace in Constantinople, where in 561 a street fight took place that lasted two days and saw fires set along the city's main thoroughfare and at warehouses at the docks.[23]

Although we do not have such rich sources for other families elsewhere in the eastern empire, we can surmise from what survives that the Apions were representative of an elite that benefited from state service, rising to high office and securing for themselves the resources to buy and profit from land. They were international landholders and aristocrats, standing above the gentry that comprised most *curia*. But all curial families invested heavily in their various homes, building mansions and palaces in city centres and villas in suburbs, coastal exurbs, and on their country estates. Such buildings were richly decorated with mosaics, many depicting aspects of city life, including the games that their owners supported and sustained. At Smirat in North Africa, in a country villa surrounded by olive groves, Magerius wished to be reminded always of the acclamations he had received in the amphitheatre for paying the performers at a *venatio*, a wild beast hunt, and buying each of the four leopards they killed. He therefore commissioned a floor mosaic for his reception hall, where the hippodrome scene was depicted: the leopards – Romanus, Victor, Crispinus and Luxurius – and their killers are portrayed and named, and the full acclamation inscribed (Plate 6). Bags of money are shown brought forth on a tray, each marked as containing 1,000 denarii, twice the sum demanded in the inscription, which concludes: 'This is what it is like to have the means, this is to have power, this is my moment.'[24]

The urban home of Synesius of Cyrene has been identified by a mosaic inscription celebrating his father, Hesychius 'the Libyarch'. The country estate of Synesius has not been identified, although we know that he preferred to think and hunt there, and that he left it quite unhappily in order to undertake his embassy to Constantinople, and with even greater regret to take up a bishopric at Ptolemais. At Carthage, a mosaic in an urban mansion portrays villa life, reminding the rich of their country estates. The mosaic of Dominus Julius, which dates to the later fourth century, shows the strong walls and towers of the central building on an estate (Plate 7). An arcade running along the length of the wall looks out over the fields, where the master and mistress are portrayed receiving the fruits of their estate through four seasons. In the uppermost of three registers, estate workers bring the seated lady olives and a brace of ducks from her left (winter), and lambs from her right, where we also see a field of golden wheat (summer); in the lowest register, she appears to the left standing to receive flowers and fish (spring), while her seated husband, to the right, wearing his robes of office, is handed a scroll marked with his name ('Master Julius') even as his workers collect grapes from his vineyard and game from the hunt (autumn). The villa dominates the middle register. It is approached from the left by the master in official dress, mounted on his horse and accompanied by his groom, even as hunters depart to the right, accompanied by their dogs. Within the walls of the villa are four domes, perhaps a bath house or a three-apsed 'triconch' dining room where all the good things of the estate will be consumed.[25]

Villas were hubs of agricultural life, and like agriculture they imposed order on nature. Estate villas emerged as the heart of working farms, blending utility with luxury. Over time, however, luxury and competitive display came to the fore, and one can quite clearly identify gradations of wealth in the number of rooms they possessed and in their decoration. The Villa del Casale near Piazza Armerina, Sicily, belonged to one of the internationally super-rich, a senator of Rome who held one of the highest offices of state. Built in the first part of the fourth century, and occupied through the sixth century, the villa has the greatest complex of floor mosaics of any building of Graeco-Roman antiquity, with forty separate mosaics covering 3,500 m². Even the villa's two large latrines, catering to the estate's extensive staff as well as guests and clients, had floor mosaics. Abundant marble was used throughout the villa, some reused, but much imported from eighteen quarries across the Mediterranean world. The greatest volume of marble was the Bianco di Luni, quarried in northern Italy, and the second greatest was the cheapest and most readily available Proconnesian. These were both white marbles. The third and fourth most abundant marbles were grey Bigio venato and yellow Greco scritto, both quarried near the North African coast to the west of Hippo Regius.

One of the rarest and more expensive marbles used at the villa, the Giallo antico, was mined at Chemtou, between Carthage and Hippo.[26]

Decurions presided over the links between their cities and the emperor by the maintenance of the imperial cult, serving as priests and conducting rituals performed according to a standard calendar. When Christianity became the state religion, powerful families came to dominate local church hierarchies, and emperors found ways to be viewed as quasi-divine, their names intoned and prayers said for their immortal souls in services in every city every week. The increased prominence of the church and its bishops, together with the emergence of monasteries, diverted imperial largesse and local wealth from civic buildings and projects, and, as significantly, away from the theatres and games, to those that benefited particular ecclesiastical and monastic foundations. The proliferation of grand and spectacularly decorated basilica churches distinguished the late antique landscape of the fifth to seventh centuries, just as the construction of oversized theatres, villas and city walls had defined earlier ages.

John Chrysostom, whom we shall meet properly shortly, encouraged the construction of churches on country estates for the spiritual needs of agricultural workers, just as he approved of the maintenance of baths and latrines at villas for their hygiene and health. Surveys have identified sixty churches in Crete that functioned in the fifth to seventh centuries, and eighty on Cyprus. Most contained mosaic floors. Two richly decorated churches in the heart of the Balkans, a far less wealthy region than Cyprus or Crete, are illustrative. Stobi sits some 150 kilometres north of Thessalonica, joined to it by a road that follows the Axius (Vardar) river. A modest city, it had a splendid theatre and public baths, to which was added a palace to accommodate a visit by Theodosius I. Towards the middle of the fourth century an artificial terrace was raised on the hill immediately above and abutting the theatre, on which was constructed a large church, decorated with floor mosaics of geometric design. A century later this was replaced by a still larger church, the episcopal basilica, which reached its fullest extent in the age of Justinian, adorned with more beautiful mosaics and wall frescoes. The mosaic floor of Stobi's baptistery (Figure 6) depicts four cantharus urns as fountains, around which are gathered deer, peacocks and water birds, thirsting for the water of life, illustrating Psalm 41 of the Greek *Septuagint* (also known as Psalm 42), 'As the hart longs for the water fountains, so longs my soul for you, O God.' The hart, the

Figure 6. Mosaic at the baptistery at Stobi, Northern Macedonia

adult male deer, was considered a symbol of the Christian soul, and this psalm was sung as part of the liturgy of baptism as Christian catechumens approached the font.[27]

At nearby Heraclea Lyncestis (modern Bitola), a large episcopal basilica was placed over a smaller, older church. It overlooked a modest theatre, which had been abandoned by the fifth century, illustrative of changed times and priorities. A series of mosaics of the later fifth to middle sixth centuries adorns the church complex. The outer narthex of the church has an exquisite mosaic depicting creation, the ocean full of fish and octopus surrounding the earth and its bounty through four seasons, spring and summer, winter and autumn. Unlike the Dominus Julius mosaic, where seasons and their bounty are illustrated literally, each season is represented by a single tree, full of life, leaves, fruit and birds, and surrounded by living creatures, notably a goat, a bull and a lion confronted, and a red dog tied to a tree. These are zodiacal references, the dog being Canis Major, whose dog star shone brightest in late summer and early autumn. The final scene, which depicts autumn, also shows death, with a hart lying on its back, its entrails ripped out by a panther. That this is an ephemeral death is suggested by the fruit in the tree: pomegranates, a symbol of resurrection. The mosaic's central motif, a cantharus urn, is approached by deer and peacocks, now sprouting a vine and grapes (Plate 8).[28]

71

Games and partisans

The wealthy staged games in their cities to court public support and as a civic duty. Some decurions, like Magerius, relished the opportunity to do so and to receive plaudits. Others, including quaestors, praetors and consuls, were obliged to stage games in large cities as a function of office. Olympiodorus, writing of senatorial wealth at Rome in the first quarter of the fifth century, reports that three sets of praetorian games cost 1,200, 2,000 and 4,000 pounds of gold, between a quarter and four-fifths of the annual income of wealthy households. The ordinary consuls, appointed by the emperors to represent Rome and Constantinople, were the most senior magistrates, who gave their names to the year. They were responsible, throughout their year of office, for putting on a series of games and contests in the hippodrome and in other theatres in the imperial capitals, including two sets of horse and chariot races, wild beast hunts and fights, and theatrical entertainments. A law of 384 established that the gifts distributed at games should only be silver coins, limited in size and quantity, and plaques or tablets 'of materials other than ivory'. Ordinary consuls alone had the right to present as gifts gold and ivory diptychs. From this time until the sixth century consuls did just that, and dozens of exquisitely carved ivory plaques have survived (see Plate 15). Competitive display was a huge burden for those who did not share the fabulous wealth of consuls, and it is worth noting that very frequently emperors served personally as ordinary consuls.

In 409, a law was drafted for Theodosius II, who served as ordinary consul on sixteen occasions in his fifty-year reign. This was intended to prevent extravagant spending on games other than those associated with high state office.

> We admonish all judges that they shall be present at the festivities of shows, as is the custom, and elicit the favour of the people by amusements, but their expenses shall not exceed the modest sum of two *solidi*. They shall not, through imprudent and insane craving for applause, destroy the resources of decurions, the fortunes of citizens, the houses of the chief decurions, the riches of the landholders, the strength of the municipality.[29]

The largest cities had circus factions known by their partisan colours as the Blues, Whites, Reds and Greens.[30] The factions were not, as once was imagined, an urban militia, nor were they political and religious partisans, the standard bearers of orthodoxy (Blues) and Monophysitism (Greens). They did not represent the upper and middle classes (Blues), or the lower classes (Greens). Rather, they were sporting associations, the rabid devotees of individual athletes and teams, who were led in their fervour by highly organised and professional cheerleaders, that is claques. Traditionally, claques had comprised bands of professionals employed by pantomimes to lead applause. Crowds are unlikely to start coherent cheers and chants spontaneously, particularly when confronted by the unfamiliar. Hence, claques were needed to initiate and sustain crowd participation. These experts were also deployed to lead the acclamation of politicians, including prefects and governors and, on occasion, emperors. However, they might also be bought by a politician to withhold applause or praise from a rival, or to signal dissent from a policy or imperial action. The claques were attached to the principal sporting teams of Constantinople, Antioch and other major cities where the Blues and Greens, Whites and Reds devoted themselves to chariot racing in the hippodromes. It was here that dissent against the emperor or provincial governor could be most easily demonstrated and crowd behaviour manipulated in a city's febrile atmosphere.

The hippodrome housed many statues with which the people interacted daily, usually political erections. The circus factions at Constantinople were permitted to celebrate the successes of their most gifted and successful charioteers by the raising of statues on the central reservation of the hippodrome, something unknown at any other city. Each statue was placed on a section of the median (*euripus*) that faced the named faction's seats, which were located in the stand facing the *kathisma* in order – Blues, Whites, Reds, then Greens – from the starting gates (*carceres*) to the curved banked end of the stadium, called the *sphendone*. Permission to erect a statue required the consent of the emperor and the agreement of both main groups of circus partisans, the Blues and the Greens, although statues were also put up for heroes of the Whites and the Reds. Statue bases dedicated to Porphyrius, who raced for both the Greens and Blues, bore inscribed acclamations from both factions. He was the only racer to receive statues during his own lifetime, and was honoured thus seven times. The *Greek Anthology*, a medieval collection of earlier poetry, records fifty-four separate epigrams, each copied down from a charioteer's statue base in the hippodrome before the end of the ninth century, and perhaps far earlier. Of these, thirty-two are dedicated to Porphyrius, two

*Figure 7. Two statue bases of the charioteer Porphyrius,
from the hippodrome at Constantinople*

of which have been preserved on one surviving statue base, and four on a
second (Figure 7).[31]

A powerful recent analysis of Roman civilisation has characterised it as
'a way of living and thinking the city that was diffused with Rome's empire'.
Cities absorbed and generated wealth and talent, were sustained by and
sustained elites, and harboured great poverty and disease. If four-fifths
of Romans lived outside cities, on the land and in villages and towns, still
their lives were defined by the needs of cities to which they were attached,
where they might on occasion consult a lawyer or marvel at a portico, and to
which they sent their produce and taxes, or their sons and, more occasionally, their daughters. The populations of cities were constantly refreshed
from without, introducing new blood and pathogens, dangerous ideas and
deviant behaviour. Immigrants learned to live as citizens, and the wealthier
learned a good deal more in the schools that cities sustained. The following
chapter is devoted to a fuller exploration of education and culture imparted
in cities, and to the communications and commerce that linked them.[32]

4

CULTURE, COMMUNICATIONS AND COMMERCE

※

The fabric of Roman cities was covered by words. Antique and contemporary inscriptions competed with graffiti, signage and public notices. Citizens were regularly confronted with documents produced by the imperial bureaucracy, including legal codes and compilations, new edicts and sealed documents of validation, tax registers and demands. Basic education was widely available and functional literacy was common in the late Roman world among merchants and traders, in the army camps, and even among household slaves. All cities, and many towns and larger villages, would have had schools to provide elementary education in Greek and Latin reading and writing. Secondary instruction was offered to boys – and rarely to girls – by a grammar teacher (*grammaticus*), who introduced them to a small canon of literature to master, composed in Latin and Greek quite different to the spoken languages. The grammarian might also provide a foundation course in rhetoric, beginning with basic rhetorical exercises, or *progymnasmata* – literally 'before exercises' – which provided pupils with practice in composition before proceeding to exercises in declamation, the delivery of the speech. After years with a grammar teacher some might proceed to higher education, to receive instruction from a rhetor. According to Diocletian's edict of maximum prices, issued in AD 301, an elementary teacher might command 50 denarii per pupil per month, whereas a teacher of rhetoric and public speaking would earn five times that amount. Higher education was very expensive but it was a key to advancement in the literate Roman world.[1]

A literary education and grounding in rhetoric might secure employment locally, in a municipal bureau or as a provincial city lawyer. Even quite trifling or mundane legal matters were addressed in a highly rhetorical fashion in late Roman documents, as we have seen in Egyptian papyri. To aspire to a career in the service of a provincial governor or administrator, or even a role in the capital itself, required higher education at an elite school. Only a few families had the means and desire to send their children further away to enrol with famous educators. However, most senior bureaucrats had acquired such an education, and despite the enormous expansion of bureaucracy that attended the establishment and development of Constantinople as the imperial capital, there were always far more qualified candidates than there were jobs.[2]

Literary culture and education

The dominance of Roman decurion families was sustained by higher learning and the cultural system it imparted, frequently called simply 'education' (*paideia* or *paideusis*). Only major cities sustained facilities for higher learning, which was the greatest asset that the wealthy could purchase other than land. Rhetoric, which we might crudely consider to be the art of persuasion, was for the Romans the means to transform speech. Romans distinguished between everyday speech and ritualised speech, which was the foundation of social order in the Roman world; it defined public life for all late Roman citizens, and skill in composition and declamation was not only highly prized, but a prerequisite for social advancement. Poets, priests, philosophers, politicians and lawyers were performers on the public stage in all the empire's cities. Many represented their cities before the governor or even the emperor. Ritualised speech even had a role in private life: orations were commissioned to celebrate major life events within families, perhaps to be declaimed within urban palaces or country villas in the presence of family and friends, employees and slaves.[3] A standard textbook by Menander Rhetor, which begins with instructions for the imperial oration, proceeds to offer notes on composing speeches to be delivered on various family, city and state occasions: on the arrival of a notable person in a city; on a birthday; to console the grieving, or at a funeral; at a coronation; and when acting as an ambassador or representative of one's city. There are even instructions on how to write a speech of encouragement for those about to have sexual intercourse for the first time, following a wedding.[4]

Only the most fortunate Roman youths, the richest if not always the most able, received higher instruction in rhetoric at the greatest schools of the ancient world at Antioch, Alexandria, Athens and, latterly, Constantinople. The *Theodosian Code*, a legal compilation completed in 438, tells us that in 425 Constantinople acquired a 'university' with five rhetors and ten grammarians for Greek, three rhetors and ten grammarians for Latin, plus a professor of philosophy and two of law. If two law professors would not suffice, those seeking to excel in forensic rhetoric for legal purposes might prefer to head to Berytus (modern Beirut). For a time, Pergamum appeared to outshine all others, even attracting the emperor Julian. Aedisius, a student of Iamblichus, trained theurgists and sophists at Pergamum, and Oribasius wrote massive medical treatises there. The conversion of Pergamum's Temple of Serapis into a mighty church signalled the triumph of Christianity, but the schools endured a while longer. Others made for Alexandria or Athens, where philosophical education flourished, at least until Justinian's reign. As we have seen, Synesius of Cyrene studied in Alexandria under Hypatia, becoming a devoted Neoplatonist versed in the Chaldaean Oracles. He also appears to have developed an interest in scientific instrumentation, allowing him to devise an astrolabe that, as we have seen, he gave as a diplomatic gift. Following his graduation, Synesius travelled widely, including to Athens, where he observed that the schools hardly compared to Hypatia's. Those who studied in Athens disagreed, including Damascius, a pupil of Isidore, who thought his master surpassed Hypatia 'as a man surpasses a woman, and as a real philosopher surpasses a mere geometrician'.[5]

As the Roman empire became, at the end of the fourth century, formally a Christian empire, one might expect to identify clear changes in the traditional system of education. There were, however, few major developments – the Syriac school of theology, philosophy and medicine, founded in Nisibis in *c.*350 and moved to Edessa in 363 is a noteworthy exception – and Christian Romans continued to receive the same 'Hellenic' (now a synonym for 'pagan') education at all levels. It was only during

Julian the Apostate's brief reign that Christians were banned from teaching 'Hellenic' literature. This much we learn from one of Eunapius's *Lives of the Sophists*. Julian studied with two 'pagan' philosophers and rhetors, Themistius and Libanius. Libanius was trained in Athens and taught there and at Constantinople and Nicomedia before Antioch. Another of Libanius's pupils was Basil of Caesarea, also known as Basil the Great. Basil's education is emblematic of that available to a wealthy youth. He was 'home-schooled' for some years on his grandmother's estate, before enrolling with a grammarian in the regional metropolis, Caesarea, of which he would later become bishop. He graduated to rhetorical training in Constantinople, then for another five years he studied in Athens, achieving the equivalent of a doctorate in philosophy. Basil, by then a masterful orator, returned home to Caesarea where he taught rhetoric – including to his younger brother Gregory, later bishop of Nyssa – and practised law. Gregory recalled his brother Basil as 'excessively puffed up by his rhetorical abilities and disdainful of all great reputations', truly an arrogant young scholar and rhetor. It was some time before Basil took holy orders.[6]

Basil wrote little about his time in Athens and his subsequent secular career, which he came to believe had kept him from his true calling as a bishop. What we know of his times as student and teacher of rhetoric comes instead from a funeral oration, delivered late in 381 by Basil's lifelong friend and fellow student, Gregory of Nazianzus. Even after he became a priest, Basil only slowly reached 'beyond the erudite, the polished, the wealthy, and the powerful, who had gained their position through access to learning and clarity of expression'. His genuine concern for the simpler among his flock was ever tinged with a patrician disdain for those who could not transcend ignorance and superstition, and to whom he might never display all the facets of his supple intellect. Still, his sermons attracted large audiences and these, as well as his many letters, were preserved as models for later generations.[7]

Bishops regularly delivered homilies and encomia for saints before their congregations, but they also performed before governors and emperors. Bishops, among the richest and best-educated citizens in most cities, were employed as envoys and petitioners to governors or to the emperors themselves. Increasingly, bishops were selected for their skills as rhetors, rather than as pastors or spiritual athletes, and not all abandoned civilian life gratefully or willingly. Synesius of Cyrene, again, may stand as an example. In 410, a decade after he had returned from his embassy to

Constantinople, Synesius was summoned from his life of leisure and philosophy to become bishop of Ptolemais. He did not accept the honour without substantial objections and he was able to extract necessary concessions from Theophilus, the powerful patriarch of Alexandria, who had just a few years earlier officiated at Synesius's wedding. Synesius was permitted to remain married, would not be challenged in his devotion to philosophy, and no popular challenges to his views on doctrine, notably resurrection, would be heard. Synesius communicated to his congregation at Ptolemais in his letter of acceptance, after six months of reflection, that he had taken the job because 'I was unable for all my strength to prevail against you and decline the bishopric'. But, it was not their wishes that impelled him to relinquish his leisure and books, nor would he do so happily.

> Rather it was a divine force that brought about both the recent delay and now has caused my acceptance. I would rather have died many deaths than taken over this religious office, for I did not consider my powers equal to the burden ... How shall I, who devoted my youth to philosophic leisure and the idle contemplation of abstract being, and have only mingled as much in the cares of the world as to be able to acquit myself of the duties to the life of the body and to show myself a citizen – how shall I ever be equal to a life of daily routine?[8]

Synesius's surviving sermons indicate that he integrated philosophy into his mission, commending reason to his parishioners for their better understanding of scripture and offering philosophical interpretations of Christian festivals. However, as he feared, the burdens of daily routine did contribute to a rapid decline in his health. He died after only three years in office.

The greatest Christian rhetor of his generation was born in Antioch in c.347. John, later known as Chrysostom, 'Golden Mouth', earned such fame from his eloquent exegeses of Christian scripture and for his moral teachings that he became patriarch of Constantinople. John preached the virtues of monasticism, but also of caring for the poor and sick. John was a fierce critic of inherited wealth, and in particular of that form of wealth that was most obviously pernicious: the owning of slaves. But his was but one voice that, in the end, caused no change to the eternal verity that those who could afford it, whatever their religion and depth of faith, would continue to own other people. Before John became either preacher or rhetor he was an ascetic in the Syrian desert, an athlete for Christ, who returned to the city only when the rigours of his asceticism brought on ill health. Those endowed with spiritual authority need not stand for decades atop pillars or confine themselves to the desert to demonstrate their suitability to the episcopate, but could do so at the lectern and ambo (pulpit). The lives of holy bishops were written by their disciples in emulation of the Evangelists, and the most renowned of the fourth and fifth centuries devoted themselves in turn to education, discipline and ministry. Among their ranks were fierce advocates for divergent doctrines about the nature of Christ.[9]

Communications

An orator addressing Constantine in 312 in Trier – a veteran delivering a speech of congratulation on behalf his own city, Autun – knew that the emperor at his imperial residence would hear his speech in the presence of countless others 'from almost every other city, who have been sent on public missions or have come as suppliants on their own behalf'. Two decades later, Constantine, now the only emperor, would have received still greater throngs of envoys at Constantinople, alongside bishops and foreign dignitaries, all vying for his attention, interest and support. Something useful might be said in a letter, but the delivery of that letter required not only handing it over, but reading it aloud and elaborating on its content, frequently spelling out what was better not written down. Delivery required eloquence, a higher education in rhetoric. It also required travel, usually at one's own expense or that of the city one represented.[10]

Many books would, at this point, simply direct the reader to a map (see Map 2), imagining that this will help her understand the nature of ancient travel, its challenges and rewards. But Roman travellers did not carry maps. They looked to the heavens and at milestones to determine direction and distance. They relied on local guides, descriptions of landmarks and written itineraries. Distances offered on later Roman itineraries were not accurate and there is little that one might identify as geographical thought, beyond commentaries on Strabo and Ptolemy, until writers of saints' lives contrived to construct a composite hagio-geography of the eastern Mediterranean. There are a few works with quite distinctive approaches in the sixth century, when it was possible to contrast the *Christian Topography* of Cosmas Indicopleustes (arguing that the earth took the form of a box, after the ark of the covenant) and Procopius's *Buildings* (a very long panegyric to Justinian). Closer to itineraries were the *Ethnica* by Stephanus of Byzantium (which an extant epitome suggests was little more than a list of ethnic names and their derivations) and the *Synecdemus* of Hierocles (a list of the six imperial dioceses with their sixty-four provinces and the 920 main cities). There is no indication that Stephanus or Hierocles produced maps, although modern scholars have produced maps based on their data.[11]

Modern maps do not easily convey the choices Roman travellers made as they encountered different landscapes and seasons, weather conditions, banditry and unrest. Each journey was a series of choices, with advice taken or ignored, prayers offered and luck ridden. Journeys were made individually and in smaller and larger groups, by delegates and traders, retinues and armies. The greatest number of travellers was always bound to and from Constaninople, which could be reached by land or sea. The Mediterranean basin stretches 3,800 kilometres (2,400 miles) east to west and from 400 to 750 kilometres (250–440 miles) north to south, the greater figure when sailing north from the Gulf of Sirte on the Libyan coast midway between the islands of Malta and Crete, through the Strait of Otranto and across the length of the Adriatic Sea. As a rule of thumb, a day's sea voyage in good weather was the equivalent of four to seven days on a paved road. It has been estimated that ships would travel at least thirty kilometres a day, but rarely more than fifty. A person wishing to travel by sea would have to find space on a commercial ship heading in the right direction, and that ship would almost never travel directly to a distant destination. Unless it was absolutely necessary to sail into open seas, ships would hug the coastline and island hop, delivering and collecting goods and provisions, and other passengers, along the way. Sea travel was more substantially affected by bad weather, and although there is little indication that the *mare clausum* – a prohibition on sailing during four winter months – was enforced, it was widely observed.[12]

When setting out from Rome for Constantinople, a traveller would most likely be directed south-east to Otranto and advised to cross the Tyrrhenian Sea, landing at Avlona. The route from Otranto via Corcyra (Corfu) became dominant rather early, allowing travellers to proceed further north, if business so required, by land. When the lands between

Rome and Otranto were unsafe, as they frequently were in our period, then passage to Ravenna by road might prove more attractive than setting sail from Otranto. Departing from Classe, Ravenna's port, one might cross the Adriatic, a gulf that stretches north from the Mediterranean Sea for around 780 kilometres, to navigate south along the eastern Adriatic coast. The eastern Adriatic coast has 300 rocky inlets, thirty-six large islands and more than 500 small islands. In contrast, the western Adriatic littoral is relatively featureless, has no large islands and very few small islands, other than those clustered in the Veneto, from the mouth of the Po to the lagoon of Grado. (That the situation of Venice shares more with the eastern Adriatic coast than the Italian coast to its south certainly had an impact on its early history.) Hugging the eastern Adriatic coast, ships would stop at the major cities of Zara (Zadar), Spalato (Split), and Ragusa (Dubrovnik), sailing through the channel formed by the string of islands from the point of the Istrian peninsula to the southern tip of the island of Korčula. Within this channel ships were protected from the worst of the weather conditions that occurred in the open waters to the south of Dubrovnik as far as Dyrrachium (Dürres) and beyond. At Dyrrachium one might sail on or continue by land along the Egnatian Way, which passed by Heraclea Lyncestis en route to Thessalonica. According to Harun Ibn Yahya, in the later ninth century the journey by land from Thessalonica to Constantinople took twelve days. There is no reason to suppose it was any different in the fifth century.[13]

From Thessalonica another road ran north, following the course of the river Axius (Vardar) through Stobi and Scupi to Naissus (Niš), where it met the military road that linked Sirmium (Sremska Mitrovica) and Singidunum (Belgrade) with Constantinople, via Serdica (Sofia), Philippopolis (Plovdiv), and Adrianople (Edirne). The Balkan land roads were mountainous and dangerous, with bandits and other hazards, not least the roads themselves, which were not all in a good state of repair. We know that the Egnatian Way was restored substantially in the sixth century as part of Justinian's commitment to renovations of older structures, and will certainly have received imperial attention at later dates. The total length of the Egnatian Way was between 750 and 800 kilometres (the exact route in all its parts has never been fully identified). These were substantial distances, but somewhat less than those by paved road from Constantinople to the eastern borders between Tarsus and Antioch.[14]

From most points to the east, and certainly outside the sailing season, the land route was considered safer. It passed through major and well-defended cities, with inns and hostelries every few miles along the major roads. According to the *Antonine Itinerary*, an early-third-century document that lists 225 routes along Roman roads and their stages across the empire, the distance from Antioch to Nicomedia was 747 Roman miles, but it was a hard route in places. (A Roman mile was 1,000 paces, estimated at about 1,481 metres or 1,620 yards.) From Antioch one must head north to the Amanus Mountains before crossing the plain of the Issus south of Tarsus. From Tarsus one headed back into the mountains, passing through the Cilician gates, a pass high in the Taurus range, and heading down to Tyana and then Archelais. The road subsequently ran across the central Anatolian plateau through southern Cappadocia, to Parnassus and then Ancyra (modern Ankara). An alternative route ran to the south of this, through Galatia to Amorium and on to Dorylaeum. A road from Ancyra also ran to Dorylaeum, but the principal road ran further to the north along the Sangarius river valley to Iuliopolis and Nicaea, before turning sharply north to Nicomedia, and west again to Chalcedon opposite Constantinople. The distance was also calculated by a pilgrim from Bordeaux, who set out for the Holy Land in AD 333 and stopped at many stages along the route, counting the miles between them: from Nicomedia to Ancyra, 258 miles; from Ancyra to Tarsus, 343 miles; from Tarsus to Antioch, 141 miles; giving a total of 742 miles.[15]

For those conducting official business, and for the delivery of letters, the land journey might be considerably quicker than for others. By the Roman *cursus publicus*, the state postal service, messages were transferred through post stations (*mutationes*) eight to ten Roman miles apart, at each of which up to forty fresh horses were kept by grooms. Every third or fourth station was more substantial, an inn (*mansio*) providing overnight accommodation not only for the horses' grooms and the couriers but also for other travellers. These were more frequent where cities were rarer. It is often imagined that a relay of riders with an urgent message might cover vast distances mounted on horses ridden at full gallop. Even Procopius, the principal historian of the sixth century, claimed that the public post once covered ten times the distance of a single traveller in any given day, which might mean up to 400 kilometres (250 miles), although at that point he is exaggerating how Justinian allowed the service to decline. For this reason, it was once believed that the journey from Antioch to Constantinople took only three or four days. However, it is more likely to have taken between ten and twelve days, depending on the weather and available light. It was not Roman practice to employ relays, so there were no spare couriers waiting at stations to take over when a messenger tired. A single courier would carry each message to its destination, and he would travel in a carriage, not on horseback. A courier would regularly travel for up to ten hours a day, averaging five or six Roman miles per hour. Consequently, even if one travelled at full speed taking new horses at every station, for as long as there was daylight and by moonlight where possible, an urgent message would travel at most seventy to eighty Roman miles (roughly 100–120 km, or 60–75 miles) per day. Difficult terrain slowed travel and many routes were extremely hilly or passed through mountains.[16]

Antioch was as close to Ctesiphon, the Persian capital, as to Constantinople, and one might sail from its port, Seleucia by the Sea, across to Salamis on Cyprus or south along the Levantine coast towards Alexandria, stopping en route at any of the hundreds of ports, docks, bays and anchorages, including the mighty harbour at Caesarea Maritima. Trade in the eastern Mediterranean region was extremely active and finding a berth on board a departing ship must always have been possible. Increasingly, while holds might be full of goods, decks were occupied by pilgrims travelling to the holy places. Although ships often stopped off on Cyprus and the southern coast of Asia Minor, one could still sail directly across the sea from Alexandria to Constantinople aboard one of the convoy of grain ships that departed each year in early summer, or from Constantinople to Alexandria as the ships returned, now with far more space for passengers. The duration of the trip might be half as long heading south as north, nine or ten days rather than three weeks, thanks to the 'Etesian' northerly winds that blew from May to September. Similarly, the trip from Rome to Alexandria might be made directly in around ten days, but sailing from Alexandria to Rome could take up to two months, which left plenty of time to encounter storms and squalls. Lucian of Samosata encountered a huge grain transport, the *Isis* out of Alexandria, blown off course as far as Athens, which he described as 120 cubits long 'and well over a quarter as wide, and from deck to bottom, where it is deepest, in the bilge, twenty-nine'.[17]

Continuing along the African coast west of Alexandria, one might disembark at Cyrene, the ports of Tripolitania, and many others as far as Carthage, Hippo and beyond to Septem (Ceuta) at the straits of Gibraltar. Trips between North Africa and the major Mediterranean islands

were frequent, but any such sea journey could be traumatic and poten-
tially fatal, as the relatively seasoned traveller Synesius of Cyrene wrote
to his brother. Returning to his home town from Alexandria, he boarded
as one of fifty passengers, a third of them women. He and his fellow passengers spent much of their time berating the captain as he fought the
elements to steer clear of rocks lining the shore, and then strayed into the
deep sea shipping lanes of the Alexandrian grain fleet, those 'double-sailed
cargo vessels whose business does not lie with our Libya'.

> A violent wind soon raised seas as high as mountains. This gust falling
> suddenly on us, drove our sail back, and made it concave in place of its
> convex form, and the ship was all but capsized by the stern. With great
> difficulty, however, we headed her in. Then Amarantus [the captain]
> thunders out, 'See what it is to be master of the art of navigation. I had
> long foreseen this storm, and that is why I sought the open. I can tack
> in now, since our sea room allows us to add to the length of our tack.
> But such a course as the one I have taken would not have been possible
> had we hugged the shore, for in that case the ship would have dashed
> on the coast'.

The storm grew worse, however, and after landing on a deserted shore-
line, they set off again on a day of inauspicious omens. 'It was the thirteenth
day of the waning moon, and a great danger was now impending, I mean
the conjunction of certain constellations and those well known chance
events which no one, they say, ever confronted at sea with impunity.' A
second storm struck the ship, driving it back towards a rocky promontory,
where a local brought the passengers to shore at a small harbour. There
they were joined by other ships, some of which had left Alexandria the day
before them. Since all passengers relied on the provisions they brought
aboard, when these were exhausted the travellers were forced to fish and
collect limpets, until locals brought them food and water. Synesius of
Cyrene, who became the bishop of Ptolemais, had no qualms in enumer-
ating the astrological omens and superstitions associated with seafaring.
Just as strikingly, he was a collector of books and the author of several,
including a treatise on making an astrolabe, but he did not carry any navi-
gational tools, guidebooks or maps. He relied, as all did, on guides and
experts like Captain Amarantus, even if he later lampooned the captain
who got him home alive.[18]

Commerce

Synesius travelled considerable distances from his native North Africa, but returned and died not far from where he was born. In contrast, a young North African woman, who died at the same time as Synesius, was buried at York in the north of England, far from where she was born. She was buried with a range of personal items that included a mirror, bangles fashioned from jet and ivory, a silver pendant, yellow glass earrings and a necklace of blue glass beads. Most precious was a carved ivory plaque discovered in her stone sarcophagus, a gift from her sister in life or in Christ, reading 'SOROR AVE VIVAS IN DEO', 'Hail sister, may you live in God'. VIVAS is the most common word in late Roman inscriptions on objects found in Britain. It conveys hopes for a happy life and the best wishes of a friend or family member, frequently in a Christian context. The 'Ivory Bangle Lady' owned precious things that had been made locally, from Whitby jet, or had travelled like her across the Roman world to the north of England, such as the blue glass beads from the Rhineland or the ivory items, which are very rare in Roman Britain, from further east. She arrived in Britain by sea having grown up in a warmer place than York, certainly coastal, possibly North Africa.[19]

Coastal North Africa between Carthage and Leptis Magna (Lebda) in Tripolitania sent goods across the Roman world and imported almost as much, including provisions transported in amphorae. So many later Roman amphorae have been found at Carthage that the major amphorae types of the fourth to sixth centuries (Late Roman, or LR types 1–7) were first classified based on those finds, identifying vessels manufactured in and exported from major production centres in Cilicia and Cyprus (LR1), the Aegean and Asia Minor (LR2a/b), western Asia Minor (LR3), Syria and Palestine (LR4; 'Gaza' type, LR5/6), and Egypt (LR7). There were also so-called 'spatheion' types, carrot-shaped amphora produced in North Africa. Amphorae types and numbers are excellent indicators of the volume and directions of trade in certain commodities over time, namely wine, olive oil and preserved fish products. Many of the amphorae discovered at Carthage that originated at sites across the Aegean and Levant contained wine. Those that originated in North Africa mostly contained olive oil and fish products, including salted fish, fish sauce (*garum*) and boney fish paste (*allec*). An archaeobotanical study of a sample from the harbour at Carthage identified some other comestibles that were probably imported to the city in around AD 600, including pomegranates, hazelnuts and walnuts. A good deal more than that, according to pollen analysis, was grown locally, including wheat, barley, grapes, olives and figs. Much made its way to the coast from inland sites along river valleys; however, much less travelled in the other direction. Very few imported amphorae reached inland sites in North Africa.[20]

Ceramics are ubiquitous in the archaeological record, but some items that were equally widely traded scarcely appear. Wild animals were bred and hunted for export, notably in North Africa. In every city of the empire, before theatrical wild beast hunts were banned, there would have been demand for large cats, and still more exotic creatures made their way to larger cities. At Constantinople's Theodosian harbour, ostrich, macaque and elephant bones have all been identified. The skeletons of exotic animals are easy for experts to distinguish from the bones of dogs or cattle, and from those of humans. The bones of unfree humans are identical to those of the free, but written sources suggest that slaves were also exported in great numbers from the ports of North Africa. Perhaps as many as 10,000 captured people were shipped each year in the first century AD, falling in number over time. Leptis appears to have been a major slave market, and on the island of Jerba (Djerba), which lies halfway between Carthage and Leptis, ceramic fragments with scratched writing – probably receipts – mention a slave trader. Textiles are also hard to iden- tify in the archaeological record, although we know these were widely traded. Jerba and its city, Meninx, grew rich on murex, the sea snail whose mucus was used to produce purple dye. Piles of crushed murex shells up to five metres deep accumulated across ten hectares of the city into the fifth century, when fishermen, weavers, dyers and merchants were employed to bring the product to market. A plentiful supply of water was ensured by four aqueducts, and wool and textiles were dyed in situ, before the purple perished, and loaded on to ships at the port, where amphorae of wine from across the eastern Mediterranean were offloaded. The stench of the processing and the relentless accumulation of shells ended in the sixth century, and later occupation at Meninx is indicated only by a few sherds of pottery.[21]

Not all regions in North Africa were substantial exporters of goods. There is ample archaeological evidence for the production of wine and olive oil in Cyrene, Ptolemais and the rest of Cyrenaica in the fifth to seventh centuries, but this was on a small scale using single, not multiple, presses. Synesius of Cyrene praised his region's olive oil for its heaviness, which made it better for lamp oil than for human consumption. The absence of any known amphora type produced in Cyrenaica in Synesius's time or later suggests that little oil or wine was ever exported. Although there is no good evidence for the export of comestibles from Cyrenaica, it must have produced salted fish and fish sauce like other regions in North Africa. Moreover, Cyrene had well-established links with Crete, which lay immediately to its north, and surely exchanged goods.[22]

Cretan wine was exported far and wide in the fifth and sixth centuries – to Cyrenaica and the rest of North Africa, across the Adriatic and Aegean, to Constantinople and beyond to the northern Black Sea region – although by the seventh century this trade had tailed off significantly, ending a millennium-long enterprise. The link between Cyrene and Crete had been enshrined in the imperial period by the linking of Cyrenaica and Crete as a double province, with the Cretan city of Gortyn as its capital. The connection was reinforced by the propagation of common myths, but it appears in the material record most clearly in the form and abundance of terracotta lamps in both places, which suggest very close connections between the lamp manufacturers and traders.[23]

Other than transport vessels, ceramics, including lamps, were rarely imported into North Africa. There was no demand for such goods, as lamps and fine wares were manufactured in huge amounts for regional use and export. North African red slip ware (ARS) was traded across the Mediterranean world in great volume between the fourth and seventh centuries, and also found substantial markets in the African coastal cities themselves. However, it was very rarely traded south towards inland Africa. ARS has been found at excavation sites in all modern countries of the eastern Mediterranean, including Greece, Turkey, Syria, Lebanon, Israel, Jordan, Egypt and Libya. The largest markets for ARS were Sicily and southern Italy, which is unsurprising given their proximity to the centres of production. ARS is found at coastal sites across Sicily, and also at inland sites in lower quantities, throughout the sixth and seventh centuries. Equivalent ceramic fine wares were produced in western Asia Minor (Phocaean red slip ware) and Cyprus (Cypriot red slip ware). As Crete enjoyed political and economic relations with Cyrene, so Cyprus was linked to Cilicia and Isauria to its north, Syria and Palestine to its east, and Egypt to its south. These long-standing connections are reflected in the material records of those regions, notably in the distribution of amphorae and other ceramics. From the fourth century to the seventh, ports on Cyprus's southern coast were visited by those transporting grain from Alexandria to Constantinople.[24]

Each summer, between June and September, the Nile flooded, covering 'the entire surface of the land, effortlessly bringing forth all the fruits of the earth'. Until the modern construction of dams at Aswan (Syene), the floodwaters would rise annually to fill the narrow valley, pouring into canals, feeding extensive and complex irrigation systems that supported the greatest agricultural complex of antiquity. Abundant wine was produced, of which very little was exported, if one is to trust the amazing dearth of Egyptian amphora types discovered elsewhere in the empire. Local wine was consumed with local bread, and perhaps with imported Spanish olive oil and pickled fish from Palestine. Grain was Egypt's principal crop, harvested in the summer and brought to granaries, from which it was transported by barge to Alexandria, much of it for export the following spring. A huge quantity of grain was shipped from Alexandria, including between 160,000 and 240,000 tonnes annually to Constantinople, the so-called 'Happy Shipment'. Around 650 ships were loaded during the sailing season and fought the northerly winds across the Mediterranean. The majority of this exported grain was levied as tax in kind, the *annona*, which was transported by private shippers, *navicularii*, with government contracts, who were paid less than a commercial shipping charge, but were compensated by generous tax concessions and state support for the construction and maintenance of their ships. Each *modius* of grain was transferred from a river barge on to a sea-going transport that might be carrying other commodities that passed through Alexandria's docks, including the purple stone known as porphyry, textiles and spices, and an abundance of papyrus. Combined shipments and mixed loads were standard, as is attested by shipwreck evidence. Ships returning to Alexandria might carry wood, including timbers of tall trees that did not grow in Egypt but were essential for construction. *Navicularii* who did not manage their business well, or whose state shipments were affected by port delays, red tape, bad omens, superstition, strong winds or bad weather, faced confiscation of the ships and fortunes.[25]

One might imagine that Captain Amarantus was used to dodging the grain ships that sailed out of Alexandria, and that his own vessel carried merchants and wares whose business did indeed lie with Synesius's Libya, stopping off at ports along the coast, including several great harbours that were improved to accommodate hundreds of vessels. Forty-five North African ports (not including those in Egypt) have been identified where the Romans certainly or possibly built artificial structures – wharves with quays, jetties, breakwaters and enclosed moorings – to support trade. Roman Carthage's harbour provided almost five kilometres of docking space, more than three times that of Punic Carthage. The port at Leptis had around 1,200 metres of docking space, and that at Hadrumetum (Sousse) about the same. Alexandria had three ports: the deep eastern Great Harbour that serviced the grain transports; the smaller western Eunostus Harbour; and an inland port on Lake Maeotis that was linked by canals to the Nile, the Mediterranean and the Red Sea. The two sea harbours were separated by a mole that joined the mainland to an island, Pharos, which offered shelter. On the island's north-eastern promontory sat the famous lighthouse.[26]

Seafarers suffered the shock of regular, destructive storms in both winter and summer, and even after reaching port they were not assured a safe harbour. Seismic activity occasionally sent huge waves across the Mediterranean. Constantinople was particularly vulnerable because it sits on a tectonic fault. Since it also sits at the end of a relatively narrow channel, the velocity and magnitude of a tsunami wave is greatly increased as it approaches the city. Excavations at the Theodosian harbour, a site today known as Yenikapı, appears to show evidence of past tsunami destruction and sediment deposition. Excavators have uncovered thirty-seven shipwrecks in the Theodosian harbour, all buried in the silt between the fifth and twelfth centuries, including small transports, medium-sized merchantmen and four galleys, which were warships. The largest merchant ship excavated at the site, Yenikapı 22 (YK22) was a little under thirty metres long, constructed principally from pine, cypress, red oak and white oak. This ship was built, according to radiocarbon dating, between AD 430 and 603, but a more precise date is given by a rare gold coin of Leo I (r. 457–74), minted at Thessalonica, which was driven into one of the hull's fragments.

Yenikapı 35 (YK35) was a large merchant ship with a high load capacity and a thick hull framed in elm and covered with Mediterranean cypress planks. Its keel was of Turkish pine. It sank in the fifth century with a cargo of amphorae, coarse ceramic transport vessels, and lamps thought to have originated in the Black Sea. Remarkably, the cargo was wholly intact. Many of the amphorae were incised with letters where the neck met the shoulder, marks of ownership and evidence that they were reused, perhaps frequently. Several of the amphorae contained tiny bones from anchovies. A rather smaller merchant ship, YK11, was only eight metres long, with a keel of oak but otherwise principally constructed in softwoods, notably pine and cypress. YK11 showed signs of frequent repairs, keeping it operating through much of the seventh century. The remains of the hull were extensively damaged by shipworms (*teredo navalis*), suggesting that it spent little time moored in the harbour where it sank, and that it was never sheathed in lead. Constantinople's only river, the Lycus, discharged into the Theodosian harbour, making the water brackish and hostile to shipworms, thus preserving the caulked hulls of many ships, which if moored in other seawater harbours would have suffered far more extensively.[27]

Amphorae discovered at Yenikapı and elsewhere in Constantinople demonstrate that the city enjoyed links with sites across the Mediterranean world, including the Black Sea, northern Adriatic and, abundantly, the Aegean islands and coastal North Africa. As patterns of trade changed and volume decreased towards the end of the sixth century – a phenomenon we shall explore in greater detail in a later chapter – changes can be seen in the fabric and manufacturing of ceramic containers. Production of all late Roman amphora types appears to have ended with the sixth century, all replaced by a singular form, globular amphorae that are found at sites across the eastern Mediterranean during the seventh and eighth centuries. Globular amphorae were formed from many different materials, proving that they were made not in a few large centres, but at many different locations in a very similar form. They were certainly made in Cyprus, as they were in Crete and across the Aegean, with examples found in Asia Minor, Syria and Egypt, and in Italy. The Yassı Ada shipwreck, discovered off the coast of Turkey at Halicarnassus (Bodrum, Map 4), and dated securely to *c.* AD 625, contained a rich variety of amphorae, around 900 of them (of which 800 were globular amphorae) from a range of locations. It has been suggested that this was an *annona* shipment, destined for Heraclius's army, then engaged with the Persians. However, there is no reason that private enterprise would not have operated on this scale at the time, purchasing goods from many local ports and reselling at larger markets and population centres. Just as at Yenikapı, many of the amphorae also show scratched marks of ownership and regular reuse.[28]

Trade in some commodities appears to have endured rather longer, for example glass, perhaps due to continued operation of specialist production facilities. Glass is formed by melting sand (silica) and soda, an alkali used as a flux to lower the sand's melting point, together with a stabiliser, typically lime. Later, medieval glass was a local product, manufactured with elements obtained locally, such as river sand and ash from the burning of plants, typically beech or bracken, mixed with a lime stabiliser. Roman glass, in contrast, was manufactured at certain sites on an industrial scale using particular materials. The key raw material used in Roman glass manufacture, besides high-quality beach sand, was natron, a salt formed from the periodic evaporation of brackish water from some lakes, notably in Egypt. Natron from Egypt's Wadi Natrun was a constant in the composition of glass through antiquity, and glass manufacturing was undertaken nearby at Beni Salama, as well as at sites with abundant high-quality sand to which natron was exported. Local impurities in sand gave distinctive qualities to late Roman glass. This, as well as the type of stabiliser used, allows identification of production sites. Several glass manufacturing facilities have been identified in the Levant, where natron could easily be exported and coastal sands were available in abundance.

Analysis of glass discovered across the eastern Mediterranean suggests abrupt compositional change in the early sixth to seventh centuries, and again in the late seventh to early eighth centuries. Glass of the sixth to seventh century, 'Levantine I', was produced with significantly less lime than both earlier and later glass at Apollonia-Arsuf (34 km south of Caesarea Maritima), where four tank furnaces have been excavated. At Gerasa, today in Jordan, sixth- to seventh-century natron glass exported from Apollonia-Arsuf is quite different to natron glass of the second to fourth centuries. The 'Levantine I' glass weathers to an opaque white, due to its low manganese content. Even as 'Levantine I' was still in production, 'Levantine II' commenced manufacture. At Hadera (Bet Eli'ezer in Israel, Map 5), to the south of Caesarea, seventeen furnace floors began producing this last generation of natron glass. Slabs of raw glass weighing around eight tonnes were formed in these tank furnaces. These enormous slabs were cut into chunks, to be melted and formed in workshops around the empire. Large chunks of unworked glass that were exported from this region have shown up in shipwrecks across the Mediterranean, and at inland sites where secondary workshops processed the raw material. From the later eighth century, plant ash glass is found for the first time in Syria and Palestine, and a century later it appears in Egypt, marking the end of natron glass production, a process that had endured for two millennia.[29]

Late Roman civilisation was defined by its cities, from York to Carthage, Cyrene to Antioch, Alexandria to Constantinople. The literary culture that sustained the elites and staffed administrations was an urban phenomenon, and its product, men like Synesius of Cyrene, travelled widely to undertake the business of politics and persuasion, to seek and secure concessions, to represent interests. Many cities of the Roman empire looked, smelled and sounded alike. Their porticoed streets were covered with writing and echoed with noise, their public waterworks gushed and trickled, splashed and plinked. While some stank year-round of fish and products drawn from the sea, others hummed with activity when grapes and olives were harvested, or when flocks and herds were driven to their markets. Most cities burned fuel in abundance to run hot baths and fire ceramics in which goods were stored and exported. Few cities manufactured fine slip wares, but all had potters and glass-blowers, smiths and metalworkers who fashioned items from imported glass and iron, lead, silver and copper. The goods for sale in shops and stalls reflected long-distance trade as much as local production, and most subjects of the Roman emperor enjoyed a common material culture as well as a similar diet. Ships, ferries, carts and pack animals brought provisions from a city's hinterland or ivory bangles as far as Britain. Wine and oil, fish sauce and anchovies all travelled in familiar containers, amphorae of clearly identifiable types. Although nothing rivalled the annual spectacle of the grain fleet leaving Alexandria, ships of all sizes, fashioned from many woods, plied the harbours of the Mediterranean and occasionally sank, providing us with singular insights into centuries of commerce and communication. In our period, all roads led to New Rome. The political narrative offered in later chapters might be read as a series of journeys that elucidate the passage of power from Rome to Constantinople, the beating heart of a new Roman empire, to and from which veins and arteries carried goods and letters and, above all, people.

5

CONSTANTINOPLE, THE NEW ROME

In around AD 1200 in the monastery at Reichenau, where the Rhine emerges from Lake Constance, a remarkable work of medieval cartography was produced. The *Tabula Peutingeriana* or Peutinger map was a faithful copy of an earlier work, itself a Carolingian copy of a graphic itinerary of the fourth century. This enormous chart comprises eleven separate sheets, perhaps of an original twelve. When placed together the sheets are almost seven metres long but the map is only thirty-three centimetres high. On it there are more than 3,000 place names, including more than 500 cities shown linked by roads. The three greatest cities of the empire can be distinguished from all other cities and from each other by their individual *tychai*, female personifications or Fortunes. These are Rome, Antioch and Constantinople. Constantinople's Fortune is seated, pointing to her city's distinguishing landmark, a tall column formed from multiple drums surmounted by a nude standing statue holding an orb and a spear. This was a statue of Constantine himself, which had stood atop the column in the fourth century, but had long since fallen by the thirteenth (Plate 9).[1]

How Constantinople became first prominent, then ascendant, and finally the empire's dominant city is central to our history. Constantinople was founded on the site of Byzantium as the victory city of Constantine the Great following his defeat in AD 324 of his last imperial rival, Licinius.

Constantinople was not conceived as a new Christian capital to replace Rome, but that is what it became, achieving a civic status that surpassed all other cities. Constantinople also grew into the most populous city in eastern Christendom, with as many citizens as Antioch and Alexandria combined. In other regards, however, it remained inferior to its two eastern rivals. In the words of Robert Taft, 'early Constantinople was known for little either culturally or ecclesiastically. It produced almost no literature of any importance, it was not a great intellectual or monastic center, and it was not the cradle of saints and martyrs. Furthermore, its homiletic and theological production was slim.' It was the infrastructure of Constantinople that bludgeoned the other great cities into submission, with its scale and magnificence, the imperial presence and patronage, the levels and levers of government, and the fact that all were required to travel to the capital, to recognise its primacy and prostrate themselves before the imperial majesty.[2]

Government at New Rome

Constantinople was a city unique among imperial foundations in being endowed with a senate. According to Zosimus, 'Constantine continued wasting revenue by unnecessary gifts to unworthy men'. His views reflect a pagan tradition that was well established by the later fourth century that Constantine had squandered money on the worst types and on indulgent building projects, extracting that wealth from the cities of the east and their temples. Eusebius, Constantine's biographer, treated the same matter with approbation. He did not record the founding of a senate, but noted that 'some received gifts in money, others in land; some obtained the praetorian prefecture, others senatorial or consular rank ... for the emperor devised new dignities that he might give as tokens of his favour to a larger number of people'. The office of ordinary consul was retained but carried no power. It was accorded, however, only to wealthy men, who already wielded power. Consuls would pay for games and have remarkable diptychs featuring their own portraits carved from ivory to distribute as gifts (see Plate 15).[3]

The senate house stood in Constantine's forum, reminding all that the senate of Constantinople was itself the creation of that emperor. The *Origin of Constantine*, an anonymous work of uncertain date, but which supplies reliable information, states plainly that in Constantinople Constantine 'founded a senate of the second order', whose members were called *clari*, 'distinguished', lower than the *clarissimi*, 'most distinguished' members of the senate of Rome. However, within a few decades this distinction was abolished, and all senators were *clarissimi*, affiliated to Rome or New Rome according to whether they lived in the eastern or western part of the empire. In 337 the senate numbered perhaps 300 in Constantinople and more than 600 in Rome. Constantius II, Constantine's son, would expand this greatly to around 2,000 in Constantinople alone, mostly to shore up his power and to provide for the increasing ceremonial needs of his court. The highest officers of state held the rank of senator, including the men in charge of the military and civilian administrations, who might thereby be recognised as consuls.[4]

The emperor as commander-in-chief stood at the head of the military administration, followed by his top generals who held the rank of 'master' (*magister*). The master of the foot (*magister peditum*) and the master of the horse (*magister equituum*) commanded the central field armies stationed near Constantinople. Together they were known as the *magistri militum praesentales*. Generals of the regional field armies were also *magistri*, attending to the frontiers in Gaul (Rhine), Illyricum (Upper Danube), Thrace (Lower Danube), and the East (Persia). The eastern command was the largest and most significant of these. The imperial guard units were separately commanded and attended to the person of the emperor. Men of provincial backgrounds who held regular commands could rise rapidly, for example Zeno the Isaurian, who became emperor in 474, had earlier been a commander of the imperial guards and, from 469, general of the eastern armies (*magister utriusque militia per orientem*). There were also at times special commands entrusted to 'barbarians' commanding their own forces in imperial service, some holding extremely elevated ranks, like Alaric, whose Goths fought on the side of Theodosius I at the battle of Frigidus in AD 394, but who went on to sack the city of Rome in 410.

Civilian administration was divided into palatine and provincial. In charge of the palatine administration was the master of offices (*magister officiorum*), who might best be described as the emperor's chief of staff. His only equal was the quaestor, the emperor's legal secretary and general counsel, who drafted imperial legislation and replied to letters and petitions. Perhaps the most famous holder of both these offices was Tribonian under Justinian. The only person closer to the emperor himself was the grand chamberlain (*praepositus sacri cubicula*), the emperor's butler who ran the private apartments and accompanied the emperor when he travelled. The grand chamberlain was usually a eunuch, for example Narses under Justinian, who later and exceptionally became a trusted general. The master of offices had the rank of count and administered three secretariats (*scrinia*), dealing with petitions, reports and the requests of embassies. The quaestor drew his principal subordinates from these offices. The master of offices also handled a cadre of roving 'agents' (*agentes in rebus*), called in Greek 'the master's men' (*magistrianoi*), whose tasks were ill defined but included running the imperial postal system (*cursus publicus*), ports and customs, ensuring coordination between disparate jurisdictions, and information and intelligence gathering. A focus on spying has led to suggestions that the 'master's men' were in fact the secret police. There were more than a thousand agents, including a certain Symeonius, who, as we have seen, invested some of the money he earned in an inscribed silver chalice that he donated in the middle years of the sixth century to his local church in a village some miles east of Antioch (Plate 2).[5]

Two counts (sing. *comes*, pl. *comites*) ran separate departments in the imperial treasury. *Comites* might also be translated as 'companions', for these men were regarded as, and in many cases were, the emperor's closest and most senior advisers forming his entourage (*comitatus*). The count of the private fortune (*comes rerum privatarum*) ran five secretariats administering the imperial estates and properties, which included market gardens and villas, pasture, agricultural land, mines and quarries. Individual estates, or groups of them, were overseen by curators who reported through senior administrators to the bureaux at Constantinople. Imperial estates might be leased out or worked directly. Little is known about the scale or functioning of the emperor's private estates, both those owned by the person who became emperor and those that were conveyed with the office of emperor. However, there is sufficient evidence to state that these were vast and grew regularly by donation, forfeiture and seizure. Anastasius found it necessary, as part of his general restructuring of the empire's finances, to establish a count of the patrimony (*comes patrimonii*), leaving the count of the private fortune to handle land acquisitions. In effect, the older role became judicial rather than financial. The larger part of the imperial treasury, comprising ten *scrinia,* was administered by the count of the sacred largesses (*comes sacrarum largitionum*). This figure oversaw the minting and distribution of currency, the collection of public revenues paid in gold and silver – notably the transaction tax known as the *chrysargyron* (since payment was originally to be made in 'gold and silver') – and other indirect taxes such as customs duties. Separate secretariats also controlled the quality of precious metals – fivefold control stamps on silver items bear the mark of one office – and the distribution of donatives to the army.

In provincial administration, the empire at the start of the fourth century was divided into twelve dioceses, each of which comprised a number of provinces that totalled a hundred or more. Each diocese was under the care of a vicar (*vicarius*). All provincial governors answered to a vicar, and the rulings of governors could be appealed to him. Above the vicars were the praetorian prefects, who stood in place of the emperor in a prefecture. In 314 there were four prefectures, each containing three dioceses. Constantine expanded this to five, granting Africa alone to a praetorian prefect. From 395, there were once again four prefectures, two in the west, being Gaul and Italy, and two in the east, being Illyricum and 'the East' (*Oriens*). The east was by far the largest prefecture, comprising four large dioceses (Syria, Palestine, Mesopotamia and Egypt), embracing sixteen, later seventeen, provinces. In regions where praetorians were present, vicars generally disappeared. However, until the reign of Anastasius, the prefect of the east was based at Constantinople and retained his five subordinates, the vicars of Thrace, Asiana, and Pontica, the count of the East (*comes orientis*), and the Augustal prefect (of Egypt).

Praetorian prefects no longer had a military role, although their title had originally designated the commander of the praetorian guard, the imperial regiment based in Rome, or the military officer immediately subordinate to an emperor. Praetorians enjoyed overall responsibility for civilian government in vast regions, most significantly in political and judicial affairs, financial matters and in the levying of taxes. According to a law issued in 331 (*CTh* 11.30.16), judgements rendered by vicars might be referred to the imperial court, but those of praetorians could not be appealed, since they 'alone may truly be said to judge in place of the emperor'. *Praefectiani*, the prefect's men, were divided into two categories: those who dealt with political and juridical matters (*schola exceptorum*), and those with financial duties (*scriniarii*). The prefects, through their agents, received direct taxes levied from the land and those who owned and worked it and distributed it to where it would be spent. An extended quotation from A. H. M. Jones captures rather well the scope of the prefects' remit at the height of their powers, although over time certain tasks or functions passed to others.

> By far the most important of the financial departments was that of the praetorian prefects. They were responsible for the rations, or later ration allowances (*annonae*), which formed the bulk of the emoluments of the army and the civil service, including the palatine ministries, from the highest ranking officers, the *magistri militum* and the prefects themselves, down to the private soldiers and the humblest clerks and attendants. They were responsible likewise for the fodder or fodder allowances (*capitus*), of officers and troopers, and of civil servants who held equivalent grades. They had to supply corn to feed the two capitals and some other leading cities of the empire, and paid the *navicularii* [who held government contracts to transport grain in their ships] the freight charges for sea transport. They had to maintain the public post, furnishing the beasts of burden and their fodder, and the rations of the staff. Down to the end of the fourth century they managed the arms factories, and they continued thereafter to supply their raw materials and rations for the workers. They were responsible for public works, in so far as these did not come under the care of the urban prefects in Rome and Constantinople or the city authorities in the provinces, or the army on the frontiers: roads, bridges, post houses and granaries were their particular care, and for them they levied the stone, timber and labour required ... they had annually to compute the needs of the state in detail and calculate the rate of indiction [general taxation rate] required.[6]

Since the praetorians controlled justice, foreign affairs, home affairs, the levying of taxes and the distribution of provisions, their activities and those of their subordinates were monitored by officials working under the master of offices and the quaestor. These were men at all ranks, including men known as *comites provinciarum*, 'counts of the provinces', whom we see working alongside vicars in their dioceses. There is no indication whether this was a permanent post attached to a particular vicar, but it is clear that the counts roved widely, ensuring that the emperor's interests were served, rooting out corruption and reporting abuse.

There was, therefore, within the administration a duplication of responsibility, which was intended to provide oversight but may also have layered bureaucracy and increased inefficiency. The system would, over time, acquire even more layers, with new offices created to perform established tasks, and surviving offices performing new tasks. Indeed, the tendency not to disband bureaux, but simply to duplicate and allow the older office to wither, led to the bureaucratic system that we call 'byzantine'. Many administrative posts were regarded by those who held them as prizes rather than jobs, to be won then exploited before moving on to the next. A successful job collector might eventually return home with sufficient funds to buy land and status for his family. Continued employment in the administration over several generations conferred substantial, entrenched privileges. Bureaucrats, and their sons, and their grandsons, were exempted from serving as decurions in their home towns or cities, and might not be required to billet troops, nor to supply materials or labour for public works. Similar exemptions were afforded to bishops.

Emperors considered bishops to be another group of subordinates, whose spiritual and pragmatic authority was not qualitatively different to their own, just less abundant. Henceforth every city had its own bishop, with a hierarchy that mirrored that of the cities themselves, each operating within their region and diocese, and with rural parishes answering to their urban bishop. The early Church councils make it clear how closely arrangements in church administration were modelled on those in provincial administration. Although all bishops were nominally equals, the importance of their sees determined precedence, and Rome, Antioch, Alexandria and Constantinople were pre-eminent. At the second ecumenical council, held at Constantinople in 381, the third canon established that 'The Bishop of Constantinople is to enjoy the privileges of honour after the bishop of Rome, because it is New Rome (*nea Rome*)'. From 451, the bishop of Constantinople enjoyed the honorific 'Patriarch'.

General taxation revenues met the burden of funding the administration, in all its complexity, but these vast costs were still lower than the expense of the army and the public post. Beyond the payment of salaries, these both consumed massive resources to sustain their infrastructure, including factories, workshops, tanneries, inns, stables and stud farms. The enormous costs of sustaining the army is well known, but less so the public post. It has been calculated that feeding the horses at a single post station (*mutatio*) might cost up to 1,500 solidi annually, to which must be added the costs of tack and stables, the grooms and farriers and their needs, and veterinary care. As we have seen, there was a station every eight to ten miles and every third or fourth station was a more substantial inn (*mansio*), which might comprise several buildings within an enclosure and might be fortified and defended.[7]

The imperial household, as distinct from the palatine administration, was funded from lands and properties owned by the emperor himself. However, it is pointless to seek to distinguish properly between public and imperial wealth, land and revenues, since 'it is undisputed that from Augustus himself onwards the emperors in effect disposed as they wished even of those funds which were clearly public'. Emperors did put great store on personal largesse and desired to demonstrate it clearly on certain occasions. None is clearer than the manner in which Constantine and his family endowed the churches of Rome as set out in the *Book of Pontiffs*, even as he offered the use of the *cursus publicus* to bishops to travel across the empire at public expense.[8]

Building a New Rome

Constantinople was defined by water. The Sea of Marmara lay to its south, the Golden Horn, an inlet, to its north. They met at the Bosphorus, the channel that separates Europe from Asia. The smell of salty water permeated the air of the late antique city of Constantinople, so distinctive and different from that of fresh water, or rainwater falling on to dry earth or sitting, stagnating, in an open cistern. Fishmongers bought the catch from fishermen and sold it raw or cooked from stalls and shops around the central fish market. Today's stink of diesel would have been unknown, but the aroma of fish roasting on grills would not, nor the smells of freshly caught fish in the morning and their remnants on a sunny afternoon, bait for the next catch. The citizens of Constantinople took to the water to swim and relax, as do small boys today off the jetties that line the shores of the Golden Horn. People travelled by water within and beyond the city, to work, to transport goods and to catch fish. For half of the sixth century a whale terrorised shipping in the narrow strait between the European and Asian shores, earning the name Porphyrius, 'Purple' or 'Royal' beast, before it beached itself while pursuing a pod of fleeing dolphins.

The city of Constantinople was set out on the model of other eastern Roman provincial capitals, such as Antioch, with a grand colonnaded central street (*embolos*), called the *mese* or 'middle street', which was punctuated by public squares and circular plazas, monuments, columns and statues. By the end of the fifth century, the essential form of the city was well established (Map 10), with the hippodrome-palace-cathedral complex linked to the fora and marketplaces, and to the land walls and sea walls by greater and lesser roads projecting from the Mese. The Mese was also the city's main processional route, running from the Golden Gate, a grand entrance, to a terminus near the hippodrome and Hagia Sophia, the city's cathedral church, typically called simply 'the great church' (*megale ekklesia*). Greatly damaged in frequent riots and fires, but always rebuilt, the city centre was a stage for imperial majesty, but also a venue for entertainment and exercise, administration and government. Bureaucrats despatched thousands of documents to the provinces, preserving sealed copies of them in archives, of which just the lead sealings have survived. Their bureaux taxed the goods that flowed back and forth within the city and those that came into it, monitored the city's ports and warehouses, tanneries and factories and their products, regulated its tradesmen, controlled weights and measures, minted coins, and much else besides. Garrisons defended the palace and the walls, their kit manufactured in imperial workshops, their horses bred on imperial stud farms. Engineers maintained the water supply of the city with aqueducts and cisterns feeding fountains and baths, all overseen by water guards and a 'Master of the Waters' who answered to the mayor of the city, the urban prefect (*eparchos*). All served the emperor, who presided over the business of government each morning and some afternoons in the imperial palace.

The Great Palace of the eastern Roman emperors at Constantinople is today notable for its absence (Map 10, detail). Its remains, between Hagia Sophia and the landscaped remains of the hippodrome, lie underground. Only a small portion of the site has ever been formally excavated and a few fifth- to seventh-century floor mosaics are accessible to the public. Rather more has been excavated privately, in the basements of bars and restaurants, hotels and shops in modern Istanbul. The palace was built for Constantine the Great, partly new and partly developed from existing structures that abutted the hippodrome. Constantine was above all a general, and therefore his palace was in its first iteration a military installation with an administrative function. The great bronze (*chalke*) gate opened on the public side to the *regia*, the royal road, bordered to the north by the Augustaeum, a large plaza named for the empresses. On the palace side the Chalke opened to the barracks of the imperial palatine guard, the *scholae palatinae*. Of these, forty *candidati* were chosen to form the emperor's personal bodyguard. By the end of the fifth century, however, the *scholae* had become bound to the palace rather than the emperor, and to court ceremonial rather than conflict. A second guard unit, the Excubitors ('Sentinels'), was created to protect the imperial person within and without the palace. Its commander was a powerful man indeed, and proximity to power from a base within the palace ensured the succession of several commanders of the Sentinels to the throne, including, in the sixth century, Justin, Tiberius and Maurice. Justinian served in the Excubitors before promotion by his uncle, Justin.

Beyond the guards' barracks lay a courtyard surrounded on three sides by a semicircular portico that gave the space the form of a horseshoe, the Onopodium, which granted access to the inner palace from the Tribunal, an open space to its north. In the centre of the outer wall was a gate, the Triple Door, and above this a gallery, the Heliacum, from which the emperor might address those gathered in the Tribunal. Beyond this stood the Daphne, comprising a southern courtyard enclosed by a porticoed villa in which the emperor's private apartments were situated, and a northern courtyard flanked to the east by a reception hall, the Augusteus, named for the Augustus, probably the first throne room, and to the west by a banquet hall, the Triclinium of the Nineteen Couches, where the emperor hosted dinners of state, reclining in the presence of high dignitaries and ambassadors. The two halls were linked by passages that met at an octagonal chamber, an imperial dressing room. Court ceremonial required frequent changes of costume. Beyond this stood the Consistorium, a large council chamber where the emperor consulted with his inner circles of advisers and where promotions took place. The number and juxtaposition of halls and rooms in this upper terrace palace complex reflects its development over the period between the fourth and sixth centuries. The palace would, in the sixth and seventh centuries, expand further to occupy a lower terrace running down to the sea walls on the southern shore of the peninsula, and over time the buildings of the upper terrace fell out of use and into dereliction.

North of the Tribunal were placed the grand imperial Baths of Zeuxippus. The northern entrance to the baths lay on the southern side of the Augustaeum. At the south-eastern end of the Augustaeum stood the Magnaura, a name derived from its being a 'great hall' (*magna aula*), a reception hall used for greeting foreign dignitaries. Proceeding westwards

Figure 8. Column of Constantine, Istanbul

from the hippodrome along the Mese, one passed several palaces belonging to rich families, including the Palace of Antiochus, adjacent to the hippodrome, and the Palace of Lausus, which was famous for its sculptural collection. A short walk took one to the circular Forum of Constantine, in the middle of which stood a tall column constructed from seven porphyry drums that together formed a purple pillar on a stepped base that supported a bronze statue some thirty-seven metres high. This was the column illustrated in the graphic itinerary that would later be a model for the Peutinger map (Plate 9). The drums of Constantine's column were brought from Mons Porphyrites (modern Jabal Abu Dukhan), the porphyry quarry in Egypt's eastern desert. The column was damaged by several earthquakes, and as early as 416 was encircled by the metal hoops that are still in place (Figure 8). The statue, of the emperor as a nude sun god, dominated the skyline of the city for centuries, although not without incident. The spear the god-emperor was holding fell in 554, to be replaced, perhaps by another spear or by a sceptre. A globe surmounted by a winged victory he was clutching was thrown from his other hand by an earthquake in 477, and again in 869, to be replaced with a *globus cruciger*, an orb surmounted by a cross. On the northern side of the forum was placed the senate house, furnished with bronze doors once given by Trajan to the Temple of Artemis at Ephesus. These were embossed with a Hellenistic depiction of the Gigantomachy, the battle between the gods and Giants.[9]

The Mese continued in a straight line from Constantine's forum to the Philadelphium, where stood the Capitol, a building associated with the imperial cult. A statue group located here was believed to show Constantine and his three sons embracing, demonstrating 'brotherly love'. The discovery of a fragment of porphyry that fits the famous statue group of Tetrarchs taken from Constantinople to Venice in 1204 suggests a different original identity for the statues. Immediately to the east of the Philadelphium was the Forum of the Bull (Tauri), which at the start of the fifth century became the Forum of Theodosius. A 'historiated' column – decorated with images in a narrative sequence – dedicated to Theodosius I was raised at its centre. There the Mese bifurcated, running to the walls to the south-west and north-west. Heading south-west one passed through two more commercial plazas, the Forum of the Ox (Bovis) and the Amastrianus, before reaching a third imperial forum, the Forum of Arcadius. Here again the road split, with one branch reaching the walls at the Rhesium Gate, while the Mese continued south-west, splitting again to arrive at both the Xylocercus Gate and the Golden Gate.[10]

Constantinople had two circuits of walls. The first, built under Constantine, were supplemented by massive new land walls, begun under Theodosius I with the erection of a triumphal arch that became the Golden Gate. The land walls were developed in earnest between 405 and 413, under the direction of Anthemius, praetorian prefect to both Arcadius and Theodosius II. Anthemius was assisted by his son Isidore, who was appointed as urban prefect. The new land walls lay two and a half kilometres west of the city's Constantinian walls and stretched almost seven kilometres from the shore of the Sea of Marmara, just south of the Golden Gate, to the banks of the Golden Horn east of Blachernae. An extension to the west, shortly after 626, incorporated Blachernae, with its palace and shrines (Figure 9). The land walls comprise an inner wall, built first, that was four and a half to five metres thick and up to thirteen metres high, with towers spaced at around 180 metres apart; an outer wall up to two metres thick and nine metres high; and beyond both a terrace and

Figure 9. Theodosian land walls of Constantinople at Blachernae

a wide ditch (perhaps never a moat) some seven metres deep supported
internally by side walls and buttresses. Badly damaged in an earthquake
in 447, the land walls were rebuilt and have survived in part until today,
with additions and accretions, conservation and over-restoration. A far
longer circuit of sea walls was completed in 439, by the praetorian prefect
Cyrus, which joined the land walls to north and south, but also supported
the redevelopment of the harbour of Eleutherius on the Sea of Marmara,
henceforth known as the Theodosian harbour, and today as Yenikapı
(Plate 10).[11]

It was the success of Constantinople, rather than its fortunate situation, that required the construction of the expanded Theodosian harbour.
Of the thirty-seven wrecked ships excavated at Yenikapı, thirty-one were
propelled by sails. The principal exceptions were war galleys, which relied
on oarsmen. Sailed ships that crossed the Black Sea, approaching the city
from the north, were more fortunate than those travelling from the south
because in the summer sailing season the prevailing winds were northerly and thus against those sailing from the Mediterranean, including the
incoming grain fleet. Throughout our period the harbour was maintained
and regular repairs and new construction were carried out. Evidence from
dendrochronology shows that new docks were constructed with wood cut
in the 520s (similar to that used in the construction of Hagia Sophia),

Figure 10. A wooden jetty at the Theodosian harbour (Yenikapı)

then repaired in the 580s, and again in the 610s. Jetties were repaired after damage that might be associated with storms, earthquakes and concomitant tsunamis (recorded in AD 447, *c.* 477, and 557), or just because submerged wood rotted over time (Figure 10). The wood, principally oak, appears to have been shipped into the harbour from the Black Sea, likely from forests in the Haemus (Balkan) Mountains.[12]

Animal bones have been unearthed in great numbers at Yenikapı, and we have mentioned earlier exotic imports such as ostrich, macaque and elephant. Native animals include hares, weasels, foxes and bears, while birds include vultures, pelicans and storks. The bones of bottle-nosed and common dolphins have been found alongside shells of both loggerhead sea turtles and tortoises. More common are the bones of donkeys, mules and horses, which will have worked at the harbour. Some bones show signs of brutal hard work, including fractured limbs and palates punctured by bits. It appears to have been established practice to throw dead dogs into the harbour. The Yenikapı excavations have uncovered the skulls of more than 500 dogs, 97 per cent of them mesocephalic breeds, meaning with middling snouts, almost all being medium-sized dogs of light or medium build such as terriers. Only 3 per cent of skulls were dolicocephalic, meaning there were only fifteen dogs with elongated snouts like that of a German Shepherd. Packs of feral dogs were a frequent threat to citizens,

Figure 11. Hunting dogs chasing hares, Great Palace mosaic, Istanbul

but many of the Yenikapı dogs died at an advanced age, suggesting that they were well kept and fed, both as pets and working dogs. Among the many animals depicted in the Great Palace mosaics are hunting dogs chasing hares (Figure 11). John Malalas, who offers some wonderful quirky insights into life in the city, writes about one remarkable dog that could identify coins by their emperors, and could also tell 'pregnant women, brothel-keepers, adulterers, misers and the magnanimous'. A far later work suggests the use of guide dogs by the blind. More commonly, dogs were used to guard property in the city, and, outside the walls, to herd flocks and for hunting. Superstition held that a dog seen early in the morning presaged bad luck.[13]

Within Constantinople's land and sea walls, existing public spaces were renovated and new plazas were added at the start of our period, all adorned with monuments. Perhaps the most famous today is the Theodosian Obelisk (Figure 12), also known as the Egyptian Obelisk. A granite monolith carved for Pharaoh Thutmose III (r. 1479–25 BC), the obelisk is one of two raised at the Temple of Amun at Karnak. Constantine wanted it for his city, but it lay on a dock at Alexandria for some decades until it was brought there and raised under Theodosius I to mark his victory over

Figure 12. Theodosian Obelisk and masonry obelisk in the hippodrome at Constantinople

Magnus Maximus in 388. Inscriptions on two faces of the lower marble base upon which the obelisk sits attest to its erection after the defeat of Maximus; that the raising took thirty (in Latin), or thirty-two (in Greek) days; and that it was the work of Proculus, who was then urban prefect. Since Proculus was put to death in 392, there is only a short period within which the obelisk could have been raised. On the third, north-eastern face of the lower marble base, the obelisk is shown on its side, being raised; on the fourth, south-east side, the obelisk is shown in place in the hippodrome, where a chariot race is taking place. The inscriptions on the marble base do not reveal that the obelisk was broken during its erection, so that its height

Figure 13. Theodosian Obelisk, imperial scene on upper base.
Obelisk above is supported on bronze cubes

is 19.6 m (64' 3") rather than the original 34.8 m (127' 6"). A second marble base was carved, therefore, to accommodate the smaller footprint of the shorter obelisk, which was then placed upon the original inscribed marble base, supported by four bronze cubes at its corners. This ad hoc measure has been invaluable to historians and art historians, for on the four carved faces of the upper base the emperor and his entourage are depicted in the *kathisma*, the imperial box in the hippodrome (Figure 13).[14]

To further commemorate Theodosius's victories, and to mark his own succession, Arcadius completed his father's redesign of the Forum Tauri in Constantinople. In the centre of what was now called the Forum of Theodosius stood the Column of Theodosius, a tall drum carved with a spiralling narrative scene of Theodosius's victories that was surmounted by a statue of the eponymous emperor. The forum and the column emulated those of Trajan in Rome, since Theodosius falsely claimed descent from Trajan. However, the design also echoed and surpassed in scale and splendour Constantine's forum, column and statue. A second statue of Theodosius astride a rearing horse stood nearby on a lower rectangular marble base, which was possibly brought from Antioch, a reused statue of a Hellenistic king. Its new identity was revealed in an epigrammatic inscription recorded in the *Greek Anthology* (XVI. 65).[15]

Figure 14. Knotted columns, alluding to the club of Hercules, raised at the Forum of Theodosius on the Mese, now Divan Yolu, Istanbul

You sprang from the East to mid heaven, gentle hearted Theodosius, a second sun, giver of light to mortals, with ocean at your feet as well as the boundless land, resplendent on all sides, helmeted, reining in easily, O great-hearted emperor, your magnificent horse, although he strives to break away.

Later sources suggest that there were also equestrian statues of Arcadius, Honorius and Theodosius II in the forum, as well as two monumental arches, each constructed of four columns carved in the shape of knotted wood, perhaps alluding to Hercules's club. Several of these knotted columns can be seen today in pieces along Divan Yolu, the street the follows the path of the Mese in modern Istanbul (Figure 14). Theodosius did not, in fact, 'spring from the East', but was born in Spain, so the design may refer to the 'Pillars of Hercules', which flank the Straits of Gibraltar.[16]

Further along the Mese, a forum was established and named for Arcadius himself, at the centre of which stood another narrative 'historiated' column celebrating that emperor's victory over the Goths led by Gainas (Plate 13). We shall turn to this episode in the next chapter. Constantinople gained further imperial fora in the fifth century. A forum of Theodosius

Figure 15. Column of Marcian, Istanbul

II was constructed in 435, according to Marcellinus Comes, 'in the place called Heliane', which is to say the former Palace of Helenianae on the southern branch of the Mese as it approached the Golden Gate. This included a semicircular portico called the Sigma, which was the name by which the forum was commonly known. There a statue of Theodosius II was erected on a high column. Marcian's forum lay immediately to the north-west of the Forum of Theodosius I, on the northern branch of the Mese. It is identified today by a column, which at around ten metres (33') tall is far smaller and less impressive than those that dominated the city's other imperial plazas (Figure 15). Its extant inscription, once in bronze, indicates that the prefect responsible was Tatianus, and that the column was once surmounted by a statue of Marcian. The location of Leo's forum is far less certain, although it has been suggested that it stood in the vicinity of the extant Church of St Irene, in what is today the second courtyard of the Topkapı Palace (Figure 16).[17]

There were frequent renovations and restorations of buildings and monuments damaged by fires and earthquakes. The fifth century is full

Figure 16. Hagia Irene and Hagia Sophia, Istanbul

of examples. In 404, a fire engulfed the great church of Constantinople and the nearby senate house. At this time a new great church was built that would later burn again, in 446. In 406, a fire burned the hippodrome gates, as well as the city's textile market (the 'Prandiara') and sections of the porticoes lining the main street. In 407, the city was struck by a devastating earthquake and associated tsunami. The seventh-century *Easter Chronicle* (often called the *Paschal Chronicle* or *Chronicon Paschale*) reports that 'the bronze tiles of the Forum of Theodosius were swept away to Kainoupolis, and Christ's emblem on the Capitol fell down, and many ships foundered and a considerable number of bodies were cast ashore at the Hebdomon'. Kainoupolis was the 'new city', an area of reclaimed land between the Forum of Theodosius and the new Theodosian harbour. When, during an earthquake in 420, stones began shearing away from the Column of Constantine, the iron bands that still bind the drums were added (see Figure 8, p. 104). In August 433, according to Marcellinus Comes, 'most of the northern part of the imperial city was ablaze for three successive days and collapsed to the ground'. An earthquake in 447 brought down sections of the new land walls, which were promptly rebuilt by Constantine the praetorian prefect. The greatest fire to date, which began on 2 September 464, burned the easternmost eight of the city's fourteen administrative districts. A second fire in 469 burned much that had been rebuilt, but it was built up again, before being burned again in 476. Every time the city burned toxic fumes will have filled the air for weeks. Aerosols from the lead used in construction will have settled and poisoned the earth in and around the city.[18]

Constantinople was supplied with drinking water brought from far beyond the city's walls. Water and its conspicuous consumption demonstrated power, and Constantinople lacked what Rome had in the Tiber and Antioch had in the Orontes, a symbol for the city and an abundant natural supply of potable water. In Constantinople, there is only a single modest river, the Lycus, which runs part of its length underground. It is uncertain whether Constantine planned to supplement the single Aqueduct of Hadrian, which served the needs of the palace complex and the neighouring Baths of Zeuxippus. The baths have generally been attributed to Septimius Severus. However, the juxtaposition of the baths with an imperial palace and hippodrome recalls Tetrarchic developments elsewhere, notably Diocletian's on the Orontes Island at Antioch. It is possible that the Baths of Zeuxippus were developed by one of Constantine's rivals – Diocletian, Galerius and Licinius are all candidates – and appropriated by Constantine, who preferred to attribute the foundation to a great emperor he had not recently murdered.[19]

Certainly, it was at Constantine's initiative that the baths gained a remarkable sculpture collection, with around eighty statues brought from across the Roman world. Isocrates, Demosthenes and Aeschines, the greatest Greek orators of classical antiquity, reminded visitors of the instructional function of baths, alongside statues of seventy-seven more writers and fighters, prophets and heroes, goddesses (including three Aphrodites) and gods (including three Apollos). Love and war, therefore, were well represented, but could not approach the written and oral arts: practitioners of history and song, philosophy and science, oratory and epic. There were many more Greek thinkers than Roman statesmen – only, perhaps, Julius Caesar and Pompey – in this collection of 'mute bronze', which was given voice in around 500 by the poet Christodorus of Coptus, three decades before the baths and its statues were destroyed by a fire that engulfed Justinian's Constantinople in 532. So complete was the destruction that it was thought Christodorus may have invented the statues, until two marble plinths were excavated from the site of the baths, one of which was inscribed 'Hecuba' (EKABH, Figure 17).[20]

A second, higher waterline entered the city in 373, called the Aqueduct of Valens. Flowing from Thrace, the water passed through channels and aqueducts that crossed the countryside circuitously, with channels totalling some 268 kilometres (167 miles). It terminated in the famous viaduct, today's Bozdoğan Kemeri, which surmounts the valley between the city's fourth and third hills. The waters flowed thence into a new Great

Figure 17. Column base inscribed 'Hecuba' (EKABH), once at the Baths of Zeuxippus

Nymphaeum at the Forum Tauri. The completion of the high waterline, able to supply locations at an elevation of *c.*56 metres (184') or lower, was reported upon succinctly by Jerome: 'Clearchus, eparch of the city of Constantinople, is well known, by whom the necessary water which was daily awaited with vows is brought to the community.' The orator Themistius was on hand to praise the undertaking more lavishly, alluding to the arrival of water nymphs, whose 'names are Thracian and manly, but their beauty and splendour are extremely delicate'. Eventually this high line comprised almost 600 kilometres (373 miles) of channels.[21]

The Great Nymphaeum, which collected and distributed water from the new Aqueduct of Valens, was joined before 425 by three other large public fountains. This and other unique information about the early fifth-century city is preserved in a document of 425 known as the *Notitia Urbis Constantinopolitanae*, which lists structures to be found within each of the city's administrative districts. The Great Nymphaeum lay at the south-eastern edge of region X, a large district that stretched from the Constantinian walls to the Forum of Theodosius, bordered on the north by the Golden Horn. Regions V and IV, which also had public fountains, sat to the north of the new imperial complex, whose major landmarks were the Great Palace (region I) and Hippodrome (region III), and west of the acropolis and Hagia Sophia (region II). Region XIV, where the fourth public fountain was located, may have comprised the large area now between the Constantinian and Theodosian walls. The *Notitia Urbis* also records, in addition to eight large public baths (*thermae*), 153 private baths (*balnea privata*), with at least five in each of the fourteen regions of the city, attesting to the increased capacity and reach afforded by the expanded water delivery network (Map 10).[22]

An imperial edict of *c.*440, preserved in Justinian's *Code*, specifies that water from the older, Hadrianic line should henceforth be used only for the public baths and imperial buildings, suggesting that the Valens line was to supply most other demands. The city's waterlines also supplied several large outdoor reservoirs and many more underground cisterns, of which more than 150 have been identified. The oldest large cistern, named Modestus for the prefect of Constantinople who oversaw its construction in 363–69, predated the Valens line. However, three vast open-air reservoirs were added in the fifth century, those of Aetius (420/1, which is today a football stadium, Figure 18), Aspar (459) and Mocius (499/500, or possibly 514/15). These were constructed outside the walls established by Constantine but within those added by Theodosius II and supplied the majority of the city's inhabitants. Water stored in the open-air cisterns was topped off with rainwater and generally was far less clean. In normal circumstances, therefore, it was probably destined for agricultural or industrial uses, or for display purposes. Fresher water suitable for drinking would be stored underground, where it could be kept cooler and sufficiently aerated. Anastasius, during whose reign the Mocius reservoir was built, also had a cold cistern installed in the vaults of the hippodrome, beneath the so-called *sphendone*. In the sixth century, the city's most famous surviving cisterns were built, the Binbirdirek Cistern, supplied by the Valens line, and the Basilica Cistern, today's Yerebatan Sarnici (Plate 11), which was at a low enough elevation to be supplied by the Hadrianic line. Many additional cisterns were built, occasionally converting older buildings for the storage of water.[23]

Exactly how water entered and left the cisterns remains unclear, although it did so in abundance to supply nearby buildings and fountains through terracotta and lead pipes. Large marble water pipes were

*Figure 18. Cistern of Aetius, a large open-air water reservoir
that is today a football stadium, Istanbul*

discovered in 1964 beneath what was the Mese, which had once carried
the water at pressure between the Forum Tauri and the Forum of Constantine. Officials known as 'water guards' (*hydrophylakes*) were charged
with overseeing and maintaining the flow of water to and through the city,
and those through whose land the water channels passed were obliged by
law to keep them clean and maintained, in return for which they were
spared any further financial imposition by the state. Officials policed
encroachments on the water supply. Although a pipe of a certain diameter
could be purchased, any unsanctioned private diversions of water were to
be resisted and reversed. An edict of Zeno commanded the urban prefect
to undertake a 'careful investigation of fountains which were originally
public, as well as those which, derived from private sources, later became
public and then have been converted to the use of private individuals'.
The water supply could also be threatened from outside the city. A brief
reference by John Malalas suggests that an aqueduct bringing water to
Constantinople was cut by Theoderic in 487, but quickly restored. When
this happened again later, during the great Avar siege of 626, the line
remained cut for over a century.[24]

The imperial city

Theodosius I was the first emperor after Constantine to base himself principally in Constantinople, and was there more than anywhere else from the time he first entered the city in triumph in November 380 to his final departure from it in August 394. Five imperial children were born there in the Great Palace, namely Honorius (384), Pulcheria (385), Gratian (388), Galla Placidia (392) and John (394). Imperial celebrations brought tens of thousands on to the city's streets and into the hippodrome. The crowded state of the city at that time is well described by Gregory of Nazianzus, the city's archbishop. Every imperial victory was also marked by processions and games.

After Theodosius, emperors were resident for much of the year in Constantinople. They moved around it, frequented its summer palaces and hunting lodges, and built their own. Beyond its walls, the city was surrounded by the lands and villas of those who could afford a suburban idyll, including emperors and their families. Basiliscus, who usurped Zeno and reigned very briefly, found refuge at the imperial suburban palace at Hebdomon, seven Roman miles outside the walls on the southern shore. Justinian, who reigned for almost four decades but travelled only rarely beyond the hinterland of Constantinople, had the Hebdomon renovated or extended and its harbour dredged, so it could be reached more easily by imperial barge. It appears to have been renamed, briefly, the Secundianae (or Jucundiana). There were also imperial villas across the Bosphorus at Chalcedon (Kadiköy) and Hieria (Fenerbahçe), where Theodora, Justinian's consort, preferred to summer. Paul the Silentiary described it in a poem. 'The sea washes the abode of the earth, and the navigable expanse of the dry land blooms with marine groves. How skilled was he who mingled the deep with the land, the seaweed with gardens, the floods of the Nereids with the streams of the Naiads.' Theodora also used an urban mansion immediately south of the Great Palace, the so-called House of Hormisdas – also called the Palace of Hormisdas, and later the Bucoleon Palace – as a refuge for Monophysite holy men. John of Ephesus recounts how she collected in this 'great and marvelous desert of solitaries' no fewer than 500, who 'under the constraint of persecution' by the Chalcedonians, had 'come down from columns and been ejected from places of seclusion' to occupy this palatial monastery dedicated to St Sergius.[25]

Justinian built a pavilion at Pege, a holy healing spring immediately beyond the city's walls, which he and subsequent emperors visited at least annually on Ascension Day, the Thursday that falls forty days after Easter Sunday. An account of the annual visit is preserved in the *Book of Ceremonies*, which has the emperor setting out from the Great Palace when it was still dark, travelling aboard the imperial barge (*chelandion*) as far as the jetty at the Golden Gate, then riding through the fields beside the land walls as far as 'the gate that leads out opposite the spring'. The emperor is not alone, of course, but accompanied by his entire court, which then processes to the shrine. 'The patricians and the whole senate proceed on foot, and the emperor, only he on horseback, is escorted by them.' Having arrived at the church and changed costume, the emperor emerges to acclamations by representatives of the city's circus factions, the Blues and Greens, who hand the emperor 'rose-entwined crosses and sweet-smelling flowers', and awaits the arrival of a procession of candle-bearers led by the patriarch of Constantinople.[26]

There were other regular imperial processions that linked the suburbs to the city and employed the cityscape and its colonnades and plazas as sites for acclamations, prayers and costume changes. These intersected with a calendar of stational liturgies presided over by the patriarch of Constantinople. We know that stational liturgies, always led by the named bishop of a city, took place in many cities, including Antioch and Jerusalem, Alexandria and Oxyrhynchus, Rome and Milan, where they were linked to an urban calendar of commemoration, including the anniversaries of earthquakes or the death dates of martyrs. The dates of processions that marked key events in the life of Constantinople included: 1 September, annual indiction and great fire of 464; 25 September, great earthquake of 447; and 11 May, dedication of the city in 330. It was during one of Constantinople's urban processions that the Trisagion was first shouted: 'Holy God, Holy Strong, Holy Immortal, have mercy on us!' Litanies of the fourth and fifth centuries preceded the development of imperial shrines and grand churches and to some extent defined their form and architecture. Processions ended at churches, and became a defining feature of the liturgy later performed within them.[27]

Ceremonial journeys and arrivals could also be made by water. From the foundation of the city, both Sykae (modern Galata), across the Golden Horn, and Chalcedon (modern Kadiköy), across the Bosphorus, were parts of Constantinople's cityscape and its waterscape. A regular procession by water took place each year on 29 June, when the faithful visited the

Church of St Peter and St Paul near the Palace of Rufinus, which gave its name to a suburb of Chalcedon, Rufinianae. The same group might have travelled on several of the regular ferries to Sykae to celebrate Ascension Day at the Church of the Maccabees at Elaia ('Olives'), a hill above the suburb. The name of the hill echoed the site of Christ's Ascension at the Mount of Olives in Jerusalem.[28]

The sacred topography of the city was transformed by the acquisition of relics, gathered from across the empire and brought into the new churches of Constantinople. Theodosius I set the standard when, in 391, he personally carried the head of John the Baptist to the Hebdomon Palace and installed it in a newly dedicated church. Shortly before 400, the body of St Phocas was conveyed to the city by sea, and preached about by John Chrysostom. In 406, the relics of the Old Testament prophet Samuel were 'conveyed to Constantinople by way of the Chalcedonian jetty', which is to say across the Bosphorus, and paraded with great ceremony by Emperor Arcadius, his praetorian and urban prefects, and the whole senate. In 411, Samuel's remains were transferred to a newly built sanctuary at the Hebdomon. In 415, the bodies of Joseph, son of the prophet Jacob, and Zachariah, father of John the Baptist, were brought into the great church on 2 October, eight days before the church itself was rededicated. In that same year the newly discovered relics of St Stephen the Protomartyr were installed in the Church of St Sion at Jerusalem, receiving a homily from the local presbyter Hesychius. Jockeying began immediately to secure the relics for Constantinople. Aurelian, the praetorian prefect, built a chapel, but failed to secure even a single bone. Pulcheria, the emperor's sister, was more successful, and in 421 received an arm, which was placed in a chapel she had constructed within the imperial palace. In return, she sent alms and a large golden, gem-studded cross to be erected on Golgotha. The rest of Stephen arrived in 439, at the initiative of Empress Eudocia, to be deposited in the Church of St Laurence.[29]

The arrival of Stephen's relics was akin to an imperial *adventus*, a ceremonial entry marked by solemn ceremony and cheering crowds. A scene very like this, although now understood to depict a rather later episode, is captured in a carved ivory plaque, today known as the Trier ivory. A casket is shown in the clutch of two men, bishops seated on a carriage drawn by two mules, which is passing the gate to the imperial palace on which an image of Christ is prominently displayed. They are preceded by a procession of figures holding scrolls, who approach a standing figure in

*Figure 19. Trier ivory depicting the arrival of holy relics
at the Great Palace of Constantinople*

imperial costume, and the empress holding a large cross. They all stand
before a chapel, newly built, roofers still struggling to place the final tiles.
All are watched from a three-storey arcade by figures dangling censers or
bells, perhaps a choir, and above them many more figures gazing down
(Figure 19).[30]

Shortly before Stephen, Constantinople had celebrated the return
from exile of John Chrysostom, whose relics entered the city with great
celebration on 27 January 438. John was interred at the Church of the
Holy Apostles, which was added to a mausoleum built for Constantine
and was otherwise reserved for the tombs of members of the imperial
family. Theodosius I was buried in the original mausoleum with Constantine and Constantius II, probably at the expense of Jovian, whose
remains were removed to another part of the complex, the north stoa.
Arcadius, Eudoxia and their son Theodosius II were buried together in
a new cruciform structure, the south stoa. The transformation of Holy
Apostles into a grand imperial mausoleum was accelerated by the decision to disinter imperial bodies buried elsewhere and bring them to the
building, notably empresses including Helena, mother of Constantine,
and both wives of Theodosius I. Quite remarkably the body of Julian
the Apostate was brought from Tarsus to join Jovian in the north stoa, a
new imperial tradition apparently overriding other considerations. Space
was later made to accommodate imperial couples. Marcian and Pulcheria, Leo and Verina, Zeno, Anastasius and Ariadne were all inserted into

Constantine's mausoleum, which in later centuries became even more crowded. Later descriptive accounts report on the many coloured marble sarcophagi, of Egyptian porphyry and green Thessalian stone, and several purple and white marble tombs have survived, including at Hagia Irene at the Topkapı Palace and in the grounds of the Istanbul Archaeological Museums (Plate 12).[31]

The image of the emperor at New Rome

The dominant image of the emperor in imperial art of the third and fourth centuries is as a commander and victor, a vanquisher of barbarians, and a beneficiary of foreign tribute and divine favour. The first duty of each Roman emperor was to command the army, the Roman people at war. It is for this reason that many generals of the fourth century became emperors, acclaimed by their own troops, including Constantius II and Constantine I, Jovian, Valentinian II and Theodosius I. Others acclaimed in the same way are remembered as usurpers and pretenders, meaning only that they failed to hold the throne, including Magnentius, Magnus Maximus and Eugenius. Three fourth-century emperors died commanding their troops, namely the pagan Julian and the Arians Constantine II and Valens. On each occasion this was explained by religious opponents as God's judgement. When faithful commanders lost battles whitewashing was necessary, and there were many gifted writers willing to offer their pens. They understood that God offered indications of his favour and displeasure. It was the emperor's duty to earn favour and fight those who enjoyed the patronage of fallen angels and demons, like Julian. Just and victorious emperors ensured the safety of Christians and protected the church. The mighty right arm of the emperor wielded the sword of the Old Testament God, 'his fierce, great, and powerful sword' (Isaiah 27:1). The New Testament, a far more pacific collection, generally cast Christ in a gentler role. He gained the ultimate victory by knowingly granting an ephemeral win to his adversary, Satan, and allowing Judas to betray him. Christ knew 'that through death He might destroy him that had power over death' (Hebrews 2:14). Christ's triumph through defeat was the message disseminated in the Pauline epistles. It was the foundation for calls to martyrdom within Christian communities. It provided comfort to those who would be martyred and to their families, and an explanation for persecution. This Christ was of little value to emperors who yearned for earthly victories and triumphs in this world. For them, the New Testament offered Christ of Revelation, who returned in majesty seated on a white horse, 'just in judgement and just in war', a sword projecting from his mouth to smite the nations. 'For he it is who shall rule them with an iron rod and tread the winepress of the wrath and retribution of God.'

Although Revelation, the Greek apocalypse by John the Divine of Patmos, stood apart from the Orthodox lectionary, it survives in around 300 discrete Greek manuscripts, suggesting that it was very widely read and copied. The oldest fragments date to the second century and the oldest complete text to the fourth. In John's apocalypse, Christ takes Yahweh's role as cosmic conqueror, as foretold in Psalm 90 (91):13 – 'You shall tread upon the lion and adder: The young lion and the serpent you shall trample under foot' – and remembered at Luke 10:19 – 'Behold I give unto you the power to tread on serpents and scorpions and the power to overcome all the power of the enemy, nothing will harm you.' A rare image of this militant Christ has been preserved in a sixth-century mosaic at Ravenna's Archbishop's Chapel, which clearly emulates earlier imperial art. Christ is dressed as a general, with a cross slung over his shoulder in the manner an emperor might be shown with a spear or, occasionally, a trophy (Figure 20). His tunic billows out like a commander's cape and is fixed at one shoulder with an imperial brooch. Under his right foot is a cowed, crouching lion, and under his left a rather modest snake. Christ holds open a book at a Latin verse from John 14:6: *Ego sum via veritas et vita*, 'I am the path, the truth, and the life, nobody comes to the Father but through me.' This is a gloss on the psalm, a prayer-poem to the protection offered only by the Almighty.

The paradox of victory through defeat remained available as an explanation for an unexpected loss, and the ubiquity of sin allowed God to punish his elect. Still, emperors did all they could to demonstrate their piety and humility. According to Rufinus of Aquileia, Theodosius I 'prepared for war by arming himself not so much with weapons as with prayers and fasts, guarded not so much by the nightwatch but by nightly prayer vigils'. In contrast, he claims that Eugenius examined entrails and indulged in pagan sacrifice. As matters went against Theodosius at the Battle of Frigidus, he lay prostrate and prayed, shouting that his campaign had been undertaken for Christ. Consequently, a wind blew up of such strength and direction that it blew the arrows unleashed by Eugenius's

Figure 20. Militant Christ, Archbishop's Chapel, Ravenna

archers back against them. A strikingly similar message was communicated by John Chrysostom, preaching (probably) in Constantinople on the fourth anniversary of Theodosius's death, 17 January 399.

> When the two armies were drawn up facing each other, and clouds of spears were hurled, and his own troops were being turned back by the violent onslaught of the enemy, Theodosius leapt from his horse, threw his shield on the ground, and fell to his knees calling for help from heaven, transforming the site of the battle into a church, fighting with tears and prayers not arrows and javelins and spears. At this a sudden wind arose, and the spears of the enemy were blown back on those who had thrown them. Seeing this, the enemy, who till then had been breathing fury and slaughter, changed tack, acclaiming Theodosius as their emperor ... Unlike other emperors, his soldiers did not share victory with him. On the contrary, it belonged to him alone, and to his faith.

Certainly, Theodosius did not fight alone and there is no evidence that Eugenius was a pagan. These were fictions propagated by those who wished to cast Theodosius as a righteous warrior against the last generation of pagans, the heir to Constantine's singular legacy and unified empire, rather than as one of several emperors to have ruled together, and fought each other, over the past two decades. Civil war had become increasingly imbued with sacred meaning, as is clear from Rufinus's account of the public liturgical events that Theodosius had staged before he left Constantinople.[32]

Theodosius I's image as a ruler protected and rewarded by God was reinforced after his death by writers anxious to influence his heirs. Ambrose of Milan presided at a memorial service for Theodosius, offering an oration in praise of the emperor's piety. For Ambrose, Theodosius was a merciful ruler, a Christian soldier 'whom power impelled toward vengeance, but whom compassion recalled from taking vengeance'. We might have ended this first section of the book, therefore, where we began, at an imperial burial presided over by St Ambrose. Others, however, were less flattering of Theodosius, and when they looked to the heavens they did not see the sun's sanctifying beams energising a devout imperial family. Philostorgius considered the last years of Theodosius and the reign of his son Arcadius to be a nadir for the empire, 'the loss of human life was so great that never did any age know the like since time began'. He imagined this had been foretold by a sword-shaped star that appeared to Theodosius following his victory over Maximus, the inverse of the sign of victory and divine favour that had appeared to Constantine. Likewise, devastation in the reign of Theodosius II was foretold by the appearance of a peculiar solar eclipse, so dark that the stars were visible, when appeared a 'cone-shaped light, which some out of ignorance called a comet'. It traversed the sky for several months before it disappeared. This is surely the solar eclipse and 'star from the East' identified by Marcellinus Comes in 418. Let us turn, then, to a study of power and politics in the age of the Theodosians.[33]

PART 2

POWER AND POLITICS

6

THE THEODOSIAN
AGE, AD 395–451

Theodosius 'the Great' (r. 379–95) is known as the last emperor to rule the whole Roman empire, meaning both its western and eastern portions. A reason to begin a political narrative in AD 395, therefore, is that it allows focus to fall on the empire's eastern lands as a separate entity. There was not, however, a constitutional division of the empire into east and west at Theodosius's death. By placing his older son Arcadius in command in the east and his younger son Honorius in the west, Theodosius extended a practice that had endured, with significant breaks and through several civil wars, since 285, when the Augustus Diocletian had appointed a trusted subordinate as his junior emperor or Caesar. Countless emperors had tried to ensure the succession of their son or sons, but Diocletian had no son, only a daughter, and his innovation was to invest someone from outside his family as his partner and heir. In 293, by the same principle, Diocletian established an expanded imperial college, the Tetrarchy, where power was shared by two senior and two junior emperors. Soon, however, each of these men sought to promote family interests. Through civil war, the son of one emperor, Constantine 'the Great', established himself as sovereign Augustus across the whole empire. However, even Constantine formed an imperial college with four Caesars, his three sons and a nephew.

Divisions of empire

In 335, two years before his death, Constantine divided the empire. Constantine presided at Constantinople as sole Augustus, while his oldest surviving son Constantine II settled in Trier in the diocese of Galliae, and governed also Viennensis, Britannia and Hispaniae (see Map 2). Constans, the youngest son, received the spoils of Constantine's first great victories, in Italy and Pannoniae, with the cities of Rome, Milan and Sirmium. Flavius Dalmatius, a nephew, was granted Moesiae and Thracia, including Naissus, Constantine's birthplace. Constantius II, the middle son, was to administer the mighty diocese of Oriens from the imperial palace at Antioch. This arrangement did not survive Constantine's death. Dalmatius was killed immediately, and in September 337, each of Constantine's sons took the title Augustus and a portion of the empire. Constantine II returned to Trier with no more land than before, whereas Constans and Constantius II shared the Balkan territories of their dead cousin. Since Constans was still only seventeen, Constantine II served as his guardian. Tensions soon emerged between Constans and Constantine II, fomented by officials on both sides seeking to advance their own interests. In 340, Constantine II was killed. Constantius and Constans now ruled as Augusti, each holding one half of the empire. Constantius devoted himself to the war with Persia, spending the better part of the following decade in a regular routine: winter in Antioch, summer campaigning. He made only a few brief trips to Constantinople. Constans travelled extensively through his western provinces, but over time he lost the support of much of his army, and at Autun, on 18 January 350, Magnentius, one of his generals, was acclaimed emperor by Constans's own bodyguard. Constans was murdered shortly afterwards.[1]

Other pretenders rose and fell besides Magnentius before Constantius emerged victorious and took his cousin Gallus as his Caesar. When Gallus was put to death, Constantius replaced him with Gallus's half-brother, Julian, known to history as 'the Apostate'. Julian was an astute ruler and was recognised as Augustus in 360, before Constantius died in November 361. The brief pagan interlude of Julian's reign served only to sharpen the Christian triumphalism that attended his death in Persia in June 363, falling in war against the infidel. Julian was buried in Tarsus by his general and successor Jovian, who reigned briefly. Neither Julian nor Jovian is known to have appointed a Caesar. In March 364, a month after the death of Jovian, the new emperor Valentinian invested his brother Valens as Augustus with command of the empire's eastern provinces. Valentinian operated in the western provinces from a base at Milan, as we noted at the start of our first chapter, rather than from Rome itself. From northern Italy he confronted a series of threats in Britain, Gaul and Germany, dispatching armies to the Danube and to North Africa. Valens's court was at Constantinople, although he lost the city for some time to the usurper Procopius, and he spent much time in the field, engaged principally in a protracted war with the Goths.[2]

There is no indication that Valentinian's division of empire, an arrangement with his own brother driven by prevailing circumstances, was intended to constitute a permanent settlement. However, it served as a template for Theodosius I, who had two sons. Valentinian died in November 375 and Valens died fighting the Goths at the battle of Adrianople in 378. After Julian and Valens, no Roman emperor would die in battle until 811. Theodosius replaced Valens as emperor in the east. His imperial colleagues were Gratian and Valentinian II, both relatives of Valens, and Theodosius's own sons Arcadius and Honorius; his rivals for authority were Magnus Maximus (d. 388) and Eugenius (d. 394), both acclaimed as emperors, with Maximus recognised for a time by Theodosius. Gratian ruled the west from 367, first under his father Valentinian and later with his half-brother Valentinian II, who joined the imperial college in 375. Arcadius, who had just turned six years old, was raised to the rank of Augustus in January 383, to mark his father's fifth year in office, his *quin-quennalia*. Gratian died in August 383, and Valentinian II in May 392, both buried in a mausoleum in Milan. For a little under a year, therefore, Arcadius was his father's only recognised imperial partner, until they were joined in January 393 by Arcadius's younger brother Honorius, who was named as Augustus and co-ruler of the empire's western lands. This was in response to the election of Eugenius as emperor in the west, in August 392, and in preparation for Theodosius's campaign against him, which culminated in September 394 at the battle of Frigidus.

The battle of Frigidus was portrayed by ecclesiastical writers as a struggle between a Christian emperor, Theodosius, and a pagan usurper, Eugenius. But there is strong evidence that Eugenius was in fact a Christian. The empire was now predominantly Christian and the rate of conversion appears to have accelerated in the last decades of the fourth century in part because of new legal penalties. There was, however, no unified

Christian community across the empire that challenged and defeated a last pagan generation. Rather, there was a patchwork of rival communities that professed their faith competitively. Theological writings that have been preserved are frequently polemical and suited later tastes for reasons of style and theology. Works by devout Christians who were judged to have been heretics were not copied, so they appear only in refutations. The canons of Church Councils are where we look for dogmatic compromise and condemnation, each instance generating a substantial volume of dissenting literature, or provoking further dangerous and divergent thinking. A good example, and the state of play at the start of our study, is found in the canons of the second ecumenical council, convened by Theodosius at Constantinople in 381. There are, according to which side one chose, four or seven canons. Several of these are administrative, restricting the authority of metropolitan bishops to their own dioceses (canon 2), and asserting the primacy of the bishop of Constantinople over all others bar the bishop of Rome (canon 3). The first canon, however, is dogmatic, a denunciation of several Christological heresies, notably Arianism, Macedonianism and Apollinarianism, which challenged the position established at the first ecumenical council, convened at Nicaea in 325. Principal among these was Arianism.[3]

Arianism holds that Christ was created by God the Father, and therefore that Christ's nature, in its humanity and divinity, is distinct from and subordinate to that of the Father. Macedonianism, considered a semi-Arian heresy, held that the Son created the Holy Spirit, which therefore is subordinate to both the Son and the Father. Apollinarianism held that Christ had a human body, but his mind or spirit was divine, the Logos. Macedonianism and Apollinarianism arose after 325, but Arianism had been the principal matter before the bishops who assembled at Nicaea. There was a large Arian community in Constantinople, which Theodosius drove outside the walls to worship. Most of these, known as Homoians, were stridently opposed to another powerful Christian sect in Constantinople, the Homoiousians, who held that God the Father and the Son were of similar, but not the same, substance. Small distinctions were vital, because salvation could only be guaranteed by correct belief, in Greek *orthodoxia*. Orthodoxy and the issue of salvation were not trivial matters. There were also composite groups identified in our sources as 'barbarians', families of different origins who had migrated into the empire, peoples formed and held together by success in war and loyalty to their leaders, and increasingly by a common Christian faith. We know them as Goths, Huns and Vandals, and they were mostly Arians.

We may, perhaps, overstate how this divisiveness affected the lives of many Christians, since we read frequently of disputes and dogmatic pronouncements in our sources. Most Christians in Constantinople, Antioch and Alexandria knew little of the issues that drove discourse in Milan and Rome, and vice versa. Those who lived in smaller cities, or villages, had quite different priorities to highly educated writers. However, all might be swayed by the rhetoric of powerful preachers and the threat that heresy posed to their salvation and that of their families. As Gregory of Nyssa famously, and rhetorically, wrote of those gathered in Constantinople at the time of the second ecumenical council,

> The whole city is full of it, the squares, the market places, the cross-roads, the alleyways; old-clothes men, money-changers, food-sellers: they are all busy arguing. If you ask someone to give you change, he philosophises about the begotten and the unbegotten; if you inquire about the price of a loaf, you are told by way of reply that the Father is greater and the Son inferior; if you ask 'Is my bath ready?' the attendant answers that the Son was made out of nothing.[4]

Secular historical works of this period are far fewer than theological writings and some are deliberately opaque, including the poems of Claudian, written at the court of Honorius, and the letters and allegorical essays by Synesius of Cyrene, some of which were composed at Arcadius's court. We do not have the twenty-two books by Olympiodorus of Thebes or the eight-volume history written by Priscus of Panium, both courtiers of Theodosius II, although fragments of both are preserved in later texts. The pagan historian Zosimus, writing in Greek in the sixth century, uses Eunapius of Sardis for the period to 404, and Olympiodorus from 407. Unique information is also preserved by Procopius of Caesarea and John Malalas of Antioch, both writing under Justinian, who drew from sources now lost to us. All extant chronicles were compiled far later, including the sixth-century *Chronicle* of Marcellinus Comes, the seventh-century *Easter Chronicle* and the ninth-century *Chronicle* of Theophanes Confessor. Laws issued in this period are preserved in the *Theodosian Code* and as novels ('new laws').

Our fullest narrative sources are church histories by Theodoret of

Cyrrhus, who wrote a lucid and staunchly anti-Arian work that ends in 429, and by Socrates Scholasticus and Sozomen, which both end in 439. Socrates completed his work in that year and Sozomen about a decade later. Sozomen used Socrates as a source but employed more elevated language. A further, still longer church history in twenty-four books by Philip of Side was completed at about the same time but is now lost. Unlike the others, this was not presented as a continuation of Eusebius's church history, but it started, rather, with creation. Photius, writing in the ninth century, offered terse judgement on Philip of Side's verbosity. He also supplied a full epitome of a lost church history by Philostorgius, an Arian, originally published in twelve books and ending in 425. The extant histories share common sources but present material differently and omit key events and figures that might complicate their interpretations or offend their writers' quite different patrons. Theodoret of Cyrrhus also wrote many letters, of which around two hundred survive. The sermons and letters of John Chrysostom present the views of a key protagonist in the lives of the cities of Antioch and Constantinople.

Arcadius

Towards the end of 394, while he was in Rome, Theodosius appointed his *magister militum* Stilicho to serve as regent to Honorius, who had just turned ten. It was Theodosius's intention to return to Constantinople, where Arcadius and Honorius still resided, in order to dispatch Honorius to Milan, then the site of the western court. In his father's absence, Arcadius had occupied the imperial throne, but authority in all matters of state had been entrusted to the praetorian prefect of the east, Rufinus. Stilicho was married to Serena, Theodosius's niece and adopted daughter, and Rufinus hoped to marry his daughter to Arcadius. While Theodosius lived, he trusted Stilicho and Rufinus to act for him and on behalf of his inexperienced sons. When Theodosius died, on 17 January 395, Stilicho claimed that on his deathbed the emperor had commended both sons to his care, a claim repeated at Theodosius's memorial service by Ambrose of Milan. Whether Theodosius made this commendation, and what it meant, was disputed. It can hardly have meant, as Stilicho claimed, that he was to act as ruler on behalf of both monarchs. It was the task of Claudian, Stilicho's court panegyrist, to propagate that fiction, supporting Stilicho as the power behind the western throne until he was killed in August 408. In contrast, in the east, Rufinus was murdered before the end of 395.

In the summer of 394, Alaric, whose men had formed the principal part of the Gothic army that defeated Valens at Adrianople in 378, had marched a Gothic army to the walls of Constantinople to fight on the side of Theodosius at Frigidus. Following the emperor's death, they turned to plundering the Balkans. In September 395, Stilicho was still in command of both the western and eastern imperial armies, men who had faced each other across a battlefield following Eugenius or Theodosius. He marched at the head of both armies, barely containing the enmity between them, but failed to engage Alaric's Goths. It seems likely that Stilicho feared a second Adrianople, the disastrous battle where Valens had died, and did not trust his own Goths to remain loyal. Instead, he sent them, under Gainas, with the rest of the eastern army back to Constantinople, where they were met by Arcadius and, at this side, Rufinus. Gainas promptly killed Rufinus. His head and right hand were removed and paraded through the city.[5]

Rufinus was replaced as the power behind the throne by the eunuch grand chamberlain Eutropius, who was no more favourably disposed to Stilicho. Eutropius relied on the support of the master of offices, Hosius, rumoured once to have been a cook. Claudian celebrated this pithily: 'Here they sit, twin rulers of the east, a cook and a pimp.' Threatened from the west and fearing a military coup, Eutropius engineered the dismissal of two generals, Timasius and Abundantius, both former consuls and veterans of Frigidus. The atmosphere at court appears to be captured in a law issued in Arcadius's name in 397, which condemned to death anyone who had 'conspired with soldiers ... including barbarians' against those illustrious men 'who belong to our consistory or offer us counsel'. Abundantius's position in command of Illyricum was conveyed to Alaric, who had lately been responsible for ravaging the region. Eutropius took Timasius's role as commander-in-chief in the east for himself, even launching a successful campaign against the Huns in 398, before he was faced with a rebellion in Phrygia led by Tribigild the Goth. The commanders chosen to suppress Tribigild, Leo and Gainas, brought about Eutropius's downfall. When Leo suffered a major defeat, Gainas, possibly a relative of Tribigild, commandeered his Gothic army and demanded Eutropius's removal and his head. Eutropius sought refuge in the great church, where he received the support of the patriarch, John Chrysostom. Spared execution, Eutropius was exiled to Cyprus and a law, preserved in the *Theodosian Code*, was issued annulling his acts and erasing his deeds, removing his name from the list of consuls and from public monuments, destroying his statues, busts and portraits in marble, bronze and paint, in both public and private places, and seizing all his property for the imperial treasury. Within months, however, Eutropius was recalled and sentenced to death at the behest of a new praetorian prefect of the east, Aurelian, who was named consul for the year 400. Only this was considered sufficient to erase the pollution of his period of office, the infamy of a eunuch consul who had brought such afflictions upon the empire, according to Claudian's invective *in Eutropium*.[6]

Gainas was not, however, satisfied. In April 400, he appeared at Chalcedon, opposite Constantinople, leading his Gothic army. At a personal audience with Arcadius, he secured the removal of Aurelian, who was handed over to the Goths and sent into exile. The praetorian prefecture went to Caesarius, Aurelian's brother, who agreed to support the Goths. Gainas was admitted to the city and appointed to the consulship for 401. Goths flooded into the city, so that 'its residents were reduced to a condition equivalent to that of captives', according to Socrates Scholasticus (6.6). 'So great was the danger to the city that a comet of prodigious magnitude, reaching from heaven even to earth, such as was never before seen, gave forewarning of it.' The city erupted into violence and Gainas fought his way out while thousands of Goths were burned alive in a church adjoining the palace. As Gainas fled, he was caught and killed. His faction victorious, Aurelian was recalled to Constantinople. Synesius of Cyrene witnessed all these events as an ambassador at Constantinople and client of Aurelian. In the year 400, he wrote a work of allegory, *On providence*, also called *Egyptians*, that recounted the rivalry between two brothers, Osiris and Typhus, identified as Aurelian and Caesarius. Typhus was favoured by the barbarian Scythians (the Goths), who deposed and exiled Osiris, but were butchered by the natives of Thebes (Constantinople). Synesius appears as himself, 'nurtured by philosophy in a rather rustic manner and unacquainted with the ways of the city', writing 'both poems and speeches' in praise of Osiris (Aurelian).[7]

Instead of celebrating his consulship at the start of January 401, Gainas had his severed head paraded through the city. 'And in the same year Theodosius *nobilissimus*, son of Arcadius was born', on 10 April 401. The following January, the infant prince, not yet a year old and sickly, became Augustus. Theodosius was Arcadius's fourth child with his empress Aelia

Eudoxia, who had given birth to three girls in four years, and later bore a fourth princess. The abundance of daughters led critics to spread rumours that the boy was truly the son of a courtier. Rumours had always circulated, but Eudoxia was vulnerable to slander. She had no power base in the east, being the daughter of Bauto (d. 388), a Frank who had been *magister militum* and consul at the western court.[8]

Factionalism at Arcadius's court intersected with that enveloping the church, leading to the appointment of John Chrysostom as patriarch of Constantinople, at the suggestion of Eutropius, and later to John's exile and death. In February 398, John had been summoned from Antioch, where as a presbyter he had established his reputation as a powerful and popular preacher, and installed as patriarch. His concern for the poor had earned him a loyal following in Antioch. However, since he did not avoid difficult subjects and was unrelenting in his determination to improve others, he had also made powerful enemies. Constantinople was home to large groups of Christians opposed to John's Christology, including Arians and Homoiousians who would not willingly follow him. But he also failed to win over those who agreed with his Nicene dogma. John, a former ascetic, expected monks to devote themselves to prayer and seclusion and not be involved in the life of the city. He intervened in their affairs, clashing with Isaac, a powerful abbot and an extremely worldly former ascetic, who had risen to fame by clashing with the emperor Valens, an Arian. John also earned the hatred of many who enjoyed the benefits of high ecclesiastical office. He demanded that his subordinates emulate his own high standards in conduct and discipline, and was willing to condemn and expel clergy he considered corrupt or abusive. He inconvenienced many by seeking to end a practice that was standard in Constantinople, whereby clergy lived with female housekeepers, virgins sworn to maintain their own chastity and a priest's household affairs. He ended episcopal banquets, spending the money saved on a hospital, and sold off cathedral treasures and marble that was destined to furnish a church. Most significantly, he determined to assert the primacy of his patriarchate in matters relating to other bishoprics.

In his first year in Constantinople, following a disappointing turnout on Good Friday, John delivered a sermon, 'Against those who have abandoned the church and deserted it for hippodromes and theatres', banning from his church and communion those who were foolish enough to repeat their sinful truancy. More than anything, John raged against wealth as the root of all evil. 'Every place is brimming with turmoil and uproar, everywhere are cliffs and precipices, submerged and hidden rocks, fears and dangers and suspicions ... nobody trusts anyone, each person fears a neighbor ... What is the cause of it all? The love of money!' This is from John's sermon delivered in 400, after the exile of Aurelian but before the defeat of Gainas. John's eloquence earned him the support of the young empress, Eudoxia, whom he praised as 'mother of the churches, feeder of the monks, patroness of ascetics, supporter of beggars'. When John organised litanies and processions, aimed at countering those of the Arian communities of Constantinople, Eudoxia supplied his followers with silver crosses and lighted wax tapers and paid for all their needs. However, John also insulted the empress by delivering a sermon on the vanity and luxury of women, which Eudoxia took as a personal slight. John urged every woman to 'renounce pomp and vanity, set aside useless and superfluous adornments, and bejewel your soul with the only gems of any value, true and solid virtues'. By alienating the empress, John gave an opportunity to his enemies, both clergy and lay.[9]

In 401, John had travelled to Ephesus to convene a synod to appoint that city's new bishop. He installed one of his own deacons rather than one of two candidates favoured locally. The synod proceeded, at his instruction, to depose a number of other bishops, and on the way home John deposed the bishop of Nicomedia. John's opponents responded by bringing twenty-nine charges against him, relating to financial mismanagement, to his imperious style of leadership, and to his methods for disciplining subordinates. These charges were brought before the Synod of the Oak, named after its location at a suburban palace in Chalcedon, which was convened in July 403 by Theophilus, bishop and patriarch of Alexandria. Theophilus, an opponent of John, had the support of powerful figures at court. There appears to have been widespread concern among Constantinople's elite about how successfully John had recruited the widows of wealthy men, notably the widows, wives and daughters of 'barbarian' generals, and mobilised their wealth in support of his own agenda. At Theophilus's urging, Arcadius deposed John and sent him into exile, at first for a single day. A year later, John was exiled again. Socrates Scholasticus relates this to the erection of a silver statue of Eudoxia on a porphyry pillar, 'which is still to be seen to the south of the church', which is to say in the Augustaeum, a public square dedicated to an earlier Augusta, Constantine's mother Helena, and other empresses.

The event was celebrated there with applause and popular spectacles of dances and mimes, as was then customary on the erection of statues of the emperors. In a sermon to the people John charged that these proceedings reflected dishonour on the church. This remark recalled earlier grievances to the recollection of the empress and irritated her so exceedingly that she determined to convene another council. He did not yield, but added fuel to her indignation by still more openly declaiming against her in the church.[10]

A summary process was staged, where John was deposed and exiled to the eastern reaches of Cilicia, and from there to Pityus (modern Pitsunda in Abkhazia), an isolated spot where the Caucasus mountains reach the Black Sea coast. It was while being transported to the more isolated location that John died in 407. Eudoxia was by then also dead. She died on 6 October 404, while giving birth. A supporter of Chrysostom, known as Pseudo-Martyrius, celebrated her death and its circumstances, as well as an earlier miscarriage that she had suffered. His cruel words reveal the bitterness that John's supporters felt at his fall, and the intense hatred they felt for the young woman whom they blamed.[11]

Knowing that the root of all evil has been concealed in this woman who exercised power, [the Lord] released his hand. The arrow flew and hit the womb of the wretched woman, reminding her of the saying 'Woman, in pain you will give birth to children' (Genesis 3.16), sending them forth from your womb straight to the grave, mixing with the first swaddling clothes the final burial shroud and becoming in one instant both mother and childless.[12]

Misogyny has also been identified in John's writings. His enthusiasm for asceticism did not make him sympathetic to domestic life generally, and his sermons are severe and replete with imagery that reveals how he regarded women. For example, his sermon against the games and theatres is an archetype of the 'male gaze', focused on the presence of shameless women, performers and prostitutes. 'Often if we meet a woman in the marketplace, we are alarmed. But you sit in your upper seat ... and see a woman, a whore, entering bareheaded and with a complete lack of shame, dressed in golden garments, flirting coquettishly and singing harlots' songs', and you take her home, figuratively, 'the sin in your mind'. Yet it is clear that Chrysostom maintained friendships with several women of wealth and rank in both Antioch and Constantinople. His concern for them and their health, their spiritual lives and personal problems, was expressed in his letters, which evince a quite different tone to his invective that so insulted the empress and other powerful women in Constantinople. The difference appears to be that John's female correspondents and friends had embraced chastity.

John's most intimate relationship and expansive correspondence was with Olympias, a wealthy deaconess of the 'great church'. Olympias was the daughter of a senator and widow of Nebridius, an urban prefect of Constantinople, and had fought to gain control of her own wealth. With it she hosted clerics from across the empire at her home, and founded a nunnery adjacent to the cathedral where she enrolled fifty of her own household servants and slaves. A passageway between the nunnery and church allowed the nuns to attend John's sermons, and he alone of men was permitted on their grounds. One of the charges brought against John was that he met women individually and without a chaperone. When leaving his church to begin his exile, John spoke last of all to the deaconesses. These women had found refuge and support in the church following the death, or exile, of their husbands, fathers and brothers. By their intimacy with John they had earned the enmity of those who had bested him.[13]

On the day of John's second and final exile a fire engulfed the 'great church' and the senate house in the vicinity of the imperial palace, which Zosimus described as 'a beautiful and lavish building adorned with statues worth seeing and with marbles in colours no longer quarried', including the statues of the Muses brought for Constantine from Mount Helicon. John's supporters, including several deaconesses, were blamed and exiled. At the same time, it was learned that 'hordes of Isaurians, who live beyond Pamphylia and Cilicia in the pathless and rugged mountains of the Taurus range', were raiding neighbouring provinces, 'and although they were not strong enough to attack walled cities, they overran the unwalled villages and everything before them'.[14]

The Isaurian uprising continued until after Arcadius's death, on 1 May 408. The smoothness with which the child Theodosius succeeded is evidence for the authority exercised by the consistory, or governing council, at Constantinople. In his last years, Arcadius had been reliant on his last praetorian prefect, Anthemius, whose name is linked with several massive infrastructure projects in Constantinople, which we explored in the last chapter. In foreign policy, relations between the eastern and western Roman courts soured, as did those between Constantinople and the bishop of Rome, Pope Innocent. Innocent condemned the treatment of Chrysostom and convinced Honorius to do the same in a letter to Arcadius. When papal envoys arrived from Rome they were denied entry to Constantinople. Intending to benefit from the rift, Stilicho 'fortified the shores and harbours with numerous guards so that there should be no access to this part of the empire for any person from the eastern empire'. This situation was only reversed on 10 December 408, after the deaths of both Arcadius and Stilicho, according to a law preserved in the *Theodosian Code*, issued in the names of Honorius and Theodosius II.

Theodosius II

Literary sources for Theodosius's fifty-year reign are poorly preserved and this is especially unfortunate, because Theodosius's court was full of poets and philosophers. Among the most famous writers is Cyrus of Panopolis, who rose through the ranks at Constantinople to become both urban prefect and praetorian prefect, for almost two years holding both offices at once. In Alan Cameron's estimation, Cyrus will have written 'panegyrics on Theodosius and Eudocia, *epithalamia* and the like on lesser members of the court, epics on the emperor's wars, and perhaps the odd mythological poem ...'. Only twenty lines of poetry, fragments preserved in the *Greek Anthology*, are convincingly attributed to Cyrus. We are more than usually reliant on the church historians Socrates, Sozomen and the fragmentary Philostorgius, who all completed their works under Theodosius II, and on later chronicles that used works now lost to us. Here Theodosius is remembered as a mathematician fascinated by astronomy-astrology, a calligrapher and bibliophile, an archer and horseman who built a polo field (*tzykanisterion*) in the imperial palace, and as a pious Christian who rose early each day to pray with his virginal sisters and fasted twice each week. Rumours about his sexuality and relationships with his favourites circulated, but the image projected beyond the confines of the palace was of a devout and humble imperial family, committed to prayer for the common good. Theodosius has often been compared poorly to his grandfather and namesake, since he never led his own armies or travelled extensively across his empire. However, the ambitions of his generals were successfully contained, in part because they won no major victories. Opportunities for glory and booty were limited, and minor victories were celebrated and successful commanders generously rewarded.[15]

In contrast to literary works, the documentary records for this period are extraordinarily rich. Fergus Millar wrote that 'The bulky conciliar dossiers, together with letter collections, doctrinal tracts, and the laws collected in the *Theodosian Code* and the post-*Code Novellae* – that is, "new" laws – of Theodosius II ... offer an extraordinary resource for understanding the inner workings of imperial government'. This is testimony to the efficacy of officials who worked for the emperor and advised him, teaching him the business of government, if only as much as he was interested to learn. Sozomen observes generally of the first fifteen years of Theodosius's reign that 'the eastern lands of the empire were free from wars, and contrary to all opinion, its affairs were conducted with great order, for the ruler was still a youth'. Sozomen's heroine was Pulcheria, who pledged herself and her sisters to virginity aged fifteen, and thereby denied other powerful families a claim on the throne through marriage. It appeared to him 'that it was the design of God to show by the events of this period, that piety alone suffices for the salvation of princes; and that without piety, armies, a powerful empire, and every other resource, are of no avail'. However, Pulcheria was not a member of the emperor's consistory, his governing council, which at his accession was led by the praetorian prefect Anthemius, whom Socrates Scholasticus praised as 'the most prudent man of his time'. Other members of the consistory were the holders of high office, including the quaestor and master of offices. Those who retired from the consistory retained power and influence, and were ensured a place at imperial feasts and access to the emperor himself. Many later returned to high office and a seat on the governing council. Anthemius was replaced in 414 as praetorian by Aurelian. We do not know what happened to Anthemius, but he is likely to have retired or died. His son-in-law Procopius became *magister militum*, and his son Isidore, formerly urban prefect, became praetorian prefect of Illyricum and, in 435–6, praetorian prefect of the east.[16]

A remarkable story, reported more than a century later by Procopius of Caesarea, claims that while Theodosius was still an infant, Arcadius had arranged for the ruler of Persia, Yazdgerd I (399–420), to act as his guardian. Upon Arcadius's death, Procopius reports that a letter was sent by Yazdgerd to the senate of Constantinople threatening war if anyone should seek to depose Theodosius. Moreover, a peace between the Persians and Romans, supposed to last 100 years, was agreed in the first year after Arcadius's death. The presence at court of a Persian eunuch named Antiochus, a Christian, may support Procopius's claim. Antiochus served as Theodosius's chamberlain and tutor; he received a letter from the exiled Chrysostom, and is described in a letter written by Synesius of Cyrene as having licence to 'do whatever he wants'. The détente with Persia at this time reflects the deterioration of relations between the eastern and western Roman courts during Stilicho's supremacy, and Yazdgerd's guardianship of the Roman emperor may have been directed as much against Stilicho as anyone at the eastern court. However, matters in the west reached a crux at this time. Having learned of Arcadius's death, Honorius planned to travel to Constantinople. Stilicho sought to persuade him otherwise and volunteered to travel there instead, but listening to those at court who feared that the 'barbarian' general was conspiring with Alaric, Honorius put Stilicho to death at Ravenna on 23 August 408.[17]

The western court had, at some time before Stilicho's death, moved from Milan to Ravenna. Alaric, following the death of Stilicho, besieged Rome, and when a deal was being negotiated to relieve the city, 'almost all the slaves that were in the city poured out to join the barbarians'. Alaric pressed his siege through the winter of 409–10, summoning his cousin Ataulf. Alaric and Ataulf travelled to Ravenna to agree a truce, and according to Olympiodorus, who provides a very full account of these events, blame for the failure to reach an agreement rested with Honorius and his praetorian prefect Jovius, who rejected Alaric's 'fair and prudent proposals'. Alaric's ambition appears to have been to secure a senior generalship as well as material concessions. Early in 410, Alaric besieged Ravenna. At the same time he established his own emperor, Attalus, in Rome, and had himself named master of soldiers. Honorius might have capitulated had not a small naval force arrived at Ravenna, buying him some time and giving him some hope of escape. Moreover, two of Honorius's subordinates proved effective: Sarus drove Alaric from Ravenna, and 'Heraclian held all the ports in Africa under a strict guard so that neither corn nor oil nor any other provisions could be conveyed to Rome, and a famine struck the city'.[18]

When Attalus refused Alaric's instruction to send a force to Carthage, Alaric stripped Attalus of his rank and kept him, his son Ampelius, and Honorius's half-sister Galla Placidia as hostages, hoping to strike a deal. At this point, Zosimus's history suddenly ends, not preserving an account of how shortly afterwards, in August 410, Alaric entered the city of Rome and allowed his men to plunder. Socrates, in a somewhat garbled summary, asserts that Theodosius had sent an army to relieve Rome when Alaric withdrew. Possibly he means the army that had earlier arrived at Ravenna. In any event, Alaric did not mean to hold Rome, but instead set off for Africa to relieve the blockade. Before he could leave Italy he died and was succeeded by Ataulf. J. B. Bury called Alaric 'Moses of the Visigoths', because he 'guided them on their wanderings until they came in sight of the promised land which he was not destined to enjoy himself'.[19]

There was at this time a profusion of usurpers, which Orosius sets out clearly and which Sozomen, drawing upon Olympiodorus, addresses succinctly, attributing the failure of each one to Honorius's piety. Armies identified as Vandals, Alans and Sueves had begun to plunder extensively within the empire's frontiers. The situation in Britain was turbulent, with the threat of Saxon invasions, and from 406 the army proclaimed and killed emperors in quick succession, first Mark, then Gratian, before choosing Constantine, who rallied the army in Gaul to his cause and secured the Rhine frontier. Soon he held sway as far as the Cottian Alps, sending his son Constans to secure Spain. However, when Constantine sought to cross the Alps into Italy to challenge Honorius he was driven back to Arles and besieged there by Constantius, Honorius's capable general. Orosius (7.42) reports that 'Honorius, seeing that nothing could be done against the barbarians when so many usurpers were opposed to him, ordered that the usurpers themselves should first be destroyed'. At this time, the Britons were told by Honorius in a letter to look to their own defences, traditionally interpreted as the end of Roman rule in Britain.[20]

At Mainz, Jovinus, yet another pretender, was raised up, now by the Alans. Remarkably, he was put down by Ataulf, who sided with Honorius because Sarus supported Jovinus. Heraclian, count of Africa, also chose this moment to invade Italy and seek the throne, having sustained the grain and oil blockade against Honorius himself. Like Jovinus he failed. Ataulf, however, gained in strength, taking his hostage Galla Placidia as his wife and setting them up in Narbo (today Narbonne) as a regal pair. They had a son, named for his grandfather, Theodosius, who died in infancy. Ataulf's attempts to gain recognition from Honorius, however, were in vain. Constantius was sent against him and drove Ataulf south, towards Barcelona. After Ataulf was killed in autumn 415, Galla Placidia was returned to Honorius. Attalus was also handed over to Honorius and sent into exile. Galla Placidia, meanwhile, was offered no reprieve. She was promptly married to Honorius's lieutenant Constantius, with whom she had two children, Honoria (in 418) and Valentinian (419).[21]

These events were followed closely in Constantinople, where prayers were offered and victories celebrated. There is no indication, however, that substantial military assistance was offered to Honorius after the detachment of around 4,000 men sent to Ravenna in 410. As we have noted, Sozomen observed that at this time 'the eastern lands of the empire were free from wars'. When news reached Constantinople in 415 that Ataulf had been killed, there was a solemn procession with lamps, but also chariot races, a blend of celebrations of which Chrysostom would have heartily disapproved. But when news reached Constantinople that Galla Placidia's second husband, Constantius, was to be elevated, Theodosius refused to recognise him as Augustus and her as Augusta. Their status was proclaimed unilaterally at the western court in 421. Philostorgius relates that 'Constantius was making ready to go to war over the insult when death intervened and relieved him of life and his cares after he had reigned for six months'.[22]

Shortly before this, in autumn 420, the Persian ruler Yazdgerd died, precipitating a power struggle from which Yazdgerd's son, Bahram V, emerged victorious. News came out that Christians in Persia and Persian-controlled Armenia, Persarmenia, were being persecuted and the conversion to Zoroastrianism of some men of high rank was being demanded, provoked in part by fears that Persian Christians, including some commanders, would side with the Romans if war came. Christians fled into Roman territory, and a Persian embassy was depatched demanding their return. At the same time Roman gold miners who had been contracted to the Persians were not permitted to leave, and merchandise was seized from Roman traders. Theodosius responded by sending an army under Anatolius into Armenia, the location of the gold mines. A second army under Ardabur crossed the frontier at Amida, laid waste to Persian territory, engaged a Persian force and won a minor battle. The Persians withdrew and Ardabur proceeded into Mesopotamia, laying siege to the city of Nisibis. Both sides prepared to dispatch additional forces and when the Romans 'heard that the ruler of the Persians was leading a multitude of elephants against them, they grew afraid', burned their siege engines and withdrew into their own territory, to Theodosiopolis (Erzurum), where they in turn were besieged. Helio, the master of offices, travelled to Mesopotamia to negotiate a truce, which was helped by the arrival from Armenia of Anatolius's troops, now under the command of Procopius, who won a second minor victory. A story circulated that Areobindus, a Gothic commander, covered himself in glory on that occasion by accepting a Persian challenge to decide the fate of the battle in single combat.

The events of 421–2 offered great propaganda value and Theodosius's literary circle celebrated the emperor and his generals who hitherto had been denied the opportunity to fight the old enemy, the Persians. The reputations of Ardabur and Areobindus, Anatolius and Procopius, were all established or burnished, and those men remained prominent for years or even decades afterwards. The last three were awarded the highest honorary rank, patrician (*patricius*), which was a distinction awarded only to some former consuls, prefects and *magistri*. Ardabur may have died before he was thus honoured, but his son Aspar became a patrician and even named a son Patricius. Socrates drew upon the panegyrical orations delivered before the emperor when he wrote about the end to persecution of Christians, and how Theodosius, at the urging of the pious Pulcheria, had sent a gem-studded cross of gold to be raised on Golgotha. Golden coins were minted in Constantinople with a novel reverse image of Victory standing holding a large jewelled cross. The stories even reached North Africa, where Quodvultdeus of Carthage wrote of an emperor, whom he misidentifies as Arcadius, who 'went to war with the Persians, assured of victory in advance by a sign: bronze crosses which appeared on the cloaks of his soldiers as they went into battle. For this reason, when he had won victory the emperor ordered also that gold coins be struck with the same sign of the cross'.[23]

In August 423, Honorius died. Immediately, Theodosius sent an army to capture the city of Salona as a forward base for western operations, and only then announced the news of his brother's death. Shortly before, Galla Placidia had departed Italy with her children for Constantinople, where she was now recognised as Augusta. Her son Valentinian was made Caesar and betrothed to Licinia Eudoxia, Theodosius's two-year-old daughter. A new western emperor was raised up, an imperial secretary called John. He sent an embassy to Constantinople seeking recognition of his elevation, but Theodosius arrested the ambassadors and sent a fleet from Salona. When its commander, Ardabur, was captured, his son Aspar marched against Ravenna and took the city by surprise. According to Marcellinus Comes, 'John was killed by treachery, rather than the manliness of Ardabur and Aspar'. When news of John's death reached Theodosius he was attending the games, but immediately departed the hippodrome for the great church to give thanks, bringing the whole crowd with him. Valentinian was then recognised as emperor and Theodosius seemed intent upon escorting him personally to the west, travelling as far as Thessalonica before ill health obliged him to turn back. Valentinian and Galla Placidia continued together to Ravenna under the care of Helio, then Theodosius's master of offices. Ardabur and Aspar were also on hand, as was a new warlord who had appeared on the scene, Aetius, whose 6,000 Huns supported the installation of Valentinian III as western emperor.[24]

Almost everything we know of the politics of the period from the death of Arcadius to the accession of Valentinian III derives from a long and detailed work by Olympiodorus, which has survived only in a few fragments, in the works of others who used his narrative, and in a summary produced in the ninth century by Photius. Opinions on Olympiodorus and his work differ. Since he wrote at Constantinople, where he was an influential courtier and an occasional diplomat, it is remarkable that he focused almost exclusively on affairs in the empire's western lands. It has been suggested that his purpose in writing was to contrast the western court with its eastern counterpart. The west was riven by factionalism, the court dominated by barbarians, a hotbed of sexual scandal and impiety, where intrigue frequently gave way to outright civil war. The eastern court was presented as stable, pious and unsullied by scandal, able when the time was right to assert its dominance. Alternatively, his work is seen not as a formal history, but rather as a form of semi-confidential report, directed to Theodosius and his immediate circle. Olympiodorus presented his work as 'materials for a history' rather than as a polished work, perhaps by way of an excuse for its style.[25]

Since Olympiodorus and those who followed him focused on the western lands of the empire, to reconstruct the political history of Theodosius's reign in the east we must draw on later chronicles that preserve a pool of dramatic and salacious stories that jar with the impression of a monastic court painted by the church historians. Each version of a story provides slightly different details or an interpretation that encourages conjecture. One story, told by Malalas in the sixth century, in the *Easter Chronicle* at the start of the seventh, by John of Nikiu at the end of the seventh, and by Theophanes in the ninth century, is probably from Priscus. This relates how Theodosius, aged nineteen, came to marry Athenaïs, a beautiful and eloquent virgin, who was the daughter of the Athenian sophist Leontius. She was identified as a suitable consort, according to the story, by Pulcheria and Paulinus, Theodosius's sister and his closest friend, when she appeared at court to petition the empress. Athenaïs was baptised with the name Aelia Eudocia and married the emperor on 7 June. A year later a princess was born, Licinia Eudoxia. Eudocia would bear Theodosius two more children, a second daughter named Flacilla, who died in infancy in 431, and a son named Arcadius. Arcadius's name was carefully removed from contemporary sources. This seems to relate to suspicions that he was no longer regarded as Theodosius's son.[26]

Only a few paragraphs after he has introduced Athenaïs, Malalas recounts how, 'some time later', she and Paulinus were suspected of conducting a secret affair, because the empress gave Paulinus, by now the master of offices, an impressively large Phrygian apple, a gift Theodosius had given to her. Paulinus then gave the apple back to the emperor, unaware that it had been his gift to the empress. Questioned about the fate of the apple, now back in Theodosius's possession, Eudocia claimed that she had eaten it. Consequently, the emperor 'became angry, suspecting she was in love with Paulinus'. Paulinus was put to death and Eudocia despatched to the Holy Land. The *Easter Chronicle* offers the same explanation for Paulinus's demise and Eudocia's retirement to Jerusalem, but places them more than twenty years later, in 444. Theophanes Confessor also picks up on the magnificent apple, placing the episode in 447–8 and adding the information that Paulinus was exiled to Cappadocia, where he was executed. Confirming the location, Marcellinus Comes records tersely in the year 440 that 'Paulinus, the master of offices, was killed at Caesarea in Cappadocia on the order of the emperor Theodosius'. At around this time Pulcheria appears to have removed herself from the court, setting up her own household at the Hebdomon Palace. In autumn 441, Cyrus of Panopolis was removed as praetorian prefect and sent away to become bishop of Cotyaeum in Phrygia. Eudocia returned to Jerusalem around the same time.[27]

Great influence at court was now wielded by Chrysaphius, a eunuch chamberlain, who had the emperor's ear. Malalas (presumably copying Priscus) observed that 'Theodosius was passionately in love with the chamberlain Chrysaphius', who was 'extremely comely', showering him with gifts and offering him 'unfettered access' (*parresia*). Paulinus is earlier described in exactly the same terms as Chrysaphius, as having *parresia* with the emperor, from which one can infer a range of meanings other than 'the opportunity to talk freely'. Paulinus is described as 'an extremely handsome young man' and as Theodosius's *paranymphos*, which means literally that he was part of the imperial wedding party, 'standing behind the bride'. He had opportunity to visit both the emperor and Eudocia in the palace without others present. The apple in the story may have been real, as gifts were exchanged at Epiphany, but it is also a metaphor for the complicated relationship between the three intimates. Yet the fact that Paulinus was put to death, apparently uniquely among holders of high office during Theodosius's reign, indicates the need to end a genuine threat, rather than to repay an insult or betrayal, however deeply felt. Instead of imagining Theodosius as naïve and vindictive, one might read Malalas, or Priscus, as implying that Theodosius had been aware all along, and accepted, that Paulinus and Eudocia were lovers, but widespread rumours that his heir was in fact the son of Paulinus had to be addressed and a challenge to his power quashed. His actions suggest that Theodosius was quite certain that Arcadius was not his own son. His certainty suggests that Theodosius was no longer intimate with his wife and had not been for some years. We do not know what happened to Arcadius, and it may even be the case that he was executed, although it was not unusual, as we have seen many times, for imperial children to die. His name was removed from all official records and, although he lived for perhaps ten years, we know of his existence from a single inscription preserved at Ravenna.

If a coup seeking the succession of Arcadius, and perhaps the regency of Eudocia and Paulinus, was foiled, this was obscured by John of Nikiu (18.1–23), writing in the later seventh century, who also records the story of the apple. It begins: 'during the childhood of the emperor Theodosius ... he had with him a child named Paulinus ... who learnt with him and they grew up together. And the emperor loved him and appointed him' to high office. After the apple exchange, 'Eudocia was greatly put to shame, and a sense of pain and offence existed for a long time between' the imperial couple. Shortly afterwards, the emperor was told that Paulinus was preparing a revolt, presumably in Cappadocia. 'Accordingly he had him executed.' John concludes the episode thus: 'Lying historians who are heretics and abide not by the truth' wrote that 'Paulinus was put to death because of the empress Eudocia. But the empress Eudocia was wise and chaste, spotless and perfect in all her conduct.' John immediately turns to the fact that Theodosius 'had no male offspring to succeed him on the throne', and observes that on the advice of Egyptian monks, who warned that such a son would become wicked, Theodosius and Eudocia 'abandoned all conjugal intercourse and lived, by mutual consent, in befitting chastity'. Reordering these observations, reanimating Arcadius, and relieving Eudocia of the burden of perfection, leaves us with a story quite similar to that we have already reconstructed.[28]

Only a little more can be said about Eudocia's two decades at court. The empress's elevation cemented the authority of her maternal uncle, Asclepiodotus, who would become praetorian prefect of the east and consul in 423. Somewhat later her brother Valerius was named as ordinary consul in 432 and in 435 as master of offices. Educated by her father, Eudocia spent a good deal of time engaged in intellectual games and writing, including some remarkable poems known as centos, which told Old and New Testament stories using lines and half-lines of poetry drawn wholly from Homer. Some 900 lines have survived of a prose hagiography of St Cyprian of Antioch that Eudocia rewrote in hexametric verse. She also wrote a verse panegyric to celebrate the Persian campaigns of 421–2. She celebrated the marriage at Constantinople of her daughter Eudoxia to Valentinian in 437, before setting off on a pilgrimage to the Holy Land, in emulation of Helena, returning in 439 with relics of St Stephen the Protomartyr.

Since the birth of her first child, Eudocia had shared the title Augusta with Pulcheria, who had earned it not through the birth of an imperial child, as was standard at the time, but following her pledge to remain a virgin, a novelty. The true nature of the relationship between Eudocia and Pulcheria cannot be extracted from surviving sources, and the diametrically opposed impressions we are given of Pulcheria relate to her interactions with Nestorius, patriarch of Constantinople (428–31), who was later reviled by the bishops of the Chalcedonian Church but revered by the Syriac fathers. In his inaugural sermon as patriarch, Nestorius had advocated the eradication of heresy and encouraged those who set about it with zeal. Like his predecessor, Chrysostom, Nestorius was an Antiochene. He was equally opposed to the circus and theatre that threatened the morals of his flock, and had firm views on monastic life that earned him the enmity of powerful monastic leaders. Nestorius celebrated the memory of John, and in his strictness, like John, Nestorius fell foul of an empress. He is said to have denied Pulcheria entrance to the sanctuary of the great church to receive communion during Easter celebrations, a privilege afforded only to priests and to the emperor, but one which Pulcheria had enjoyed under the previous patriarch. Her access barred, Pulcheria is said to have proclaimed 'Let me enter according to my custom ... Did I not give birth to God?' Nestorius replied, 'No, you gave birth to Satan!' Pulcheria's claim to have given birth to God was in reference to the virginity of Mary, whose status as *Theotokos*, 'Mother of God' was at that moment controversial. The dominant interpretation in Constantinople, which Pulcheria upheld, was that Mary gave birth to Christ's divine nature as well as his human nature form. Nestorius espoused a view prevalent in Antioch that Mary, as a woman, could not have given birth to God. Rather, she should be known as *Christotokos*, 'Mother of Christ'. At the heart of Nestorius's Christology was a denial of the 'hypostatic union' between Christ's two natures, the human and divine, which sparked the so-called 'Two Christs' controversy. The matter would be resolved, and Nestorius condemned, at the Council of Ephesus, remembered as the third ecumenical council.[29]

Just as Theophilus, patriarch of Alexandria, had led the opposition to Chrysostom, Theophilus's nephew and successor as bishop, Cyril of Alexandria, led the opposition to Nestorius. Cyril convened a synod of Egyptian bishops to condemn Nestorius and appealed directly to Rome, securing the support of Pope Celestine. He did not until this time consult the emperor or consistory, and although Theodosius may not have paid the greatest attention to secular legislation issued in his name, his cultivated reputation as a secular monastic and a model of piety demanded that he be seen to act in mediating the dispute. Moreover, a letter he sent to Cyril displays genuine irritation at the bishop's efforts to work around him in deposing Nestorius, berating Cyril for his insolence and deceptions, but forgiving him for the sake of 'sacred peace' pending a final resolution by a council of bishops. At the same time Theodosius issued an official document in more temperate language, a *sacra* convening a council at Ephesus the following Pentecost, 7 June 431. The preamble sets out the emperor's right as divinely appointed ruler and his duty as mediator between the sacred and profane realms to establish tranquillity and concord in the universal church. However, Theodosius did not participate in the council, and nor did his nominated representative, Candidianus, commander of the palace guards, whose role was to prevent disruption to clerical deliberations, to ensure all participants were able to express their opinions freely, and to prevent the group from splintering before the proceedings were closed.

A goal of the council, from the imperial perspective, was for the bishops to act harmoniously, for all to participate in deliberations and accept majority decisions, and for this concord to be publicly demonstrated and witnessed. It was not auspicious, therefore, when Cyril and the Egyptian bishops insisted that the council commence on 22 June, two weeks after the prescribed start but still before the late arrival of delegates from Rome and Antioch. In this he had the support of the local bishop, Memnon of Ephesus, and Candidianus's attempt to stop proceedings was unsuccessful. At this first session, which Nestorius and his supporters refused to attend, the papal condemnation of Nestorius was ratified. When the Antiochenes arrived, their bishop John convened a competing council at which Cyril and Memnon were condemned. Upon its arrival the papal delegation joined the larger group of 210 in support of Cyril, and John and thirty-four of his bishops were condemned for convening a separate council and excommunicated. Theodosius despatched a letter reiterating earlier instructions on concord and declared decisions taken by only some of the bishops to be invalid. However, when the decisions of the competing councils reached Theodosius he demonstrated his displeasure by ratifying both, confirming the excommunications of Nestorius, Cyril and Memnon, and detaining all three. Nestorius was confined to a monastery and later exiled to Egypt, but Cyril made it home to Alexandria and was restored to his position, as was Memnon in Ephesus. A couple of years later Cyril was reconciled with John of Antioch, who recognised Nestorius's deposition, and communion between the churches of Alexandria and Antioch was restored.[30]

As these disputes played out in Ephesus and Constantinople, a still greater threat to the empire's harmony and integrity emerged. It began in 428 with the arrival from Spain of Geiseric (Genseric) and his army of Vandals and Alans, plus their families, which a later source numbers at 80,000. Hippo Regius, where St Augustine was then bishop, was put under siege. Augustine died in August 430 and soon afterwards his city fell. An unlikely story that circulated later, recorded by Procopius of Caesarea, held that Geiseric's invasion was the result of an invitation from Boniface, count of Africa, who was feuding with Aetius. Clearly, Boniface was unable to check the Vandal advance, even when he allied his forces with those ferried from Constantinople under the command of Aspar. This joint force was defeated by Geiseric, who took many captives, among them the future emperor Marcian, who was then part of Aspar's retinue. In the same year there was famine in Constantinople and Theodosius was pelted with stones as he processed to the granaries. Unable to remove the Vandals, an agreement was reached as attention turned to the Huns, who appeared as a threat at exactly this time. Through the 430s, an increasingly large sum of gold was paid annually to the Huns, whose leaders followed affairs to the south closely and advanced to extract further concessions when the eastern armies were occupied elsewhere. The fall of Carthage saw Aspar sent once again towards North Africa, now with an even larger fleet of perhaps 1,100 ships, which docked at Sicily but made it no further. In the spring of 442, the fleet was recalled, again in the face of an opportunistic threat from the Huns. In the same year, Valentinian agreed a treaty with Geiseric and betrothed his daughter Eudocia to the Vandal's son Huneric. North African bishops suffered the confiscation of their church treasures and the use of their buildings by the Arian Vandals. Quodvultdeus, the bishop of Carthage, fled to Italy, and others headed east. The letters of Theodoret of Cyrrhus from this period contain appeals for wealthy Carthaginians fleeing the Vandals, owners of rich villas now abandoned to new masters, 'dignified members of her far-famed senate who wander around the world', who were surviving on handouts to support their families and slaves. Most who disposed of less wealth, whose savings or takings comprised copper coins, were left behind. The fall of Carthage to the Vandals is signalled by two hoards of copper coins, one of 172 coins stretching back almost a century to the reign of Constans I, and a second of 3,942 coins covering most reigns between 337 and 439.[31]

Through the ninety years, until 533, that the Vandals held Carthage, parts of the city were repurposed. Graveyards were established within the walls, including within the ancient theatre and an area within the port. At a private villa that had been abandoned, three rooms were used to bury the dead. The bones of thirty different people, including children and infants, were discovered in a pit in one room, suggesting rapid interment following a disaster or epidemic. However, it would be wrong to imagine that Carthage fell into terminal decline during the fifth century. In the view of those who frequented the Vandal court, such as the Latin poet Florentinus in praising the Vandal king Thrasamund (d. 523), it was magnificent, a royal capital to rival Ravenna, a seat of higher learning, endowed with great houses, strong walls and wealthy suburbs. In the suburb of Alianae, Thrasamund built a new bath house, although within the city the Antonine baths, which had fallen out of use before the Vandal hegemony, were not restored until after 533. We shall return to this in a later chapter.[32]

It is possible that the loss of Carthage in 439 had precipitated a coup in Constantinople, which we have seen led to the execution of Paulinus, the promotion of Chrysaphius, and a reconfiguration of power at court. Theodosius remained in charge, but militarily matters spiralled out of control, with waves of invasions in 440–41 reported by Marcellinus Comes: 'The Persians, Saracens, Tzanni, Isaurians and Huns all left their own territories and plundered the lands of the Romans ... The Hun kings with countless thousands of their men invaded Illyricum; they destroyed Naissus, Singidunum, and very many other towns and cities.' The Hun kings, the brothers Attila and Bleda, returned the following year, moving through Illyricum as far as Thrace. The annual tribute payment was increased and when it was withheld by the eastern court the situation deteriorated rapidly. A concatenation of events seems to have precipitated a turn from cold to hot war with the Huns. In 445–6, famines and 'a plague' struck in Constantinople and the great church was burned. In January 447 a huge earthquake hit the city, bringing down many towers of the land walls. These were rapidly rebuilt, but not before Theodosius had walked in bare feet from the Great Palace to the Hebdomon, his head bowed as he passed through the ruined walls. As the Huns advanced towards the city, Aspar and Areobindus were sent against them but suffered several defeats. Anatolius was sent out to negotiate a peace, and the annual tribute payment was trebled to 2,100 pounds of gold. These were desperate days in Constantinople.[33]

At this time an Isaurian warlord who had taken the Roman name Flavius Zeno came to the fore as commander of the eastern armies and defender of Constantinople. Consequently, he was honoured as consul in 448, and therefore received a flattering letter from Theodoret: 'Your fortitude rouses universal admiration, tempered as it is by gentleness and meekness, and exhibited to your household in kindliness, to your foes in boldness. These qualities indicate an admirable general.' Zeno did not, in fact, demonstrate meekness, and appears to have become a threat to Theodosius in his last years. It may even be that Zeno's marriage, to a certain Paulina, had connected him to the house of the disgraced Paulinus.

Theodosius's last decade in power was quite different from the decades after the death of his father, which recent studies have shown to be a period of stable and well-ordered government by a cadre of powerful administrators. Administration was more efficient without the interference of an emperor with a strong agenda, and the lack of interest that Theodosius showed into his twenties is captured by reports of how he failed to read official papers, once signing a document – presented to him to prove a point – that enslaved his own wife. That is not to say, however, that the consistory did not respond promptly and decisively to pressing situations and petitions. The activity of a single season, autumn 415, is captured in a series of laws directed against other faiths issued in Theodosius's name, although at that time the emperor was only fourteen years old. In October 415, a law degraded the Jewish patriarchate. A further law was issued, forbidding the trampling of Jews or the destruction of their temples and houses. In November 415, laws directed against heretics, Montanists and Eunomians, were enacted. In December 415, a law banned pagans from holding public office. In this atmosphere, anti-Jewish riots took place and, as we have seen, the pagan philosopher Hypatia was murdered in Alexandria.[34]

The most enduring mandarin of this period was Helio, who was *magister officiorum* for more than a dozen years between 414 and *c.*427. Helio had responsibility for the cadre of *agentes in rebus*, roving agents including postal workers and spies who criss-crossed the empire. He was authorised to expel and punish those unworthy of office, so that 'the approved imperial service may be restored to its pristine status and may be thronged with an assemblage of responsible men'. Paulinus, a successor as master of offices, undertook a survey of the service and established a formal register of 1,174 agents, which became the statutory number. This appears to have been a further necessary cull and part of a broader survey of the imperial administration undertaken in the later 420s and 430s. For example, in 428, the praetorian prefect of the east, Florentius, published a report that named and shamed decurions who had sought office to avoid 'the necessary performance of their compulsory public services'.[35]

In 425, a range of teaching posts was established in Constantinople, supported from state funds and provided with teaching rooms at the Capitol. There were fifteen posts in Greek (five rhetors and ten grammarians) and thirteen in Latin (three rhetors and ten grammarians), plus two professors of law and one of philosophy. Professors who stayed in post for twenty years were to receive an honour and an associated pension. However, the posts did not attract superstar private practitioners from better-established schools. Additional state funding for education is probably best explained by the need to provide Latin instruction for lawyers in the Greek-speaking city and the rhetorical education required of all in a fast-growing bureaucracy. A parallel project, the collation of existing law, was started in 429. Delivered in 437, the laws preserved in the *Theodosian Code* are a series of imperial responses to petitions, letters in reply to individual enquiries addressing particular situations that were stripped of 'copious inane verbiage' (*inanem verborum copiam*) and codified. Each of these rulings was addressed to a specified official who was charged with carrying out the instruction. Such laws were frequently modified or recapitulated following further petitions.

There was frequent turnover in senior staff in this period, with many men serving short periods as praetorian prefect of the east, the emperor's effective deputy and head of government. Following Asclepiodotus, Hierius held the office for three years (425–8), followed by Florentius (428–30), Antiochus Chuzon (430–31), Rufinus (431–2) and Flavius Taurus (the son of Aurelian, 433–4). Hierius and Florentius held the office again in the 430s, and Florentius and Taurus held it again in the 440s. Clearly, these powerful men were not leaving office in disgrace, but rather the consistory was circulating senior positions, and on occasion reinstating the most experienced former office holders. Moreover, the sons of the highest office holders were succeeding to positions their fathers had held, demonstrating the entrenched power of an administrative elite. The forced retirement of Cyrus of Panopolis, despatched to a distant bishopric late in 441, is all the more striking in this context, and it cannot have been the case that Cyrus was implicated in Paulinus's plot, as then he certainly would not have been permitted to enter into the ordinary consulship in January 441. His demise is most easily explained by the new political climate following the fall of Carthage, which put Theodosius under pressures he had previously not experienced. A story, preserved by Malalas, suggests that Cyrus had become a little too popular, not remaining behind the scenes, but riding around the city in his carriage, 'supervising building operations and reconstructing the whole of Constantinople', apparently also introducing street lighting.

The Byzantines chanted about him in the hippodrome all day while Theodosius was watching the races as follows: 'Constantine built, Cyrus rebuilt, put them as the same level, Augustus!' Cyrus was amazed and commented 'I do not like Fortune when she smiles too sweetly' ... So a plot was made against Cyrus and he was charged with being a Hellene [pagan], his property was confiscated and he was stripped of office.[36]

The emperor appears to have withdrawn still further from active government in the 440s, doubling down on his pious image. Power remained with the consistory and even more authority devolved to a panel including the praetorian prefect and quaestor, which sat in place of the emperor as supreme judge. The role of the senate in law-making was formalised, establishing a parliamentary process whereby a 'bill' (as it were) that emerged from the consistory (generally a response to a particular query or petition) was read and discussed by the senate, then returned to the quaestor for redrafting before formal submission to both the senate and consistory for further discussion and amendment before it received an imperial signature.[37]

The power of the administrative elite was not challenged by the military, and certain families supplied sons to both branches of administration, military and civilian, even as more high commands were awarded to barbarians. Procopius, who became *magister militum* in 421, was married to the daughter of the late civilian mandarin Anthemius, and therefore was brother-in-law to Isidore, who held both high civilian office (urban prefect) and a military command (praetorian prefect of Illyricum). Procopius's son Anthemius would also have a successful military career before becoming western emperor for five years (467–72). The most senior commanders were members of the consistory, and still more were senators, although regular active participation in those bodies was perhaps restricted to the *magistri militum praesentales*, who commanded the infantry and cavalry stationed at Constantinople, and therefore were most frequently present at court. Sozomen states that Plinta, a Goth, was particularly powerful at court early in Theodosius's reign, being general of both foot and horse and even able to unify feuding factions of Arians in the capital. Theodosius's generals were not very successful and seven of them are only names to us – Agintheus, Florentius, Hypatius, Lupianus, Macedonius, Sapricius, Theodulus – each recorded once in a law or a fragment of the lost work by Priscus. Those who flourished were kept in line with high honours, ranks and offices that carried large stipends and opportunities to amass considerable wealth. Several generals were awarded the highest honour, the ordinary consulate, by which they gave their name to the year. Plinta was ordinary consul in 419, having quashed a rebellion in Palestine. Ardabur, an Alan, was consul in 427, a reward for the defeat of the usurper John. Aspar, Ardabur's son, held the office in 434 alongside Areobindus, a Goth. Anatolius, a Roman, held the consulate in 440, and Zeno, an Isaurian, in 448.[38]

The unity of the empire

At the start of the fifth century the *Notitia Dignitatum* was produced. A document of singular importance and disputed date and provenance, it listed dignities, ranks and offices in the *pars Occidens* and *pars Oriens*, the western and eastern halves of the empire. As a description of the civilian and military administrative structure of the empire in around 400, the document is both invaluable and deeply flawed. The version that has come down to us, through many copies, is a western copy that was updated into the 420s. It clearly demonstrates a desire in both parts of the empire to project an image of harmony and cohesion, but it is also accurate in some details for the east, for example the Egyptian military lists, parts of which are corroborated in extant papyri.[39]

Imperial harmony was projected vigorously in all media. Laws were issued in the names of all ruling emperors and had notional validity across the empire, even as each was drafted in response to a particular petition and implementation entrusted to a named official. The emperors of west and east minted coins for each other. After the death of his father, Arcadius struck coins for Honorius at Constantinople, Alexandria and Antioch that circulated throughout the eastern provinces. Likewise, coins of Arcadius were struck at all western mints, including Milan, Ravenna and Rome. A gold coin minted in large numbers at Constantinople between *c.*397 and *c.*402 depicts Arcadius on the obverse ('heads') and Honorius on the reverse ('tails') with the legend CONCORDIA AUGG, 'harmony between the emperors'. The association of Theodosius II, Arcadius's son, in the imperial college in January 402 saw the legend altered to CONCORDIA AUGGG, with a third G. Similar coins were struck at other mints, including Thessalonica and Rome, and bronze issues projected the same message of imperial harmony, unaffected by the actual state of relations between the eastern and western courts.[40]

On bronze coins, the bust of each emperor was usually shown individually on the obverse, but certain issues also depicted the emperors together on the reverse. For example, a bronze issue of Honorius minted in Alexandria in AD 406–8 featured Arcadius and Honorius together with Theodosius II. Occasionally, the imperial college was shown together on the obverse, for example two bronze issues from Antioch of around the same time, one showing Moneta, the divine protector of funds, and another showing the Tyche of Antioch on the reverse. Bronze coins had a far greater currency than gold, passing through many more hands, and evidence from excavations demonstrates clearly that after the death of Arcadius and throughout the fifth century far fewer new bronze coins were put into circulation, and the quality of their fabric and engraving declined markedly. Older coins remained in circulation longer. Copper coins of Theodosius I, Arcadius and Honorius all circulated for more than a century, projecting an increasingly worn image of imperial harmony.[41]

New statues were erected, and older groups refreshed, to highlight current imperial constellations. An inscribed base from Smyrna shows that an imperial statue group once dedicated to the family of Constantine was replaced with one of Gratian, Valentinian II, Theodosius I and Arcadius. Perhaps only the heads were re-carved. A record of a lost inscription indicates that an imperial group including Valentinian, Theodosius and his two sons once stood at Side, in Pamphylia. At Heraclea Pontica, in Bithynia, a group statue was erected slightly earlier, depicting the same men without Honorius. In 387, a tax demand provoked a riot in Antioch that saw painted images and bronze statues of Theodosius and his family thrown down and destroyed. Theodoret tells us that new images were rededicated very soon afterwards as the city's leaders and most prominent orators, among them Libanius and John Chrysostom, orchestrated widespread public lamentation intended to appease the emperor, whose clemency was won by a monk known as Gubba. A fragmentary plaque shows that a group with Theodosius, Arcadius and Honorius was raised at Corinth in *c.*395. Later sources report that Theodosius I had raised equestrian statues of both his sons at his new forum in Constantinople, along with one of himself, and that later a similar statue of Theodosius II was added to the group. An inscription discovered at the Porta Portuensis (Porta Portense, Trastevere) indicates that statues to Arcadius and Honorius were dedicated there in 401–2 by the senate and people of Rome

(SPQR). At Erythrae in Ionia an inscribed statue base, now lost, indicated a dedication to Honorius and Theodosius II together.[42]

Statues of Honorius were certainly raised in several cities in the east, for example at Ephesus and Aphrodisias. It is likely that Arcadius and Honorius stood as a pair at Aphrodisias, and they are certainly portrayed together on the base of the Column of Arcadius in Constantinople, centrepiece of a new forum dedicated to Arcadius. The column's narrative frieze, celebrating the defeat of Gainas, formed a single spiral from base to summit. Although the frieze is lost it is preserved in four detailed sketches produced in 1574, which also show how badly cracked the column had become (Plate 13). It was damaged in several earthquakes, and the column itself was taken down in the seventeenth century. Only part of the badly abraded base of the column survives today, proving that it had a central staircase. The sketches from 1574 show that the three faces of the column's base were carved in four registers. Each carving conveyed a generic message that complemented the spiralling narrative of victory on the column. On the eastern face the emperors, Arcadius and Honorius, appeared together, accompanied by the Fortunes (personifications) and senators of Constantinople and Rome. On the southern face ambassadors were depicted offering gifts and tribute to the emperors. On the western face barbarian soldiers bowed to a trophy, the symbol of Roman victory, and to the emperors who secured it. On two faces, the lowest register shows discarded arms, armour and military standards, as if strewn on a battlefield. This is the top register on the western face, above a pair of winged victories holding a wreath within which is placed a chi-rho flanked by the letters A and Ω (alpha and omega, the first and last). The uppermost register of the eastern and southern faces contains similar winged victories holding a cross above the victorious emperors. On the eastern face the emperors may appear again in miniature flanking the cross. The central message of the monument is clearly 'victory', but the means of achieving victory and its benefit is imperial harmony.[43]

In reality, victory and harmony were increasingly hard won. The ability of the Roman empire to absorb new peoples and ideas was sorely challenged in the fifth century. If a clash between pagans and Christians has been overstated, then the competition between varieties of Christian intensified. Moreover, the enduring polarity between Roman and barbarian entered a new phase with the rise of strongmen, Goths and Alans, Huns and Vandals, who wielded power behind the throne even as new threats appeared on all fronts. It is hardly surprising, therefore, with so much threatened, that a good deal was done in the reign of Theodosius II to insist upon the unity of the Roman empire and its unchanging dominion. The elevation of Valentinian III in 425 was confirmation that the dynastic division of empire instituted in 395 would remain in force, even as it was a demonstration of the power of the eastern court to effect its will in the west at moments of weakness. This was expressed in the eastern reaches of the empire in traditional fashion. Malalas reports that statues of both Theodosius and Valentinian were raised above a basilica in Antioch, erected under the supervision of Anatolius after an earthquake and inscribed in gold mosaic. When North Africa came under threat from the Vandals, eastern armies were sent west and their commander, Aspar, was named western consul in 434. In 435, Theodosius commissioned a map. We do not know where it was placed and no trace of it has survived, except for a short dedicatory poem, which emphasises that it embraced the whole world and followed the work of ancient cartographers.[44]

> This famous work – including the entire world,
> seas, mountains, rivers, harbours, straits and towns,
> uncharted areas – so that all might know,
> our famous, pious Theodosius
> most venerably ordered when the year
> was opened by his fifteenth consulship.
> We servants of the emperor (as one wrote,
> the other painted), following the work
> of ancient mappers, in not many months
> revised and bettered theirs, within a short space
> embracing the entire world. Your wisdom, sire,
> it was which taught us to achieve the task.

More substantively, just three years later, in 438, the *Theodosian Code* was completed and delivered to both Theodosius in Constantinople and, somewhat later, to Valentinian in Ravenna. Written in Latin, the legal language of the unified empire, its delivery to the west was timed to coincide with the marriage of Valentinian to Licinia Eudoxia. Valentinian's consistory was never consulted on the laws, which were a gift and imposition from the east.

Increased emphasis upon symbols of unity could not, however, mask the new reality, that east and west were no longer so closely bound together legally or politically. Theodosius did not send copies of his novels, new laws issued after 438, to the west until 447, and there is no indication that Valentinian ever sent copies of his laws to Constantinople. Similarly, whereas occasional updates were made in the west to the relevant portion of the *Notitia Dignitatum*, no updates appear to have been made to the eastern part after its initial publication. The administrative system of the *pars Oriens* was frozen in around 400, just as the map of 435 does not appear to have recognised that any lands had been lost to barbarians. In 439, this situation dramatically worsened with the Vandal capture of Carthage, and with it the conquest of North Africa, which E. A. Thompson judged to be the decisive political event of the fifth century, which 'split the Mediterranean into two halves'. Certainly, we now see clearly in the words of Priscus of Panium recognition that the western and eastern halves of the empire were politically separate and compelled to conduct diplomacy as discrete governments with distinct, often competing interests. At the court of Attila the Hun in 449, Priscus, serving as an envoy despatched by Theodosius from Constantinople, encountered another delegation that he described as 'western Romans' representing 'Aetius and the emperor of the western Romans', Valentinian.[45]

In July 450, Theodosius II died after falling from his horse while hunting. At this time Pulcheria, who had spent the 440s at the Hebdomon Palace, re-emerged into public life, renounced her pledge to remain a virgin, and married a new emperor, Marcian. Pulcheria had always retained a voice at court, through her patronage of men who became powerful officials. Those operating at a distance recognised her influence; for example, in 446, Theodoret wrote to her about the wretched state of his city, Cyrrhus in Syria, and its hinterland. However, the bishop wrote on the same matter to others in high positions, including the praetorian prefect Constantius, to whom he provided greater detail. Indeed, he wrote to anyone and everyone who might offer him help, notably the general Anatolius, who was charged with confining the bishop to Cyrrhus. Theodoret, who was aggressively anti-Arian, even wrote to Arian generals, to Aspar thanking him for help, and to Areobindus requesting his leniency in demanding provisioning for his troops in a lean year. From his correspondence, we gain a sense of who held power at the end of the Theodosian age. Increasingly, it seems, this was wielded by generals, and above all of them stood Aspar.[46]

7

SOLDIERS AND CIVILIANS, AD 451–527

❦

Aspar, the Alan general who served Theodosius II, marked his ordinary consulate in 434 by commissioning a large silver plate, a *missorium*, forty-two centimetres in diameter and weighing almost eight pounds (Figure 21). The composition is in three registers, or horizontal bands, and Aspar dominates the middle register, seated on a bench supported by wild cats, a version of the curule seat (*sella curulis*). He is wearing a toga and in his right hand he holds aloft the *mappa*, a cloth or flag used to start the races, the name of which became synonymous with chariot racing. Beside Aspar stands his son, Ardabur junior, named after his paternal grandfather. Ardabur junior is identified as a 'pretor' (*praetor*), the young man's first public office requiring that he also stage games. Ardabur junior would become consul in the year 447. For now, however, Ardabur's *mappa* is not held aloft, but draped over his left wrist. This is not his celebration, as the inscription around the plate makes clear, but that of *Flavius Ardabur Aspar, vir illustris, comes, magister militum et consul ordinarius*. In his left hand, Count Aspar the general holds the consular sceptre, since he is shown presiding at his consular games, a whole calendar of races and contests for which he paid. Immediately above Aspar, in the plate's upper register, is an image of his father, Ardabur (senior), who had been consul in 427, and holds his consular sceptre. Above Ardabur junior is Plinta, consul in 419 and the boy's maternal grandfather, also holding his consular

Figure 21. Silver consular dish (missorium) *of Aspar*

sceptre. Both ex-consuls were dead, so they are suspended in a heavenly band, depicted within medallions. To the left and right of Aspar and his son stand Fortunes, personifications of the city of Rome and a second city, possibly Constantinople, possibly Carthage, where Aspar spent his consular year engaged with the Vandals. The plate's lowest register features a few discarded shields and arrowheads, the arms and armour of enemies the family of Aspar had defeated for New Rome.

This is the only surviving example of a consular *missorium*, although such plates may have been produced in greater quantities than consular ivories, which were carved as gifts for supporters and colleagues to commemorate the staging of games. There are references to silver bowls being sent with ivory diptychs to the same friends, and silver spoons were made for at least one consul. Silver was far easier to work than ivory, and it was easily melted down and reused. This may explain why only one consular plate has survived, compared with around fifty known ivories. The closest iconographical parallel to Aspar's plate is to be found on the far heavier and more splendid *missorium* of Theodosius I, produced in AD 388 to mark the emperor's tenth year in power, his *decennalia*. Theodosius is shown seated, facing forward and larger than his co-emperors, Valentinian II and Arcadius, who flank him. It is the latest of nineteen extant imperial *missoria*, all silver and all dating from the fourth century. These were distributed as gifts to the emperor's companions, notably his companions in arms. It is possible that the Ardabur family possessed such a fourth-century plate, and if they were still produced by emperors in the fifth century they will have received several.[1]

Marcian and Leo

Aspar made Marcian emperor. He did so without consulting the western emperor, Valentinian III, who as senior Augustus had the right to appoint his colleague. But Valentinian held no sway in Constantinople and was too weak at home to contemplate confronting Aspar. Marcian had served in Aspar's household and seemed a suitable figurehead for the interests he represented. Flavius Zeno, who may have intervened or even sought to set himself on the throne, was bought off with the rank of patrician. Pulcheria was brought into the arrangement and agreed to marry Marcian, for which she was condemned by her opponents. In the *Life of Dioscorus*, a familiar story is transferred to Pulcheria, who is said to have 'burned with adulterous love' for Marcian, a 'Nestorian'. When Theodosius was given an enormous apple, he gifted it to his beloved sister. But Satan tempted her to give it to Marcian. Theodosius banished Marcian to the Thebaid, but he was recalled at the emperor's death and made emperor by Pulcheria, who willingly surrendered her hard-won virginity.[2]

According to Malalas, the new emperor promptly beheaded Chrysaphius, 'the beloved of the previous emperor, and confiscated his property'. Marcellinus Comes suggests that this was at Pulcheria's command. If Chrysaphius had advocated the appeasement of the Huns, Marcian ended the policy, surely at Aspar's insistence. Attila had turned his attention to the western empire and in 452 sacked Aquileia, but died in 453, ending for a while the Hunnic threat. However, matters at the western court were not peaceful. Valentinian murdered Aetius with his own hand in September 454; and, in the following March, 455, Valentinian was murdered by Aetius's loyalists. A senior senator, Petronius Maximus, appears to have been implicated in both murders and promptly took the throne, but survived only until the end of May, when he too died as he sought to flee Rome. Like Hypatia in Alexandria four decades earlier, he was killed with roof tiles. Days later, on 2 June 455, Geiseric's Vandals sacked Rome and took captive the empress Licinia Eudoxia and her daughters, Eudocia and Placidia, then aged around seventeen and fifteen. The imperial women were taken away to Carthage, where Eudocia was married to Geiseric's son Huneric (they had been betrothed thirteen years earlier to seal a treaty). The second daughter, Placidia, had been betrothed to a Roman senator, Anicius Olybrius, and this marriage appears to have gone ahead also as part of negotiations between Geiseric and Marcian, whereby Eudoxia and Placidia were sent to Constantinople. Olybrius had fled to Constantinople after the Vandal attack, where he visited Daniel, the Stylite who had predicted the return of the imperial women to Olybrius.[3]

The *Life of Daniel the Stylite* is such a compelling source precisely because we lack a contemporary secular history for the reigns of Marcian and Leo. Useful information is recorded briefly in the chronicle of Hydatius and that of Marcellinus Comes, and at greater length in fragments preserved from the histories of Malchus and Priscus. Sources that are now lost were employed by several later authors, including John Malalas, and Evagrius, who completed an *Ecclesiastical History* in 593. Evagrius, who greatly favoured Marcian, recounts that he 'was of Thracian descent, the son of a military man', who was marked by signs that foretold his accession. He was 'pious in divine matters and just in matters relating to his subjects'. Marcian relied on the support of a group of like-minded generals and senators, whose interests he promoted. Legislation of 452 limited the sum consuls were expected, or entitled, to spend on disbursements and gifts, directing their largesse instead to repairing Constantinople's aqueducts. Marriage restrictions were lifted and senators were now permitted to marry most freeborn women, even those who were relatively poor, although not, of course, the truly indigent. Most significantly, the *collatio glebalis*, a tax levied on senatorial land holdings, was abolished. A detailed senatorial property register was kept and, since 371, registration of lands liable to this extra tax had been a necessary step in the elevation of a decurion to the senate.[4]

John of Nikiu, in contrast to Evagrius, and writing far later in Egypt, reports that gloom covered the earth at Marcian's accession. John alludes to the Council of Chalcedon, the fourth ecumenical council, convened on 8 October 451, which broke open cracks papered over in 433 with the reconciliation of Alexandria and Antioch. Matters had deteriorated rapidly since 448, catalysed by the preaching of Eutyches, abbot of a Constantinopolitan monastery who had opposed Nestorius. Early in

448, Theodosius II had deposed Irenaeus, bishop of Tyre, a Nestorian, and renewed a ban on his teachings. He had then confined Theodoret, the prolific letter writer, to his see at Cyrrhus, a suffragan bishopric of Antioch. Theodoret had refused to condemn Nestorius in 431, and subsequently rejected the reconciliation of Alexandria and Antioch. He had contrived in the elevation of Irenaeus and wrote a satirical work, *Eranistes*, a dialogue between an intelligent Antiochene named 'Orthodox' and an itinerant beggar and collector of heresies, the eponymous Eranistes.

The views of Eranistes appear to be similar to those espoused by Eutyches, who became aware of the work and wrote to the pope, Leo, that Nestorianism was being revived. However, it transpired that Eutyches's staunchly anti-Nestorian Christology was not precisely that articulated by others in Constantinople. An accusation was lodged against him with the patriarch of Constantinople, Flavian, which resulted in Eutyches being put on trial and excommunicated. Precisely what Eutyches believed was not drawn out at the trial, which focused instead on what he was unwilling to admit, which is the 'double consubstantiality' of Christ. His condemnation led to wider conflict, requiring a further council, the second council of Ephesus, in the summer of 449, which was chaired by Dioscurus, patriarch of Alexandria. At this council, Flavian was deposed and exiled, as was Theodoret, and also Domnus, patriarch of Antioch. It was to address this situation that the council of Chalcedon was convened, early in the reign of Marcian.[5]

The Council of Chalcedon reversed the ruling of the 449 council, stating that 'while Christ is a single, undivided person, He is not only from two natures, but in two natures ... unconfusedly, unchangeably, indivisibly, inseparably'. The recognition of Christ's two natures was considered essential to Chalcedonian orthodoxy, since otherwise Christ would share neither God's divinity nor mankind's humanity, therefore being different to and distinct from both. Miaphysites did not recognise this as a problem, believing that Christ's single nature was not a blend of distinct natures but a combination. Chalcedon did not end the discussion but instead provoked adherents to defend more vociferously variants of Miaphysitism, notably that which its opponents called Monophysitism (and which its adherents, like John of Nikiu, called orthodoxy). Miaphysitism remained the dominant dogma among Christians east of Constantinople, even as Arianism – which holds that Christ was created by God the Father, and therefore that Christ's nature is distinct from and subordinate to that of the Father – came to dominate among the Goths and related 'barbarians' who settled in the western provinces of the empire.

Marcian fell ill on 26 January 457, while retracing Theodosius's barefoot seven-mile march to the Hebdomon. He was leading an annual liturgical procession that commemorated the great earthquake that struck exactly one decade earlier. It seems that he suffered from gout and his feet were badly inflamed, forcing him to hobble the whole distance as he distributed alms. He died the following day of gangrene or sepsis. Pulcheria had died two years earlier, but Aspar still lived and six years after he had chosen Marcian, still serving as *magister militum praesentalis* at Constantinople, the Alan general orchestrated the acclamation of Leo.

At the time of Marcian's death there was no western emperor in Rome. Petronius Maximus had been replaced by Avitus, who had lasted less than two years. According to Malalas, Marcian had crowned Anthemius, who was married to his daughter from an earlier marriage. Leo, however, did not immediately select a western partner, but instead promoted and rewarded the two men who had ousted Avitus, the generals Ricimer and Majorian. Ricimer, who was married to the daughter of Anthemius, Marcian's granddaughter, became a patrician and *magister militum* in command of the western armies. Majorian, Ricimer's immediate subordinate, led a successful campaign against the Alamanni, for which he was acclaimed emperor by his army. Ricimer recognised the value of retaining a trusted colleague on the throne, rather than representing the interests of the distant Leo, and further victories for Majorian in Gaul and Spain cemented his authority. The emperor left Ricimer in Rome while he led his own armies, which proved his undoing. Following a defeat by Geiseric's Vandals, Ricimer had Majorian detained, deposed and decapitated. Ricimer remained the power behind the throne of Majorian's successor Libius Severus (461–5) and the aforementioned Anthemius, his father-in-law, who reigned from 467 until 472, and ruled in his own right without the title emperor in the interregnum (465–7).[6]

Ricimer's counterpart in Constantinople remained the ageing Aspar, supported by his powerful sons, Ardabur junior, Patricius and Hermanaric. The son's names reflected the scope of Aspar's power: Ardabur had the Alan name of his grandfather; Patricius was given a Roman name that now reflected his rank; and Hermanaric bore a Gothic name that demonstrated his descent from Plinta and the more recent Gothic marriages of his thrice-married father. Aspar was now also related to Theoderic, nephew of his latest wife, who was leader of a Gothic army recently settled within the empire. This Theoderic would shortly become a key player in events. Priscus tells us that 'Ardabur, the son of Aspar, a man of noble spirit who stoutly drove back the barbarians who frequently overran Thrace' was made commander of the eastern armies by Marcian and retained that rank under Leo. Ardabur junior was responsible for bringing the body of St Symeon the Stylite into Antioch and constructing his shrine there. Upon his accession, Leo pledged his oldest daughter Ariadne in marriage to Patricius, who would be consul in 459. Hermanaric was consul in 465. However, in that very year, Leo moved against Aspar and his clan.[7]

According to Priscus, Ardabur junior had 'turned to self-indulgence and effeminate leisure', and 'amused himself with mimes and conjurors and stage spectacles' in Antioch. Far worse, in Constantinople, an Isaurian named Tarasis, son of Codissa, informed Leo that Ardabur junior was conspiring with the Persians. Ardabur was removed from office and Leo broke openly with Aspar, who acquiesced and protected his son from further punishment. Isaurians had arrived in Constantinople in greater numbers from around 460, many joining the Excubitors, a new imperial bodyguard formed by Leo, which was perhaps conceived as a unit loyal entirely to him, an effective counterweight to Aspar's Goths and Alans. Tarasis was a commander of the Excubitors, and our sources mention several other notable Isaurians by name, including Illus, Illus's brother Trocundes, and Indacus Cottunes. Each powerful Isaurian had slaves, freedmen, and many relatives by blood and marriage, many with the same names. Besides surrounding himself with Isaurians, Leo had moved to distinguish his rule by his personal piety, cultivating the support of key constituencies in Constantinople, including the monks. According to Malalas, he moved against a prominent pagan, the former quaestor Isocasius, obliging him to convert to Christianity; and he 'ordered that Sundays should be days of rest, promulgating a sacred law on the subject to the effect that neither flute nor lyre nor any other musical instrument should be played on a Sunday'. By some reports, this was unpopular in a city used to music and games, sports and festivities, although it seems only to have enforced a law already on the books. Far worse for the restive citizens were the huge fires that swept through Constantinople in both 464 and 469. Eight regions of the city were burned and countless homes and businesses destroyed. For a while, the destruction displaced the emperor to a suburban palace at St Mamas, to the north of the city on the Bosphorus.[8]

Leo promoted his brother-in-law Basiliscus as *magister militum praesentalis* and aligned himself more fully with the Isaurians. In 466, Leo's daughter Ariadne was married to Tarasis, who took the name Zeno (after Flavius Zeno, the Isaurian who had defended Constantinople against Attila). One source reports that Tarasis had an earlier wife and a son, also named Zeno, the date of whose deaths we do not know. Tarasis Zeno was appointed to command the army in Thrace, but proved unpopular and suffered a mutiny. His fortunes recovered, however, when in 469 he became father to a son, named Leo after his grandfather. Tarasis Zeno was rewarded with the ordinary consulship and made commander of the eastern armies. In that capacity, he drove his compatriot, Indacus Cottunes, from the fortress of Papirium. Indacus had been launching raids across Asia Minor from this base, which Zeno appears to have taken for himself to fortify further. *Castellum Papirium*, often called Papyrius (or Papirius), has been identified with two sites known today as Çandır Kalesi and Bağdat Kırı. Both would suit the description given by Joshua the Stylite of an 'impregnable fortress in his [Zeno's] own country'. Papirium features frequently in political intrigues and factional warfare during the following two decades.[9]

Taking an interest in western affairs, and offering support to Ricimer, in 468 Leo sent Basiliscus, his wife Verina's brother, in command of an expeditionary force against the Vandals. At the same time a new western emperor, Anthemius, allegedly chosen far earlier by Marcian, was now sent to Rome. Anthemius's five-year reign would end badly, with his murder by Ricimir. He was replaced by the still more ephemeral Anicius Olybrius, husband of Placidia, making the eastern princess very briefly empress in the west. Basiliscus's expedition proved disastrous, both militarily and financially. According to Malalas, a tremendous sea battle was fought, but 'Basiliscus accepted bribes from Geiseric, king of the Vandals, and betrayed his ships', his own vessel being the only one to escape. 'All the rest of the ships and the army perished, sunk at sea.'[10]

With the defeat of Basiliscus and Zeno's departure for the east, Leo was forced to reconcile with Aspar, briefly, marrying his second daughter Leontia to Patricius, who was named Caesar. But in 471, Aspar and his sons were murdered. The bloody end of Aspar's long supremacy is recorded succinctly by Marcellinus Comes: 'Aspar, chief of the patricians, with his sons Ardabur and Patricius ... the latter designated caesar and son-in-law of emperor Leo, an Arian, together with his Arian family, was wounded in the palace by the eunuchs' swords and perished.' Malalas, followed by the *Easter Chronicle*, offers more detail, stating that the father was murdered in the palace and the sons, both senators, were killed separately, during a meeting of the senate. The given reason was a plot to usurp the throne, presumably to install Patricius. Aspar's followers rioted and a band of Goths, led by a certain Ostrys, entered the palace and fought a pitched battle with the Excubitors. Ostrys escaped with Aspar's concubine, earning the esteem of the crowd who cheered 'the dead man has no friends, except Ostrys'. It was for this reason that Leo was given the epithet 'the Butcher'.[11]

Leontia, Leo's second daughter, was remarried to Marcian, named for his grandfather the emperor. Leo appears to have had a son, who died as an infant, whose name has not been preserved, and whose memory has been preserved only in a horoscope and an allusion in the *Life of Daniel the Stylite* (38). His designated heir, therefore, was his grandson and namesake, Leo, son of Ariadne and Zeno, who was named Caesar. When Leo senior fell ill, in November 473 Leo junior was raised from Caesar to Augustus. Remarkably, the protocols by which the young emperor was 'created by an emperor' are preserved in a tenth-century source, the *Book of Ceremonies*. 'The people and the ambassadors came together in the hippodrome ... and they cried out, the people in Greek, the soldiers in Latin, urging the emperor to go up. The emperor went up with the senate' to where Leo I was seated with the patriarch Acacius, and 'the people cried out Augustus!' The bishop offered a prayer and Leo placed a crown on the young man's head, wishing him 'Good fortune!' three times. The two emperors sat down together to receive the acclamations of their assembled constituents.[12]

In his last years, Leo's piety hardened into opposition to Arianism, the faith of the late Aspar. Arians were prevented from congregating in and possessing churches, where services had been conducted openly for decades in Gothic. Naturally, Leo participated in the fifth-century hunt for precious relics. According to another tenth-century manuscript, in around 473 a veil worn by the Virgin Mary was brought to Constantinople and installed in a shrine constructed by Leo to receive it. The shrine, called the Holy Soros (reliquary) was built just outside the northern land walls of Constantinople at Blachernae. Within the shrine a ciborium, or canopy, was set up over the casket in which the veil (called the *omophorion* or *pallium*) was placed. The ceiling of the ciborium, facing down upon the reliquary, was decorated with a shimmering mosaic 'all of gold and precious stones' depicting Mary on a throne flanked by the imperial family, Leo I and Verina, the latter holding the young emperor Leo, 'and also their daughter, Ariadne. That image has stood from that time onward above the bema of the holy reliquary.' Zeno was not depicted, for he was not at that time destined to become emperor.[13]

According to Joshua the Stylite, the palace officials hated Zeno 'because he was an Isaurian by birth'. This reflects a degree of metropolitan disdain for barbarous mountaineers, certainly, but also a realisation that the interests of the Isaurians did not correspond exactly with those of the Romans. Isaurians were regarded as ruggedly effective by those who deployed them in war, and barbaric by others who encountered them in books or the streets of fifth-century Constantinople. To the Roman ear and eye, there was little difference between Odoacer in Old Rome and Tarasis in New Rome. Nevertheless, when Leo I died, the child Leo took his father Zeno as co-emperor in February 474. Leo then died at the age of seven in November of that year, having reigned in total for one year and twenty-three days.[14]

Zeno (and Basiliscus)

At the death of his son, Zeno recognised that he was in danger. He departed Constantinople in January 475, taking a good deal of wealth to the fastnesses of Isauria. Basiliscus, brother of the dowager empress Verina, took the throne and Illus and Trocundes were sent to attack Zeno, taking with them many of their compatriots. In their absence, Basiliscus slaughtered those relatively few Isaurians who had not yet left Constantinople. Learning this, the Isaurian generals made common cause and seized the throne. Zeno entrusted Illus with the further fortification of Papirium. He then returned to Constantinople and was welcomed back into the city. Basiliscus was killed, but Verina survived, having come to support Zeno's restoration. The *Life of Daniel the Stylite* suggests that 'the blessed empress Verina' was involved in the plot against Zeno, but swiftly saw her error and the wickedness of her brother, taking herself away from the palace to the shrine of St Mary at Blachernae, which her father Leo had built, where she remained until Basiliscus's death. According to Candidus, a courtier and writer who is hostile to the empress, and John of Antioch, Verina's motives were less pure. She had planned to install her lover Patricius, son of Aspar, on the throne and serve as his empress, but had been thwarted by Basiliscus.[15]

Palace intrigue was not the reason for Basiliscus's demise, nor did the senate and people abandon him for his purge of the Isaurians, which many would have welcomed. Basiliscus's undoing was his promotion of Mia-physite (Monophysite) Christology, which lost him the support of the city. According to the *Life of Daniel the Stylite* (69), the pillar saint had urged Zeno to flee, but at the same time correctly predicted his restoration and the demise of 'Basiliscus – a name of ill omen – [who] made an attack on the churches of God [and wished] to deny the incarnate dispensation of God'. Basiliscus had issued an encyclical letter, sent to all churches in his realm, supporting a position articulated by Cyril of Alexandria (d. 444) that Christ's divinity and humanity are united in 'one nature of the Word of God incarnate'. This contradicted Chalcedon.[16]

The demise of Basiliscus is predicted in a remarkable horoscope, pre-served only in a ninth-century Arabic translation from a Greek original that was written shortly after 9 a.m. on 12 January 475, the time of Basilis-cus's inauguration as emperor.

> Look at the planets of this king at the hour (at which he began to) rule. How did it escape the notice of those astrologers who looked for him? Because Venus is in mid-heaven and is the lord of the Lot of Fortune and the lord of the triplicity of the Sun, the luminary of the day, and Jupiter aspects the Sun from quartile. This indicates power and emi-nence and splendour. But they did not look at its (the Sun's) aspectors, nor at the fact that the soundness of rulership is from the Sun. But the Sun is in the twelfth place from the ascendant with the descending node, and Mars aspects it from the quartile. Because of this he will fall from rulership, splendour, and so on.[17]

Further comments on Saturn, Mars, the Moon and Mercury, confirm the prophecy: 'All of this indicates poverty, the loss of rulership and its diminution, and strife and evil and misery.' We should not doubt that Zeno consulted both the holy man, Daniel, and his own court astrolo-ger, whose reading of the heavens proved more accurate than those of Basiliscus's astrologers. In his exile, Zeno had spent time in a fortress that locals called 'Constantinople', to which his response was to observe that

'Mankind is indeed the plaything of God, if the divinity delights in toying even with me in this way, for seers foretold, maintaining it stoutly, that I would spend July in Constantinople.' Perhaps it did not escape the notice of Basiliscus's seers that the splendour of the sun was compromised by its position relative to other celestial bodies, notably Mars, at the time of Basiliscus's accession, but they did not consider it beneficial to point this out.

The disposition of celestial bodies did not expressly predict the fire that swept through Constantinople in 476, and which was described in detail by Malchus, a learned man who was greatly moved by the destruction of the ancient statues curated in the Palace of Lausus and at the Augustaeum, a public square adjacent to the palace gates and hippodrome, and also the burning of the capital's great library housed in the basilica. Writing in the twelfth century, Zonaras (14.2) preserves Malchus's observation that 120,000 ('twelve myriad') books were destroyed, including a roll 120 feet long, a serpent's innards, on which were written the *Iliad* and *Odyssey* in golden ink.[18]

Malchus and Candidus were courtiers who witnessed the intrigues and coups that dominate their histories, which have been preserved only in summaries distilled in the ninth century, by the patriarch Photius, and in fragments collated in the middle of the tenth for the emperor Constantine Porphyrogenitus. Candidus 'the Isaurian' was greatly supportive of Zeno, and opposed to those he portrayed as enemies, including the dowager empress Verina, her brother Basiliscus and, eventually, Illus, who before his fall had 'contributed greatly to the Roman Empire through his manly virtues in war and his philanthropy and just actions in the city' of New Rome. Malchus, perhaps born in Syrian Philadelphia but writing in Constantinople, was highly critical of the emperor, highlighting Zeno's ignorance and cowardice, although he portrayed him in a gentler light than his predecessor, whom he named Leo 'the Butcher'. Malchus, praised as a sophist by those who preserved his work, wrote seven books of fine history on the years 474–80 entitled *Byzantiaca*, meaning 'Byzantine History', in other words a history of the empire ruled from Constantinople. Malchus always called the city 'Byzantium'. *Byzantiaca* appears to have comprised an epitome of events from the city's refoundation and dedication by Constantine in 330 until the reign of Leo, followed by a detailed account of the reign of Zeno. Other than these two works, which later historians used, an eclectic range of sources provides details of Zeno's seventeen years in, and briefly out, of power, including the aforementioned series of political horoscopes preserved in a ninth-century Arabic astrological compendium, the Syriac *Chronicle of Pseudo-Joshua the Stylite*, the fragmentary history of John of Antioch, and a lacunose hexametric imperial encomium (*basilikos logos*).[19]

Illus had hedged his bet on Zeno by taking prisoner the emperor's brother, Longinus, whom he kept as a hostage in Isauria. In this way he was able to influence politics in Constantinople to his advantage, maintaining support for Zeno so long as it suited him. Longinus's imprisonment did not check Verina, however, and she was alleged to be behind two plots on Illus's life, in 477 and 478. Verina was sent to a nunnery in Tarsus, Cilicia, and from there to Papirium, never returning to Constantinople. Illus, although far from reconciled with the emperor, came to Zeno's rescue in 479, when he was trapped in the palace by his brother-in-law Marcian, and Marcian's brothers Procopius and Romulus. Marcian, who was married to Ariadne's sister Leontia, was despatched to Papirium to join Verina. Joshua the Stylite blames Zeno for the plots against Illus, and although this may be because he omits a good deal of detail provided by others, he may thereby have got to the heart of the matter: Zeno and Illus were plotting against each other, spreading rumours that have since become history. With Verina removed, it was Ariadne's turn to plot against Illus, who was attacked by a soldier as he ascended the *cochlis*, a spiral staircase that linked the palace and the hippodrome. Illus lost part of an ear and the soldier his head before he could reveal who had sent him. Illus withdrew to Antioch and there, having command of the eastern armies, he ended the proxy war and resisted Zeno openly. When the general Leontius was sent to put him down, Illus persuaded him to rebel and raised him up as a pretender. Verina, by now under Illus's control, was brought from Papirium to Tarsus to give her sanction to the proclamation of Leontius as Augustus, but the rebels could not hold Antioch and fled to Papirium, resisting until 488, when Illus and Leontius were captured and beheaded. John of Antioch offers a great deal of sympathetic detail, noting that the executioner lost his words and wits at the moment of their deaths, and lightning, thunder and hail arose.[20]

An imperial panegyric has survived from around this time and has been interpreted by its editor as celebrating Zeno's defeat of Illus and Leontius and marking his promotion of Longinus, Zeno's brother, to be his successor. In lacunose form, the oration is less remarkable for the information it supplies than for the slavish manner with which it follows the rules for composing imperial encomia, as prescribed by Menander Rhetor. More interesting to the historian are unique details of the battle at Antioch and the subsequent siege of Papirium provided by Joshua the Stylite, who insists that Illus had made contact with the Persians, sending an embassy and large quantity of gold in Leontius's name seeking a military alliance. Joshua also places blame on 'a sorcerer, a false fellow named Pamprepius', who had cast a horoscope predicting the success of the revolt and was put to death shortly before the capture of the main protagonists.[21]

Pamprepius was, in fact, among the most learned men of his age, about which he wrote a lost historical work, *Isaurica*. A Neoplatonist and poet, born in Egypt, he taught for many years at Athens before moving to Constantinople, where he became attached to Illus. Pamprepius saw in Illus's rebellion an opportunity to promote the fortunes of Hellenism, which its opponents called paganism, and he visited Alexandria to promote his oracles that predicted success. According to one witness, Paralius, who was later a Christian convert, Pamprepius's message found favour with many.[22]

How many sacrifices we offered to the gods when we were all pagans in Caria, dissecting livers and examining them by means of magic arts when we were putting questions to the so-called gods in order to learn whether all of us would defeat the emperor Zeno, who has since met a pious end; [all of us with] Leontius, Illus, Pamprepius, all those who joined the revolt; and how many thousands of oracles and promises we received, whose combined force the emperor could not possibly survive, but the time had come when Christianity would collapse and pass away while pagan worship would take hold again.[23]

A consequence of Pamprepius's activities, which saw his posthumous reputation suffer by the pen of those whom he had wished to revive, was the persecution of philosophers in Alexandria by the local patriarch. For this reason, Damascius, who studied at Alexandria at around the time of the rebellion, branded Pamprepius 'a beast more contorted and rabid than Typhon himself'.

A Christian theologian writing at the same time, perhaps in Alexandria, composed a theosophy, a treatise in seven books about 'correct faith' supplemented by four more books that extended the Christian apologist tradition, establishing a model that would be borrowed into the Syriac tradition and persist in Greek for centuries. The theosophy contains a compilation of, and commentary on, ancient oracles intended to demonstrate 'that the oracles of the pagan gods and the so-called divine predictions of the Greek and Egyptian sages, indeed also those of the famous Sibyls, were concordant with the aim of divine scripture and expressed either the primordial cause of all things or the Holy Trinity contained in a single godhead'. Why, if the power of prophecy in their religion was recognised and shown to predict the fundamental tenets of Christian monotheism, would pagans like Paralius resist conversion? The theosophist ended his work with a reason why conversion was urgent: he offered a prediction that the world would end in AD 508. As Psalm 90 (91) observes, a thousand years are like a day in the sight of God. According to the Alexandrian calendar, which was adopted by the emperors of New Rome, the world was created in 5508 BC. Consequently, the 508th anniversary of the Incarnation was the year 6000, the end of the sixth day since Creation and the beginning of the seventh.[24]

The conversion of the Parthenon, the great temple dedicated to Athena on the Athenian acropolis, can also be dated to this time. The same theosophy records an oracle of Apollo for the citizens of Athens.

The citizens had asked Apollo in this way, 'Prophesy to us, o prophet, Titan Phoebus Apollo, to whom should this house belong?' And the oracle responded as follows, 'Do whatever calls forth virtue and order. For my part, I proclaim a single triune God ruling on high, whose imperishable Word will be conceived by a virgin. Like a fiery arrow he will streak through the middle of the world, capture everything and offer it as a gift to the Father. This house will be hers. She is named Mary.'[25]

This oracle, allegedly revealed to Athenians as they built the Parthenon in the fifth century before Christ, was discovered miraculously carved into the side of the temple by their successors in the fifth century after the Incarnation. Observing that the oracle of Apollo had recognised the true nature of God, the Holy Trinity, and the Virgin birth, they fulfilled the prophecy and gave the temple to Mary. An inscribed oracle of this nature cannot, of course, have been prominently displayed on the Parthenon for a thousand years. Exactly the same oracle, however, has survived in an inscription carved in the fifth or sixth century AD, by Christians on the island of Icaria. One can imagine an Athenian version was knocked up towards the end of the fifth century and 'affixed to the side of the temple' to justify the removal of the cult statue of Athena and the conversion of the temple to a church dedicated to Mary.[26]

The conversion of temples and attacks on philosophers in Athens and Alexandria could not hope to eliminate Hellenic culture and surely was not intended to do so. However, those who attacked pagans may have wished to remind Christians that they had common enemies who denied their faith. Zeno felt an urgent need to unite Christian communities across the east that had turned against each other since the Council of Chalcedon. In 482, 'the pious, victorious, supreme, ever-worshipful Augustus' wrote to 'the very reverend bishops and clergy, and the monks and people throughout Alexandria, Egypt, Libya and the Pentapolis' setting out a compromise that he imagined would paper over the cracks.

We confess that the only-begotten son of God, himself God, who truly took upon himself manhood, our Lord Jesus Christ, who in respect of his godhead is consubstantial with that Father, and consubstantial with us in respect of his manhood; we confess that he, having come down and been made incarnate of the Holy Spirit and the Virgin Mary, the God-bearer (*Theotokos*), is one not two; for we assert that both his miracles and also the sufferings which he, of his own will, endured in the flesh, belong to one single person; we in no way recognise them that make a division or confusion, or bring in a phantom, seeing that his truly sinless Incarnation from the God-bearer did not bring about the addition of a Son, for the Holy Trinity existed as a Trinity even when one member, God the Word, became incarnate.[27]

This 'Edict of Reunion', or *Henoticon*, made no mention of Chalcedon, but implied that its rulings were of no consequence. It recognised the definition of faith codified at the Council of Nicaea (held in 325) by '318 holy fathers', which '150 holy fathers confirmed' at the Council of Constantinople (AD 381), and which was 'followed by all the holy fathers at the Council of Ephesus' (AD 431). Some holy fathers to whom this edict was sent may have been appalled that the emperor sought to dictate to them on doctrine. More, however, will have recognised behind the emperor's edict the thought and ambition of Acacius, patriarch of Constantinople, who saw an opportunity to strengthen the authority of his see in relation to Alexandria, Antioch and Rome, all three of which claimed apostolic foundation. It has been suggested that both Zeno and Acacius had strong sympathy for Monophysite Christology, but neither, following the suppression of Basiliscus, was able to express this openly. The patriarch of Alexandria, John Talaia, a Chalcedonian, refused to accept the *Henoticon*, and consequently was replaced by Peter Mongus, 'the Fuller', a Monophysite who had earlier served as patriarch and accepted the edict. Talaia retired to Rome, where the pope, Simplicius, recognised his complaint. Simplicius wrote to complain to Zeno and summoned Acacius to answer for his support for Mongus and his promotion of 'heresy'. Simplicius's successor, Felix III, anathematised Acacius, initiating the so-called Acacian Schism, which lasted until 519.[28]

Surviving fragments of Malchus dwell expansively on Zeno's regular struggles with the two Theoderics: Theoderic Strabo, commander of the Goths in Thrace, and Theoderic son of Theodemir, commander of the Goths in Pannonia. Malchus implies that they regarded the Isaurian as a fellow warlord, hardly even first among equals, whom they could threaten in order to extract concessions, including land, wealth and titles. Theoderic Strabo, 'the squinter', had been a creature of Aspar and the nephew of his wife. His demands of Zeno frequently related to the family of Aspar or concessions secured under Leo. The younger Theoderic, heir to the royal Amal line, had spent a decade in Constantinople as a hostage, standing surety for his uncle Valamir's behaviour after 459. He would later claim that during this period he had learned to rule justly over Romans. Zeno's ambition was to turn the Goths against each other or otherwise find a means to neutralise them as a threat to the Romans, which he did by granting concessions and high office alternately to prevent them allying against him.

Military engagements took place across the Balkan peninsula between 477 and 481, and Zeno's own subjects seemed to suspect his motives, often for good reason, as Goths were permitted to settle on their lands and seize their crops, threatening several cities of great antiquity, sacking Stobi and Scampia. The wealthy of Lynchnidus, later called Ohrid, and the citizens of Heraclea Lyncestis, near modern Bitola, suffered the army of Goths on their lands, while 'the son of Valamir' – Greek sources refer to Theoderic consistently as the son of Valamir – 'restrained them from arson and murder'. On the same occasion, the citizens of Thessalonica overthrew statues of Zeno and threatened to burn the city prefect's palace, suspecting that the younger Theoderic was to be granted their city. In fact, he marched west to occupy Dyrrachium and neighbouring parts of Epirus, which he held for a number of years. Strabo, meanwhile, sought to benefit from the rebellion of Marcian, marching his Goths to the walls of Constantinople 'in support' of Zeno, retreating only when given 'promises and large sums of money that affect the greedy nature of the barbarian mind'. Malchus claims that the Isaurians were prepared to burn down the city rather than cede it to the Goths.[29]

The death of Theoderic Strabo in 481 simplified matters, although 'the son of Valamir' vacillated between peace and war for the next six years. In 484, as consul for the year, 'Flavius Amalus Theodoricus' was permitted to kill Strabo's son Recitach, while his Goths marched to war with Illus and Leontius. Eventually, in 488, Theoderic set out for Italy. Several reports attribute this move to Zeno, but he was dead by the time Theoderic secured Ravenna in March 493, murdering Odoacer in violation of a treaty. Little is known of Zeno's last three years and reports of his death are confused, stating it was from dysentery (of which his son and namesake had died some years earlier) or epilepsy. A tradition later emerged that Zeno was buried alive for three days in a sarcophagus, which Ariadne refused to open despite his screams.[30]

Anastasius

Learning of Zeno's death, a Persian embassy bringing a large elephant and a demand for gold to Constantinople halted at Antioch. John of Antioch implies that only the demand for gold was sent on, although the famous Barberini ivory of the period (Plate 14) shows an embassy from the east delivering an elephant to the triumphal emperor of the Romans. The Persian demand for gold, if not yet an elephant, was delivered to Anastasius, a palace official already aged sixty, who ascended the throne.[31]

Several of the major narrative sources for Anastasius's reign are the same as those for Zeno. The *Chronicle of Marcellinus Comes*, which begins with the death of Valens in 378 and offers a terse entry for each year thereafter, becomes an eyewitness account during Anastasius's reign, when its author lived in Constantinople. Malalas is an excellent source of information on life at Antioch, where that author was born around 491 and lived before he moved to Constantinople. Moreover, there are other written materials that supply insights, including two imperial panegyrics. A variety of contemporary inscriptions and poems has been preserved in the *Greek Anthology*. Military writings attributed to a certain Urbicius seem to originate in Anastasius's reign. There are also visual materials that attest to themes in later fifth-century cultural history, including exquisite imperial ivories produced in Constantinople and remarkable buildings erected in Ravenna.[32]

The *Book of Ceremonies* preserves an account of Anastasius's selection and inauguration, stating that the people of Constantinople, assembled in the hippodrome, greeted Ariadne with a ritual acclamation and a demand: 'Ariadne, Augusta, may you be victorious! Holy Lord, grant her a long life! Many years for the Augusta! An Orthodox emperor for the empire!' No longer would they tolerate to be ruled by a Miaphysite. Ariadne replied that she would consult and 'choose a man who is Christian, Roman, and endowed with every imperial virtue', and asked for time to bury Zeno. The crowd's response was once again the familiar acclamation, wishing her a long life; a topical aside, hoping the empire enjoyed a happy Easter (which fell on 14 April 491); and a veiled threat: 'May all blessings be upon you, Roman Empress, if no foreign element is added to the race of Romans!' No longer would they tolerate to be ruled by an Isaurian. With the consent of the leading men (*archontes*), the senate and the bishops, Ariadne chose Anastasius, who was brought to the palace and kept safe during Zeno's funeral. The following day Anastasius stood before the assembled worthies of the city and empire, each wearing a pure white cloak (*chlamys*), and swore an oath that he would bear no grudges against any of them and would administer the state justly. He then proceeded to the imperial box (*kathisma*) in the hippodrome, now wearing imperial clothing – including a golden tunic (*divetesion*) and purple sandals – but bare-headed. There he was greeted first by the soldiers and raised on a shield, a regular aspect of 'king-making'. The leader of a unit of the imperial bodyguard, the spear-bearers, placed his own golden torc on Anastasius's head. He was then acclaimed by the soldiers and civilians together.[33]

Re-entering the palace – for the *kathisma* was also a room in the palace complex, joining it to the hippodrome – Anastasius was joined by the patriarch of Constantinople, who vested him in the deep purple imperial robe and crowned him with the imperial crown (*stemma*), intoning a prayer. These ritual acts altogether, from shield-raising and acclamations to robing and coronation, comprised the inauguration ceremony of the Roman emperor at the end of the fifth century. Not every inauguration was identical, but each drew from a pool of ideas and ceremonies, emphasising what was most urgent or important at the time or to the investee. Anastasius returned to the *kathisma*, now emperor, to greet the people, to distribute largesse, and to receive their cheer of 'Augustus, revered one!' A version of his speech is preserved, punctuated with the ritualised interjections of the crowd. Having thanked the people, Anastasius thanked God, processing to the great church, where his crown was placed in the sanctuary while he prayed and a service was conducted. Receiving it back as he left the church, he processed back to the palace, undertook his first official duties, and dined with the *archontes*.

The account of Anastasius's inauguration preserved in the *Book of Ceremonies* projects an image of stability, indicating that the new emperor received the assent of all political and social groups. However, the events of the first years of his reign indicate, to the contrary, that the citizens of Constantinople were not universally supportive of their new emperor, and there were destructive riots. According to Marcellinus Comes, shortly after Anastasius's inauguration 'Civil strife arose among the Byzantines and much of the city and circus were burned.' John of Antioch provides corroborating testimony and a fuller articulation of how Anastasius responded. First, he declined to deploy troops against those who had 'pulled down bronze statues of the emperors from their pedestals and inflicted on them every kind of abuse'. Instead, he met their demand to replace the urban prefect, and by doing so angered the soldiers. In 493, there was again violent protest in Constantinople against the rule of Anastasius and 'statues of the emperor and empress were dragged down and through the city on ropes'. When calm returned, Anastasius was able to rebuild what had been burned and to turn the situation to his advantage, by using it as a pretext to expel Isaurians from the city. Even this action, however, was compromised, since those accused of conspiracy were not immediately deprived of their ranks or property, and were sent off to join a rebellion their compatriots had already started. Most notable among these were the master of offices, Longinus, and the general Ligninines. A second Longinus, the aforementioned brother of Zeno, was exiled to the Egyptian Thebaid, where he died eight years later.

Hardening his heart, on a single occasion the emperor sold by auction the property of the Isaurians and the possessions acquired by Zeno during his reign, even to the extent that the imperial garments were offered for sale; he also sent orders that the fort of Papirium be destroyed and abolished the levy that had supported it, which had totaled 1400 pounds of gold each year.[34]

Shortly afterwards a battle between the Isaurians, commanded by Ligninines, and the Romans took place at Cotyaeum in Phrygia, which the Roman army won, although apparently greatly outnumbered, but supported by Huns and Goths. The Isaurians withdrew and the conflict dragged on until 498, when Anastasius's generals secured victory.

The first half of a panegyrical oration, written and delivered by Priscian at an indeterminate date between 503 and 513, is devoted to the emperor's Isaurian victory, 'when the triumphant army of the unconquered emperor and his commanders, powerful in their loyalty as well as in their courage, visited [the Isaurians] with slaughter and rout and total ruin'. Anastasius's suppression of the Isaurians was linked to Pompey the Great's destruction of Cilician piracy in 67 BC, even as a fictive Roman lineage for the ageing emperor was created: 'Nor is it strange that such a man was sprung from the stock of Pompeius ... But yet O Pompeius, yield to a renowned descendant.' Pompeius appears to have been a family name for Anastasius, and perhaps even his father's name. Just as Pompey's victory brought economic relief to Roman markets, as the price of grain plunged, so Anastasius linked his victory with fiscal reforms that brought economic benefits to some traders: the abolition of the *chrysargyron* transaction tax. Still, he gained a reputation for parsimony, evidently linked to his introduction of a new system of coins.[35]

Anastasius is reputed to have melted down statues to mint his new copper coins. There appears to be evidence that this was common practice in the fifth century, or at least that a range of objects formed from copper alloys was repurposed to create coins later in the century. Analysis of some of the smallest copper coins, *nummi minimi*, shows a shift from minting in relatively pure copper ($c.96$–98.7 per cent) under Theodosius I to the use of recycled copper alloys and heavily leaded bronze. Coins struck under Leo I ($c.78$ per cent copper) have the least copper, less even than those of Anastasius (81–85 per cent copper). The presence in these coins of tin and zinc indicates that the base metal, copper, was recycled even before it was heavily debased with lead. One can imagine that many of the bronze statues badly damaged in fires were among those sources of copper. New statuary was rare at this time, and if bronze was in short supply we may have an explanation for an epigram in the *Greek Anthology* (XI. 271), which reveals that an iron statue of Anastasius was placed among the statues in Constantinople's hippodrome on its central median, the *euripos*, which was named after the narrow channel that separated Attica from Euboea. The poem alludes to an equally famous channel, the Strait of Messina: 'Near to Scylla they set up cruel Charybdis, this savage ogre Anastasius. Have fear in your heart Scylla, lest he devour you too, turning a brazen goddess into small change.'[36]

In 498, the same year as Anastasius's Isaurian victory, Kavad (also given as Kawad or Kavadh) regained the Persian throne after two years in exile. It was Kavad who had sent an embassy to Anastasius in 491 demanding gold, described as a tribute payment in recognition of the Persians' defence of the Caspian Gates – today the Dariali Gorge in the republic of Georgia – a heavily fortified mountain pass that defended Transcaucasia and the lands to its west, Persian and Roman, against 'barbarians' from the Eurasian steppe (Map 6). A renewed demand for this Caspian Gates subsidy was refused and Kavad punished Anastasius by allowing his Arab clients, the Lakhmids, to attack Roman territory. Kavad's need for funds, however, resulted in a further escalation, and in 502 he led a Persian army through Armenia, capturing Theodosiopolis (Erzurum) and then Amida (Diyarbakır), which fell after a three-month siege. Anastasius's response was forceful, sending the largest Roman army against the Persians since the invasion by Julian 'the Apostate'. Nevertheless, Kavad retained the upper hand until a new treaty secured peace for the rest of Anastasius's reign in return for a large cash payment to the Persians. This resumption of hostilities between Rome and Persia, later called the 'Anastasian War', presaged greater wars to come.

Wishing to strengthen the border, Anastasius then commissioned fortifications in violation of an earlier agreement with Persia. He rebuilt the village of Dara near Amida as a great fortress, naming it Anastasiopolis (Maps 5 and 6). A counterweight to the Persian stronghold at Nisibis (Nusaybin) only eighteen kilometres to its east, Anastasiopolis enjoyed an elevated position with a good natural water supply, which was augmented by the construction of an aqueduct and large cistern adequate to support a new public baths within its strong walls. Anastasius seems also to have sponsored the strengthening of Resafa, which the Lakhmids had attacked in 498. Well to the south of the Roman–Persian frontier in Mesopotamia, Resafa was home to the shrine of St Sergius, for whom it was called Sergiopolis and later, for several decades after 553, also Anastasiopolis. In the spirit of imperial relic-collecting, Anastasius had Sergius's thumb brought to Constantinople, but this did not dent the popularity of the saint's shrine as a destination for pilgrims, despite its location in the 'barbarian plain', a zone beyond the pale.[37]

The greatest structure to bear the name of Anastasius cannot, definitively, be attributed to him. The Anastasian Wall, also called the Long Wall or Long Walls of Thrace, may have been begun somewhat earlier in the fifth century, following the disastrous battle of Adrianople, when the lands to the west of Constantinople were no longer considered inviolable (see Map 11). The presence in the Balkans of armies of Goths in imperial service proved an effective deterrent to potential invaders, but one that was removed by Theoderic's departure in 491 for Italy. At least three invasions across the Danube followed this, between 493 and 502, provoking Anastasius to look to his forward defensive line in Thrace. The wall, some sixty-five kilometres from the city and forty-five kilometres in length, would have been punctuated by regular towers and required a garrison of around 4,000 men. It also protected the water supply of Constantinople. One source that supplied the Valens line, the springs and caves at Pınarca, lay around two kilometres within the wall. The water supply provided by the older Hadrianic aqueduct was well within the new long walls.[38]

Throughout Anastasius's reign, riots were associated with public spectacles and entertainments. Popular unrest was also provoked in Constantinople, Antioch and other cities by Anastasius's increasingly open support for Monophysitism. From the time of his accession, Anastasius's orthodoxy had been questioned, and Euphemius, the patriarch who invested him with the imperial vestments, had extracted from the new emperor a written profession of his faith. As Anastasius's power grew, so did his resentment of Euphemius, who enjoyed popular support in Constantinople as well as the trust of Ariadne. The patriarch was deposed for appearing to support the Isaurian rebels, provoking yet another riot in the capital, in 496. The new patriarch, Macedonius, was required to accept Zeno's *Henoticon*, but was sympathetic to Euphemius, whom he sponsored in exile, and remained committed to reconciling Chalcedonians and Miaphysites in the face of opposition from both. Macedonius failed in an attempt to reconcile the monasteries of Constantinople, which had been divided over the *Henoticon*, and faced increasing interference from influential Miaphysites, notably Philoxenus of Hierapolis (Mabbug), who had attracted the emperor's interest. Philoxenus produced a version of the bible preferred by Syrian Miaphysites.

Macedonius was still patriarch a decade later when, in 508, a certain Severus arrived in Constantinople with 200 Monophysite monks. Severus, who would become the foremost polemicist for the Monophysites and among the greatest theologians of his age, had been a student in Alexandria in the 480s, a pagan peer of Paralius, before he proceeded to study law at Berytus (Beirut). There, by his own account, he and several colleagues were inspired by St Leontius to abandon 'their hollow education along with the affairs of life and purify their minds of the fables of the Hellenes'. Severus fell in with Zacharias, already a devout Monophysite whom he had known in Alexandria and who would later become his first biographer. While from 491 Severus lived as a monk, Zacharias took up a legal career in Constantinople, and it was there that they encountered each other again in 508, when the lawyer supported the monk's cause and secured for him an audience with the consul Clementinus, and through him access to the emperor himself. [39]

Anastasius favoured Severus, as he had Philoxenus, and the monk enjoyed influence at court for three years before he overplayed his hand. In 511, the Monophysites in Constantinople adopted a modification – perhaps invented by Peter 'the Fuller' Mongus – to the Trisagion, the 'Thrice-holy' prayer, which added three Greek words, 'crucified for us' (*stauротheis di emas*). This provoked a fight in Hagia Sophia itself, and resulted in a 'Stauroтheis riot', where the factions openly declared their support either for the emperor or the patriarch, with the patriarch clearly enjoying a substantial advantage. An account of this is preserved in Severus's own words, in a letter he sent from Constantinople to Soterichus, bishop of Caesarea.

> When the Christ-loving crowd gathered together in the Great Church of the imperial city to send up to God the hymn of the Trisagion ... that son of impiety, Macedonius, enflamed by the fire and madness of impiety, commissioned and hired slaves for money ... These inflicted great blows and other wounds which were difficult to heal upon the brothers who uttered the Trisagion, as they pulled their hair, dragged them, struck them with their hands, kicked them with their feet,

poked their eyes and plucked out their nails like carnivorous birds
... Some seized the psalm singers, confined them in dark spaces, beat
them severely and left them discarded and half-dead. Others among
the impious stood at the doors of the church calling to those passing
by in the market [of the Augustaeum], shouting 'Come and see these
impious ones who entered the church of God'. They did this for no
other reason than to incite the populace against the pious.[40]

Chastened by the support the people showed for the patriarch, Anastasius was reconciled with Macedonius, but only for as long as it took to
oust him. The patriarch was accused of fomenting rebellion and exiled to
Euchaita, to join Euphemius.[41]

A similar scene played out in Antioch, where the Chalcedonian patriarch Flavian had battled Philoxenus, but was deposed along with Elias of
Jerusalem in 512. Flavian was exiled to Petra and replaced by Severus, who
was installed as patriarch of Antioch in November 512. With three powerful Chalcedonian patriarchs removed, Anastasius was emboldened to
support the modification of the Trisagion, having it announced at Hagia
Sophia on 4 November and at other churches in the city on the following two days. This provoked a second 'Staurotheis riot', which was widely
recorded and appears to have been the worst popular unrest of Anastasius's reign. The emperor fled to Blachernae, but returned on 8 November
to appear before the people in the hippodrome, just as he had before his
inauguration almost two decades earlier, without his imperial crown.
Offering to abdicate, he calmed the crowd, but if we follow Malalas,
shortly afterwards,

> the emperor ordered that many arrests be made. Of the many brought
> into custody, he had some punished and others thrown into the Bos-
> phorus by the city prefect. They suffered this way for many days and
> after countless numbers had been executed, excellent order and no
> little fear prevailed at Constantinople and in every city in the Roman
> state.[42]

Severus proved to be a powerful and popular leader in Antioch, and
his influence reached across the eastern provinces to Alexandria. His
strict anti-Chalcedonian line limited Anastasius's scope to secure support
in the west, but still the emperor made several entreaties to the older

Rome. These were provoked in part by a fear that the Balkan provinces were looking to the west. As the bishops of Illyricum contemplated papal supervision a rebellion in Thrace threatened Constantinople itself. The insurrection was led by Vitalian, a commander of forces in Thrace, and appears to have been provoked by Anastasius's religious and fiscal policies. If Vitalian was driven by ambition for himself and the Orthodox position, his troops were perhaps more concerned by Anastasius's decision to withdraw the *annona*, a distribution of provisions levied from local landholders, to which the troops were accustomed and felt entitled. The lands of Thrace seem at this time to have been unable to support the levy. Twice Vitalian led his army to the walls of Constantinople, in 513 and 514, and on both occasions was persuaded to depart by the emperor's overtures. In 515, Vitalian's army was defeated and the rebellion ended. To celebrate, Severus of Antioch wrote a hymn *On Vitalian the Tyrant and the Victory of the Christ-loving Emperor Anastasius*.

An embassy despatched to Rome after Vitalian's defeat was led by two of Anastasius's trusted counts, and pointedly it included no clerics. It carried two letters, to the senate of Rome and to the pope, Hormisdas (514–23). The letter to senators set out the emperor's hope that they would intercede to achieve peace and end the Acacian Schism, but their reply confirmed that they would side with the pope. The papal letter from Hormisdas reiterated a distinction between papal authority and imperial power, first articulated in an earlier letter of 494 from Pope Gelasius to Anastasius. An embassy carrying letters from Hormisdas and the senate of Rome was turned away by Anastasius, who banned them from stopping at any port to distribute the letters they wished to send to eastern bishops. Anastasius's last letter to Hormisdas, dated 11 July 517, ended their dialogue. Almost exactly a year later Anastasius died.

The most substantial legacy of Anastasius's reign, and certainly his own initiative, was an extensive revision of the state economy and currency, which generated huge surpluses. At his death, Anastasius left 320,000 pounds of gold in the state coffers, amounting to more than 23 million solidi, which was between three and four times the total annual imperial budget. He abolished the unpopular transaction tax, the *chrysargyron*, and thereby appears to have stimulated an already expanding economy. The scale of this tax relief is captured by the record of a single city, Edessa, which paid every fourth year a total of 140 pounds of gold or 10,080 solidi. The response of the Edessenes was to 'dress up in white, from the greatest to the least, and carrying lighted candles and burning incense, march out to the shrine of St Sergius and St Simon, praising God and the emperor'. The emperor had found other ways to generate revenue, commuting payments made in kind to gold, and appointing state officials to oversee the payment of taxes that had earlier fallen to localities and municipalities. Trade was further stimulated and facilitated by currency reform, first in 498, and subsequently again in 512. In 498, a whole new set of copper coins was introduced, each with a fixed weight and valued at forty, twenty and ten nummi (which we might translate as 'pence'). The forty-nummia piece was the follis and 420 folles bought a single gold coin, the solidus. In 512 the weight of each was doubled – in the case of the follis, from nine to eighteen grams – and a five-nummia coin introduced. The new tariff was 210 folles to the solidus. Marcellinus Comes reports that the currency reform was popular, but still Anastasius earned a reputation for parsimony. Although the same charges were levelled at most emperors, including Zeno and Basiliscus, Anastasius does appear to have deserved his reputation in the eyes of the poor. In contrast, senators and other large landholders did rather well and praised the emperor.[43]

Justin

Anastasius died on 9 July 518, aged at least eighty-seven. According to several reports, a 'bearded comet' appeared in the East, which caused widespread fear. The Monophysites came to realise that 'it portended apostasy, destruction, and the ruin of the Church', when the learned emperor was succeeded by an illiterate guardsman called Justin, by then already in his seventh decade. Justin is alleged to have stolen money entrusted to him to ensure the succession of another candidate to buy his own accession, killing the rival and his patron. Swiftly, Justin reversed a number of Anastasius's positions, showing himself to be a staunch Chalcedonian. By March 519, Justin had ended the Acacian Schism, receiving a large embassy sent by Pope Hormisdas and agreeing the terms that they offered. Negotiations with Pope Hormisdas were conducted in part by Justin's two nephews, Germanus and Petrus. Germanus served as his *magister militum* in Thrace, where he won a famous victory over the Antae, and Petrus was enrolled as commander of the Excubitors.

Justin made peace with powerful men whom Anastasius had exiled, recalling them and entrusting them with high command. Foremost among them was Vitalian, who had marched against the city. He was summoned to Constantinople and promoted to master of cavalry and infantry at court, and consul for 520. Justin sought to replace the Monophysite bishops whom Anastasius had supported, driving many into exile, including Philoxenus of Hierapolis and Severus of Antioch. Severus was summoned to Constantinople, where Vitalian demanded that his tongue be cut out, a cruelly fitting retribution for Severus's hymn to Vitalian's earlier defeat. In September 518, Severus fled to Alexandria, where he was free to write for far longer than Vitalian prevailed at court. According to Marcellinus Comes, 'in the seventh month of his consulship ... Vitalian died in the palace ... stabbed sixteen times'. The new patriarch of Antioch was Paul, a fierce Chalcedonian called 'the Jew' by his enemies. His appointment followed nomination by the papal legates, and he received 1,000 pounds of gold as an imperial gift.[44] Rapprochement between the old and new Romes did not suit a now aged Theoderic, who feared conspiracy between the pope, his senate and the eastern emperor. It was at this time that Theoderic imprisoned Boethius, a senator and father to the two consuls of 522, then put him to death with Symmachus, Boethius's father-in-law. Following Theoderic's death in 526, the eastern emperor had both a justification and an opportunity to reassert his authority in the older Rome. Before he could do so, however, on 1 August 527 Justin died.[45]

As the fifth century became the sixth, imperial power had passed from a series of soldiers to a civilian and back to a soldier. However, it would be wrong to imagine that much had really changed in Constantinople. The levers of government remained in the hands of an entrenched elite, with senior offices of state passing from father to son to grandson, senators all. There were new additions to this elite, such as the Ardaburs, established by the 'barbarian' warlords Ardabur and Plinta, and entrenched by Aspar and his sons. From the time of Aspar's consulship, celebrated with games and the distribution of sumptuous gifts to his peers, Aspar and his family wielded influence as great as any emperor. But no Ardabur ever sat on the throne and each scion satisfied himself with the curule chair. Marcian and Leo 'the Butcher', both creatures of Aspar, wore the crown before Zeno, an Isaurian warlord, rose to power. During their reigns, Anastasius became a bureaucratic mandarin, then emperor himself, raised up by leading men (*archontes*), whom Anastasius would enrich. Zeno had encouraged his praetorian prefect Sebastianus to sell offices, using much of the cash raised to pay his compatriots and tribute demanded by the Goths. Anastasius, to the contrary, undertook currency and taxation reforms that long outlasted him, supporting the state economy and sustaining the palace, the bureaucracy and the army.

Anastasius's selection as successor to Zeno was remarkable and it is hard to escape the conclusion that this was the determination of the established families, the senate of Constantinople. Anastasius remained a creature of the senate, where members of his immediate family were installed, flourishing under imperial patronage. Many close relatives held the consulships of east and west, including Flavius Paulus, Anastasius's brother, who was eastern consul in 496, a year before the emperor's own second consulship, and Flavius Hypatius, Anastasius's nephew, who was western consul in 500. The complexity of the name of the eastern consul of 517 reveals the degree of intermarriage at the heart of the Anastasian court: Flavius Anastasius Paulus Probus Sabinianus Pompeius was probably Anastasius's great-nephew and brother of Flavius Anastasius Paulus Probus Moschianus Probus Magnus, consul in 518. Despite this, none of Anastasius's closest relatives, his three nephews Hypatius, Pompeius and Probus, succeeded him.

While Anastasius proved to be fiscally and administratively competent, he struggled to steer such a steady course politically. His attempts to suppress opposition were calculated but not always successful and they became increasingly harsh. His sympathy for Monophysitism provoked a strong reaction later in his reign, although unlike Basiliscus he survived the backlash in Constantinople. Efforts to find compromise, such as Zeno's *Henoticon*, which Anastasius continued to promote, had quite the opposite effect and the fractures in the eastern Church, widened by the Council of Chalcedon, persisted into the reign of Justin. Justin secured support in Constantinople by his ostentatious support of Chalcedonian 'orthodoxy'. His successor would elevate demonstrations of imperial orthodoxy and piety to a higher register.

8

THE AGE OF
JUSTINIAN, AD 527–602

Petrus Sabbatius was born near Tauresium, a small town in the northern Balkans, in around AD 482. He would rebuild the place of his birth some eight years after he took the throne and rename it Justiniana Prima. Petrus's origins were unremarkable and, for that reason, eminently imperial. For more than three centuries, from the accession of Claudius Gothicus until the death of Phocas, the majority of emperors were men of similar background, who rose through military service. With the exception of the Theodosian dynasty, and until the establishment of the Heraclian dynasty, almost all Roman emperors were men of humble origins born in Upper Moesia and Dardania, a region that stretched south from the great Danubian fortresses of Sirmium, Singidunum and Viminacium, encompassing countless small settlements and a few fortified cities that lined the great military roads to Naissus, Scupi and Serdica (Map 3).

Petrus was Justin's nephew. He had been brought to Constantinople, provided with the education his uncle lacked, and enrolled in the palace guard. One of several family members promoted by Justin, Petrus proved the wiliest and was adopted by his uncle before 520. He was promoted to master of cavalry and infantry at court, serving alongside Vitalian. In 521, Petrus celebrated his consulship with elaborate games, including beast fights, all aimed at surpassing Vitalian. A consular diptych, a work of art in ivory that was an equivalent of Aspar's silver consular dish, is the only place to preserve his full name, Flavius Petrus Sabbatius Justinianus (Plate 15).

Justinian has been variously judged a powerful leader, firm in faith and endeavour, or as a foolish cuckold, dominated by his wife Theodora, who had a daughter before she met him but failed to bear him an imperial heir. Justinian is held to have established enduring Orthodox values and promoted morality, but also to have fractured permanently the eastern Church and persecuted those who resisted his directives. Justinian either laid the foundations for the long existence of a mighty empire, or revived obsolete institutions that wasted precious resources at a time of great need, using campaigns of reconquest and monumental building programmes to mask the underlying weakness of his domestic position. The contradictions in our understanding of Justinian and his achievements derive from the earliest and most artful literary accounts, written by Procopius of Caesarea. Procopius was author of an extensive history of Justinian's *Wars* that has generally been considered panegyrical. However, careful readers, including Edward Gibbon, have identified them as complementary to the scurrilous vituperation called *Anecdota*, 'Unpublished', because it was for a millennium. It is known today as the *Secret History*. Procopius was an eyewitness to the events he described, travelling from Persia to North Africa to Italy and back to the capital as a secretary in the company of Justinian's greatest general, Belisarius. His first seven books, of eight, on the *Wars* ended in 551, probably the very year the *Secret History* was written. Gibbon observed that 'According to the vicissitudes of courage or servitude, of favour or disgrace, Procopius successively composed the *history*, the *panegyric* and the *satire* of his own times'. In fact, the panegyric came last, being the expansive catalogue of construction works known as *Buildings*.[1]

Procopius is the most valuable commentator on Justinian, the chief protagonist in an age the emperor recognised as his own, but a familiar voice, that of John Malalas, becomes more urgent in the course of his reign. Malalas, whose name in Syriac means 'rhetor' or 'scholar', was based for much of his life in his home town of Antioch. Quite differently to Procopius, who employed classical language and allusions, Malalas wrote in a vernacular Greek, frequently using the technical language of government. It seems very likely that Syriac, not Greek, was his native language. Malalas's chronicle begins by telescoping 'sacred history', that of the Old Testament, and integrating it with elements of Greek and Roman history. He reached the age of Augustus and of Christ in book ten; Constantine in book thirteen; Anastasius, in whose reign he was born, in book sixteen; and Justin in book seventeen. By far the longest part of his work, book eighteen, covers Justinian's reign and events in Constantinople, where Malalas now lived. Malalas died after his final revision of the text, which ends with Justinian's death. Uniquely for an author of this period, Malalas cites his many sources, at least in that part of his work compiled in Antioch. For that part written in Constantinople he drew heavily on official versions of events produced by Justinian's propagandists and a city chronicle written after 532. Book eighteen begins with a sketched portrait of Justinian:

> After the reign of Justin, the most sacred Justinian ruled for 38 years, 7 months and 13 days from 1 April in the fifth indiction ... In appearance he was short, with a good chest, a good nose, fair-skinned, curly-haired, round-faced, handsome, with receding hair, a florid complexion, with his hair and beard greying. He was magnanimous and Christian. He favoured the blue faction.[2]

Justinian's reign to AD 541

Among Justinian's first tasks as sovereign was to rebuild and reorganise the eastern frontier, where a prolonged seismic event continued to cause devastation. Antioch was hit badly. Having suffered a terrible fire in October 525, it was struck by two earthquakes on 29 May 526 and on 29 November 528. Marcellinus Comes reports the devastation and the unfortunate death of the patriarch, who had joined crowds fleeing burning buildings to the open space of the hippodrome.

> A sudden earthquake struck the entire city of Antioch in Syria during meal-time. Moreover, it brought doubly fierce destruction to the great western part of the city from the easterly winds, which were soon blowing in on all sides and fanning the fires of the kitchens that were blazing at the time in the collapsing buildings. It also killed Euphrasius the bishop of the whole city when his head was crushed in a fiery grave, for an obelisk in the hippodrome was upturned and driven into the ground.[3]

Huge effort was committed to the reconstruction of Antioch, and at the same time other key cities were rebuilt, including Palmyra, where a military duke was stationed. Justinian was preparing for war with Persia.

In 522 Justin had received in Constantinople Tzath, ruler of the Caucasian region known as Lazica, ancient Colchis, which was at that time subject to the Persians (Map 6). Tzath was baptised and became the emperor's godson. Shortly afterwards, the neighbouring region of Iberia also attempted to defect to the Romans, but was brought quickly to heel by Persia. In around 525, the Persian ruler Kavad, now in his seventies, considered how best to resolve tensions with Constantinople and at the same time secure the succession of his son, Chosroes. He proposed that Justin adopt Chosroes, and sent him to the Mesopotamian frontier near Nisibis to witness negotiations. However, these faltered on the issue of Lazica and on the nature of the adoption. Chosroes departed insulted, 'praying that he might exact vengeance on the Romans'.[4]

Early in 528, the Persians attacked Lazica and Justinian committed a detachment and three generals to its defence. One battle was fought and won by the Persians, and a second by the Romans. Skirmishing also took place along the border in Mesopotamia, the land between the rivers Tigris and Euphrates, known to the Persians as Asuristan. A Roman advance on Nisibis was repelled. On the eastern frontier, therefore, Romans and Persians, their allies and proxies, were engaged in two theatres. To the south, in Mesopotamia (Asuristan), the field army of the east was commanded by Belisarius, raised in 529 from the emperor's retinue to serve as a general. To the north, the Caucasian lands were strengthened by the appointment of a *magister militum per Armeniam*, the 'warlike and very capable' Sittas, who was entrusted with four infantry units (*numeri*) of the regular frontier army (*limitanei*) and a federated force of conscripted locals (*foederati*). The Roman army at this time comprised around two-thirds frontier units, based in the regions and supported by local levies, and one-third mobile field army units (*comitatenses*). Each field army was commanded by a general, known in Latin as the *magister militum* (and in Greek as *strategos*).

In March 529, a band of Arabs, engaged by the Persians, attacked Antioch's suburbs, plundering and looting. Belisarius was sent to fortify the desert frontier at Thannuris, south of Dara, while Justinian also pursued negotiations, sending his confidants Hermogenes and Rufinus into Persia. Their approach was rejected by Kavad, who renewed the demand for a tribute payment in gold for his defence of the Caspian Gates, which he had sent both to Anastasius and Justin. He threatened war in one year if it were not to arrive. In June 530, Kavad sent armies into both Asuristan and Armenia. Procopius was at Dara with Belisarius and Hermogenes, and so provides an eyewitness account of a great Roman victory there, reporting also a second victory to the north.

Negotiations were renewed by Hermogenes through autumn and winter, but in spring 531 Kavad again sent his army deep into Roman lands. Belisarius tracked the Persians as they followed the southern shore of the Euphrates as far as Callinicum (modern Raqqa), but his attack there was repelled. Belisarius was recalled to Constantinople and Sittas took command of the eastern forces with Mundo, a Gepid based in Illyricum who had joined his forces to the Romans shortly before. Mundo's reputation as a general had been established decades earlier, when in 505 his army of Gepids, a people who had recently come to dominate the lands north of the Danube, had defeated a Roman army sent out by Anastasius. Hermogenes and Rufinus remained in charge of diplomacy.[5]

Before long, all were dealing not with Kavad, but his son Chosroes. Kavad died on 8 September 531, probably of a stroke, having ruled Persia for forty-three years. Justinian had recently survived a popular insurrection in Constantinople, the 'Nika riot', and Chosroes had pressing domestic concerns that threatened his hold on power. In spring 532, an 'eternal peace' was signed between the Romans and the Persians, which was widely commemorated, including on a column erected at Hierapolis:

> The lord Justinian alone, by divine counsel, made peace, and he put an end to the lamentations of war which had been upon the cities for three decades ... The cross extinguished the terrible roarings of war and the measureless hardships of life as if they were a rough wave or a fire. Peace was made between the Romans and Persians with the help of Holy God. Money was handed over ...[6]

It seems, after all, that Justinian had agreed to pay the new ruler of Persia the sum of gold his father had demanded for three decades.

The Nika riot in Constantinople of January 532 followed a familiar pattern. The anger of the factions was directed against the authorities and damage was inflicted upon urban structures and monuments. Matters turned when the rioters attempted to promote and acclaim a new emperor in the hippodrome. The catalyst for the riot was a decision by Eudaemon, the urban prefect, to put to death seven circus partisans, both Blues and Greens, convicted of murder. When one Blue and one Green were spared by the collapse of the hangman's scaffold, the factions engineered their escape and petitioned the emperor for mercy. They did so with orchestrated chants in the hippodrome, when next the emperor presided over the races, on Tuesday 13 January. When Justinian refused to acknowledge them, the factions united against him with their familiar cry of '*Nika*!' ('Conquer!') and began to riot. The emperor and his retinue withdrew into the palace, while the partisans descended together on the praetorium to compel the release of their men by setting fire to the building. Having secured their partisans, the factions made a more substantial demand, the deposition of the emperor's most senior administrators: the quaestor Tribonian, the praetorian prefect John the Cappadocian, and Eudaemon. Justinian conceded and removed the men from office, but this attempt to mollify the rioters emboldened them further. Forced into an action he had hitherto resisted, Justinian commanded Belisarius to march out with a detachment of Goths against the rioters. It was at this point that the factions determined to choose a new emperor, Probus, a nephew of the late emperor Anastasius. Rushing to his house, they found him absent and so turned their attention to Hypatius, another of Anastasius's nephews. Procopius reports that 'the city was set to the torch as if it had been captured by enemies. The church of Sophia, the Baths of Zeuxippus, and the imperial courtyard from the Propylaia all the way to the so-called House of Ares were burned and destroyed'.[7]

On the fifth day of rioting, Saturday 17 January, troops were brought into the city from Thrace, and the following day Justinian emerged from the palace into the imperial box in the hippodrome 'carrying the holy gospels'.

> On learning this, the crowds came in too and he made a proclamation to them on oath. Many of the people chanted for him as emperor, but others rioted, chanting for Hypatius. The people took Hypatius and led him off to the Forum of Constantine. Setting him high on the steps and bringing regalia and a golden collar out from the palace, they put these on his head. Then they took him and led him off to the hippodrome, intending to take him up into the imperial box.[8]

Justinian had withdrawn back into the palace and considered aban-doning the city on his royal barge. However, Theodora is reported to have persuaded the emperor to stay, concluding her harangue with the 'old saying, namely that kingship is a good burial shroud'. Belisarius was sent out again, now with his personal guard and palatine troops, as well as Mundo, to bring order. Some 30,000 citizens were killed in the hippodrome and surrounding streets. The following day, Justinian dealt with the nephews of Anastasius: Hypatius and his brother Pompeius were executed, Probus was exiled. 'The emperor announced his victory and the rebels' destruction to all the cities, and he undertook to rebuild the places that had been burnt.'⁹

The greatest structure that Justinian commissioned in this rebuilding programme was Hagia Sophia, the Church of Holy Wisdom. Designed and constructed by Anthemius of Tralles and Isidore of Miletus, two *mechanikoi* (architects and engineers), it took less than six years and more than 10,000 workers to build. With an interior space of more than 250,000 cubic metres, Hagia Sophia was the greatest single building ever to have been constructed. It was not a basilica, the preferred design for churches since the second century, but was domed. The dome was functional and symbolic, held to mirror heaven, with forty windows for flooding it with light (Plate 16). As a space designed for a new liturgy, which featured antiphonal chant, the dome captured sound and magnified it, creating a 'wet' sound that washed across and immersed worshippers. Reverberation of music and sound in the dome lasts up to eleven seconds, quite opposite to the effect desired and achieved in a modern concert hall, where a note reverberates only for two to three seconds.¹⁰

The magnificence of Hagia Sophia and its central dome were responses to the Church of St Polyeuctus, a colossal three-aisled basilica church that was built in Constantinople in the last years of Justin's reign. Polyeuc-tus's patron was the last Theodosian princess, Juliana Anicia, wife of the consul Areobindus and daughter of the western emperor Anicius Oly-brius (d. 472). Around the walls of that church ran a long hexametrical inscription, a panegyric to its pious founder, including the following lines: 'What choir is sufficient to sing the work of Juliana, who ... alone has con-quered time and surpassed the wisdom of renowned Solomon, raising a temple to receive God, the richly wrought and graceful splendor of which the ages cannot celebrate.' Hagia Sophia surpassed St Polyeuctus in size and splendour, such that when Justinian first entered it he is reported to have announced 'Solomon I have outdone you.'¹¹

Figure 22. Dedicatory inscription at the Church of Saints Sergius and Bacchus

Hagia Sophia was not Constantinople's first domed church. Justinian and Theodora had already built the Church of Saints Sergius and Bacchus, today known as Little Hagia Sophia, so closely does it resemble the great church (Plate 17). It was located within the Palace of Hormisdas, where Justinian and Theodora had lived as a newly married couple before 527, and which Theodora later established as a refuge for Monophysite holy men. A dedicatory inscription (Figure 22), still extant around the base of the dome, reads:

> Other sovereigns have honoured dead men whose labour was unprofitable, but our sceptered Justinian, fostering piety, honours with a splendid abode the Servant of Christ, Begetter of all things, Sergius; whom not the burning breath of fire, nor the sword, nor any other constraint of torments disturbed; but who endured to be slain for the sake of Christ, the God, gaining by his blood heaven as his home. May he in all things guard the rule of the sleepless sovereign and increase the power of the God-crowned Theodora whose mind is adorned with piety, whose constant toil lies in unsparing efforts to nourish the destitute.[12]

Among those nourished by Theodora's pious patronage were children and prostitutes. Soon after Justinian took the throne, Theodora ordered the arrest of all traffickers of child prostitutes. 'When they had been brought in with the girls, she ordered each to declare on oath what he had paid the girl's parents', then she repaid the sum and ordered the girls released, sending them away with a gold coin and a change of clothes. In fact, Malalas is rather specific about how much the girls had cost: five gold coins each, suggesting that this was a performance rather than a general attempt to ban prostitution. It is tempting to see in this act the retribution of Theodora, a woman herself coerced into public performance and, perhaps, prostitution when young, on pimps and traffickers of the young. The emperor's wife acted to defend those who could not defend themselves. Theodora's very public foray into the world of brothel-keepers can also be interpreted as publicity for the regime's crackdown on immorality, which embraced the prohibition of gambling – some gamblers known to have blasphemed had their hands cut off and were paraded through Constantinople on camels – and the imposition of harsh penalties for heresy, including 'Hellenism'. Malalas relates that in 529 'the emperor decreed that those who held Hellenic beliefs should not hold state office, whilst those who belonged to the other heresies were to disappear from the Roman state, after they had been given a period of three months to embrace the orthodox faith. This sacred decree was displayed in all provincial cities.' [13]

Immorality was blamed for natural disasters such as those that struck Antioch, which was renamed Theoupolis, 'City of God', with the expectation that its luck would change. The 'Wrath of God' is a phrase Malalas habitually uses for earthquakes. For that reason, those charged with policing sin were subject to especial censure when they transgressed, notably bishops who were 'accused of living immorally in matters of the flesh and of homosexual practices'.

Among them was Isaiah, bishop of Rhodes and formerly prefect of the watch at Constantinople, and similarly Alexander, the bishop of Diospolis in Thrace. In accordance with a sacred ordinance they were brought to Constantinople and examined and condemned by Victor the urban prefect, who punished them: he tortured Isaiah severely and exiled him, and he amputated Alexander's genitals and paraded him around on a litter. The emperor immediately decreed that those detected in pederasty should have their genitals amputated. At that time many homosexuals were arrested and died having their genitals amputated. From then on there was fear among those afflicted with homosexual lust.[14]

A great range of sins were now codified as crimes, captured in the monumental restatement and harmonisation of Roman law known as the *Corpus Iuris Civilis*. Compiled by Tribonian in three parts, the *Corpus* comprised the *Institutes*, the *Digest*, and the *Code* (*Codex Justinianus*), to which were added the emperor's 'new laws', or novels, passed after 534. An intention of the compilation project was to concentrate legislative authority in the emperor's own hands. Therefore, earlier laws were made to accord with current laws, and in the *Digest* the writings of the greatest lawyers of antiquity became concordant with the emperor's own opinions, as if they 'had been uttered from our own inspired mouth; for we ascribe everything to ourselves, since it is from us that all their authority derives'.[15]

The emperor did not see himself as bound by the law, but rather as the 'law ensouled' (*nomos empsychos*). This idea was present in Hellenistic thought, is expressed in a fifth-century compilation by John Stobaeus, and is advocated by Justinian himself in his novels. Being unbound by law, the emperor could not be indicted for such atrocities as ordering the deaths of tens of thousands of his own people, as he did in 532. He continued to issue new laws in pursuit of his own moral agenda, to serve a particular need or interest, or to advance his desire to concentrate authority in his own hands. The prefaces to these novels indicate that Justinian was frequently called upon to hear appeals and adjudicate on complex points of law.[16]

The completion of the *Digest* in December 533 was considered by Justinian an act of divine providence equal to the 'eternal peace' with Persia, and, still more recently, the defeat of the Vandals and the return of Carthage, 'nay the whole of Libya', to the Roman empire. These are the opening lines of the 'Confirmation of the Digest' written for Justinian himself. Procopius begins book three of his *Wars* with a lengthy digression on the establishment of the Vandal kingdom in Libya, and the actions of its ruler Geiseric. Shortly before his death in 477, Geiseric had agreed an enduring peace with Zeno, which lasted through the reigns of Anastasius and Justin, and those of Geiseric's son Huneric (husband of the Theodosian princess Eudocia), and Gunthamund, who was Geiseric's grandson and Huneric's nephew. Geiseric had introduced a principle of agnatic succession, meaning the oldest male member of the royal family would become ruler. In 496, Gunthamund was succeeded by his younger brother Thrasamund. In 523, Huneric's son Hilderic, first cousin to both Gunthamund and Thrasamund, succeeded. Hilderic, already in his fifties, was the son of Eudocia and grandson of Valentinian III. He was not an Arian and, favouring the dogma of his mother's line, he installed an Orthodox bishop at Carthage. Hilderic earned the enmity of his principal rivals when he became 'a close friend and ally of Justinian, who had not yet come to the throne, but was governing with full power for his uncle Justin ... Hilderic and Justinian sent large gifts of money to each other'.[17]

In 530, Hilderic was imprisoned by Gelimer, a nephew of Gunthamund and Thrasamund, who seized power. Justinian saw an opportunity, exchanging letters while the peace with Persia was being negotiated. Belisarius, recently recalled from the east and with his hands stained with the blood of rioters, was entrusted with command of an expeditionary force, the dispatch of which proved greatly unpopular at court. Memories of an earlier disastrous expedition, led by the future usurper Basiliscus, provoked John the Cappadocian, the praetorian prefect restored to power, to urge caution. The destruction of Basiliscus's fleet had proven cripplingly expensive and had ended eastern intervention in the west for two generations. However, Justinian preferred other counsel and had begun to support those fomenting rebellion against Gelimer. Tripolis rose in revolt under Pudentius, who was sent a supporting Roman army under a general named Tattimuth, even as Gelimer lost control of the island of Sardinia, where his commander Goda wished to assume power. Justinian despatched 400 soldiers to Sardinia while his generals prepared the main expeditionary force.

Procopius provides detailed coverage of these events, and setting out a narrative based on his account offers a rare opportunity to flesh out scenes and highlight particulars that afford insight into the realities of sixth-century travel and warfare, including the transportation of horses and their fodder; the difficulties of keeping a fleet together; the problems of where to land a fleet and how far and fast to march; efforts to gather information and spread disinformation; and the need to secure provisions and water and how to deny these to an enemy. All of these matters concerned Belisarius, who took command of the expeditionary force that Procopius numbers at 10,000 infantry and 5,000 cavalry, along with as many federates and sailors, all carried on 500 transport ships of various sizes and ninety-two warships (*dromones*). Room was reserved on board for horses, which were collected from imperial pastures in Thrace.

Commanding such a large fleet was a challenge in itself, and Belisarius was not an experienced admiral. His captains kept the ships together on open seas, with painted sails and signals, and kept them apart in harbours, with shouts and long poles. Weak winds made the journey protracted and both drinking water and provisions ran low. Five hundred men died from eating bad bread. Procopius himself was sent ahead on a secret mission to Syracuse on Sicily, bringing back vital intelligence that Gelimer was absent from Carthage. The fleet was able to set sail from Caucana (Punta Secca, Map 9) on Sicily's south-eastern coast and land unmolested on the Libyan coast at Caput Vada, to the south of Hadrumetum (Sousse), five days' march from Carthage (Map 8). This was around three months after its departure from Constantinople. The choice not to sail straight for Carthage was that of Belisarius. He feared a sea battle far more than the threat of a storm or a longer march.[18]

The Roman army met no resistance from locals as it advanced towards Carthage, and Belisarius forbade looting. Gelimer, who was indeed absent from Carthage, ordered Hilderic and his family be put to death there, and commanded his divided army to converge upon the Romans from three directions at Decimum, a pass 'ten miles' from the city. However, the Vandal detachments arrived at different times and the battle was lost before Gelimer arrived. Belisarius entered Carthage, marched to the palace, and sat upon Gelimer's throne and dined in his banqueting hall. Having received news that Gelimer's brother Tzazo had recovered Sardinia, Belisarius set his men to rebuilding the walls of Carthage. Gelimer, reunited with Tzazo, marched on the city and cut its aqueduct. Belisarius marched out to meet the Vandal army, which was encamped some thirty kilometres from Carthage at Tricamarum. The Roman victory was decisive: Tzazo fell fighting, Gelimer and his family fled, taking refuge on a Numidian mountain called Papua. Belisarius sent out detachments 'to recover for the Romans everything that the Vandals ruled'. Eventually, having endured hunger and a harsh winter, Gelimer surrendered.[19]

His victory seemingly complete, Belisarius sailed for Constantinople with his Vandal captives and great plunder, knowing that his regal behaviour at Carthage had made Justinian wary. Belisarius was welcomed at New Rome as a conquering hero, however, and was permitted a triumphal procession in the city. He paraded 'on foot from his own house to the hippodrome and then again from the starting gates to the imperial box' conveying the spoils of war, including Gelimer, to the emperor. Both Gelimer and Belisarius threw themselves to the ground before the emperor in deep obeisance. Belisarius was granted the consulship for the following year, 535. Gelimer was pardoned and granted imperial lands in Galatia, Asia Minor, where he lived with his family. Justinian had demonstrated admirable restraint and wiliness in his handling of a potential rival and a conquered foe.[20]

The victory celebrations in Constantinople must be read in several contexts. First, Justinian was set upon constructing a triumphal imperial image for himself, material examples of which we shall return to later. Second, by his very public obeisance, Belisarius was neutralised as an immediate threat to Justinian, for it was shown to be through the emperor that divine grace radiated and enveloped the Roman people, including his victorious general. Third, a celebration that united the citizens of Constantinople and their emperor, the general and his soldiers was a salve, of sorts, for the wound that the city had suffered just a short time earlier, culminating in the emperor's retribution through Belisarius and Mundo in the same hippodrome. Belisarius began his consular year borne aloft on the curule chair by captives of the Vandal war, a second victorious procession and a deliberate revival of an old Roman practice.

The reconquest of Libya was cast as the first act in a great restoration, the recovery of the western empire. However, North African matters were far from settled. Carthage had been entrusted to Solomon, a eunuch, who found himself engaged in protracted combat with the Berbers, called Moors (*Mauroi*), and Numidians. Having supported the Romans in their defeat of the Vandals, these native peoples had risen against their new masters. In 536, the Roman army mutinied, complaining of mistreatment of Arians and the confiscation of lands belonging to Vandal women, whom many Roman soldiers had taken as wives and concubines. There were around a thousand Arian troops including a company of Herul federates. These rebels rallied behind a certain Stotzas, who also recruited many local slaves as he marched on Carthage. The rebellion was suppressed by Belisarius, who sailed across to Carthage from Sicily, where he had recently landed en route to fight the Goths, but sailed back just as swiftly to deal with another army mutiny. Germanus, Justinian's trusted cousin, was sent to Carthage to prosecute the war against Stotzas, successfully, and later against the Berbers. Germanus won plaudits for his military strategy and fair administration before his recall in 539, to be sent on to the Persian front in 540.[21]

Procopius begins book five of his *Wars* with a discussion of how the Goths came to rule Italy and adjacent lands, a prelude to his narrative detailing Justinian's reconquest. Here we again meet Theoderic, who had made himself master of Italy during the reign of Zeno. Following Theoderic's death in 526, his eight-year-old grandson Athalaric reigned, but Amalasuntha, the boy's mother and Theoderic's daughter, ruled first as regent and then, following her son's death in 534, as queen. Amalasuntha favoured the educated Romans at court, appointing Cassiodorus as her praetorian prefect, even as her father Theoderic had appointed Cassiodorus's father and namesake to that post. Opposition to Amalasuntha and her politics was focused among the Gothic military elite, and her principal rival was her cousin Theodahad, whom she took as co-ruler upon her son's death. Theodahad's imprisonment of Amalasuntha, early in 535, and her subsequent murder, provided Justinian with both a justification and an opportunity to reassert his authority in Italy (Map 9). Both Gothic factions had been in contact with the eastern emperor, courting his approval as they witnessed the Vandal campaign. Now Theodahad saw Roman forces advance upon him from the north-west and south. Mundo marched into Dalmatia and captured Salona from the Goths, while Belisarius sailed for Syracuse on Sicily, commanding a fleet no more than half the size of that which had sailed against Carthage. This does not suggest over-confidence so much as wariness, for Belisarius was instructed to sail away to Libya in the event that he met stern opposition. In fact he did not. Syracuse and other cities surrendered quickly, and solid but brief resistance was met only at Palermo. 'Having received the dignity of the consulship because of his victory over the Vandals ... on the last day of his consulship [Belisarius] marched into Syracuse, loudly applauded by the army and the Sicilians ... after recovering the whole island for the Romans.'[22]

Theodahad considered surrender on terms that would enrich him personally, but reversed course when Mundo was killed skirmishing outside Salona. Belisarius was commanded to sail from Sicily to mainland Italy, where he met little resistance until he put in close to Naples, which had a large Gothic garrison. A siege was set, and having cut the aqueduct to deny water to those within the walls, it was discovered that troops could access the city along its length. With troops inside the city, Naples was taken by storm with much bloodshed and plundering. This loss and Theodahad's failure to march out from Rome against Belisarius during the twenty-day siege at Naples led to his murder by his own men and replacement with Witigis. Belisarius prepared to march to Rome, at the invitation of its inhabitants. As Procopius reports, for the first time in sixty years, on the ninth day of December in 536, the city became subject to the Roman emperor. Meanwhile Witigis withdrew north to deal with a threat from the Franks. Both Goths and Romans were engaged in diplomacy to secure Frankish support. Recognising that his expeditionary army, now depleted by the garrisons he had left in Sicily and Naples, was insufficient to meet the Goths in open battle, Belisarius sent troops out from Rome to secure the cities of Tuscany while preparing Rome itself for a Gothic siege. When Witigis came south he brought a mighty army. In a letter that Belisarius himself sent to Justinian, repeated by Procopius in his *Wars*, the size of the army was estimated at 150,000 men, an impossibly large number. However, by this exaggeration Belisarius's achievements are magnified, as they are by Procopius's account of the general's first encounter with the Goths at the Milvian bridge, a bridge over the Tiber made famous by an earlier battle. There Belisarius had determined to slow down the Gothic advance on Rome, and ended up fighting them personally, 'the whole battle focused on the body of one man', which was protected by his remarkable dark grey horse and the valour of his loyal retinue.[23]

However great the Gothic army was, it could not fully surround Rome, defended by its Aurelian walls and as many guards and local conscripts as Belisarius could spare. Gothic camps arranged around the city failed to sustain an adequate blockade, although cutting the aqueducts made life uncomfortable. Federate forces were sent as reinforcements from Constantinople, 1,600 Slavs and others, provoking Belisarius to attempt to break the siege early in April 537, but his whole army was driven back into the city and the siege dragged on, increasingly marked by famine and disease. Seeking to avoid another pitched battle, Belisarius sent cavalry south to harry the Gothic supply lines. Disease spread in their camps. Belisarius's secretary Procopius, our source, was sent to Naples, to gather supplies, men and information. There he met 3,000 Isaurians, who put in at the harbour, while another 1,800 cavalry had landed at Otranto. Both fleets would sail on to Rome bringing grain. Learning this, the besieging army began to withdraw and a temporary truce was signed. Both sides breached its terms. Goths were caught attempting to enter Rome along an aqueduct, and an army sent north by Belisarius succeeded in capturing

Ariminum (modern Rimini), just a day's march from Ravenna, without a fight but against orders. Another force entered Milan at the request of locals. Witigis lifted the siege of Rome and retired towards Ravenna, besieging Rimini en route and paying the Burgundians to cross the Alps and surround Milan.

Belisarius headed north from Rome against Witigis, 'at about the time of the summer solstice' of 538, capturing strongholds as he went. Further reinforcements, 5,000 imperial troops and 2,000 Herul federates, had arrived, led by Narses, a eunuch and steward of the imperial treasury. Together they relieved Rimini, although disagreement arose between the two generals and subordinates began to choose sides. This led to defeat at Medilanum (Milan), where the garrison capitulated and the Burgundi-ans massacred the locals. Narses was recalled and Belisarius proceeded to Vetus Auximum (modern Osimo), a little to the south of Ancona, which was taken after a siege through the summer of 539, and then Ravenna, further north of Ancona, which he captured in May 540. As ever, Proco-pius provides a great deal of detail about events he witnessed himself, and structures this around speeches and letters that he reports verbatim. It is for this reason that we can narrate events in such detail. At Osimo, Pro-copius reports his own words, giving advice on signalling to his general, whose commands were drowned out by the Goths beating their weapons together: 'Use the cavalry trumpets to urge the soldiers to continue fight-ing with the enemy, but call the men back with those of the infantry ... for one [sound] issues from leather and thin wood, and the other from thick brass.'[24]

Other speeches that he places in the mouth of Belisarius, or letters he attributes to his pen, Procopius may have drafted in part or in full. However, he cannot also have known the many exchanges that he reports, such as the discussion between Witigis and the Franks at Ravenna. After 541, it is likely that Procopius and Belisarius parted company. To that point, at the end of book six of his *Wars*, Belisarius is Procopius's hero, although he treats fairly his Gothic enemies, more so than his Roman rivals. Belisarius is favoured by fortune, which is to say Fortune or Tyche, and at the centrepiece of the narrative of the Gothic wars, as Belisarius rides out from Rome and risks death at the Milvian bridge, he is protected personally 'by some *tyche*' and thereby liberates her temple at Rome.[25] In a letter Belisarius sent to Justinian in 540 announcing victory, he begins by recalling events at Rome four years earlier, repeats the exaggerated number of 150,000 Goths, and notes how close he came to defeat, 'had not some *tyche* snatched us from ruin'. In his *Secret History*, Procopius relates how Fortune now turned on Belisarius, 'the man to whom shortly before *tyche* had delivered Gelimer and Witigis as captives of war'.[26] Recalled to Constantinople, the general was once again under suspicion of harbouring imperial ambitions, as he processed about town surrounded by 'a large number of Vandals, Goths and Moors', evidence of his success and great wealth.[27]

The siege of Rome and the drive to Ravenna, the years 536 to 540, coincided with the sudden onset of a period of rapid climate cooling, the consequence of a dust veil caused by a massive volcanic eruption, that resulted in widespread crop failure and famine across the northern hemisphere. Procopius, who was in Italy, reports that the sun shone very weakly 'when Justinian was in the tenth year of his reign'. Cassiodorus, who remained praetorian prefect under Theodahad and Witigis, observed that 'men are alarmed ... at the extraordinary signs in the heavens, and ask with anxious hearts what events these may portend'. Contemporary eyewitness accounts report on the same phenomenon from as far afield as Ireland, Syria and China.[28]

In 541, a deadly disease arrived at Pelusium, in the Nile delta. Soon it was reported to have spread to Alexandria, and in the spring of 542 it arrived at Constantinople, where Procopius 'happened to be at the time'.

Visions of demons taking every imaginable human form were seen by many people, and those who encountered them believed that they were being struck on some part of their body by them. The disease set in at the very moment of the vision ... Most people, however, were taken ill without the advance warning of a waking vision or a dream. They fell ill in the following way. Suddenly they became feverish, some of them when they rose from sleep ... the fever was so feeble from its beginning to the evening that it gave no cause for worry ... [but] no more than a few days later, a bubonic swelling appeared. This happened not only in that part of the body, below the abdomen, which is called the groin (*boubon*), but also inside the armpit, in some cases by the ears, while in others at various places on the thigh ... While some fell into a deep coma, others developed acute dementia, but both felt the fundamental effects of the disease.

Plate 1. Mosaic of a solar chariot, Chapel of Sant'Aquilino, Milan.

Plate 2. Silver chalice, inscribed for Symeonius, Walters Art Museum, Baltimore.

Plate 3. The stoa or embolos *at Ephesus, once a grand colonnaded street, looking towards the theatre.*

Plate 4. Mosaic showing Dionysus and Hermes, from a bath at Antioch.

Plate 5. Fortunes (tychai) of Constantinople, Antioch, Rome and
Alexandria, from the Esquiline Treasure, British Museum.

Plate 6. Mosaic of Magerius, from Smirat, now in Sousse Archaeological Museum, Tunisia.

Plate 7. Mosaic of Dominus Julius, Carthage.

Plate 8. Mosaic at the episcopal basilica, Heraclea Lyncestis, Northern Macedonia.

Plate 9. Peutinger map, location and Fortune (tyche) of Constantinople.

Plate 10. Sea walls of Constantinople, looking towards the Theodosian harbour.

Plate 11. Basilica Cistern, today Yerebatan Sarnici, Istanbul.

Plate 12. Sarcophagi of porphyry (purple stone), Istanbul Archaeological Museums.

Plate 13. Sketch of the eastern face of the Column of Arcadius, Freshfield Album, Trinity College, Cambridge.

Plate 14. Barberini ivory, Louvre Museum, Paris.

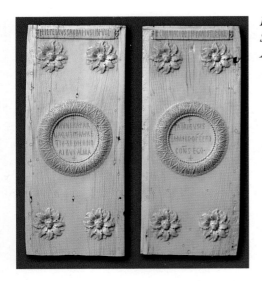

Plate 15. Ivory consular diptych of Flavius Petrus Sabbatius Justinianus, Metropolitan Museum of Art, New York.

Plate 16. Dome of Hagia Sophia.

Plate 17. Church of Saints Sergius and Bacchus, also known as Little Hagia Sophia, Istanbul.

Plate 18. The largest of nine David Plates, showing David slaying Goliath, Metropolitan Museum of Art, New York.

Plate 19. Monogrammed glass coin weight, Metropolitan Museum of Art, New York.

Plate 20. Mosaic showing St Demetrius before the walls of Thessalonica.

Plate 21. *Apse mosaic of Virgin and Christ Child flanked by angels, Kiti, Cyprus.*

Plate 22. *Fragment of mosaic from the 'House of the Bird Rinceau', Antioch, Baltimore Museum of Art.*

Plate 23. *Madaba mosaic map, showing Jerusalem and the cities of the Jordan river valley, Madaba, Jordan.*

Plate 24. *Three Fortunes (tychai) – Rome (Constantinople), Gregoria and Madaba – in a floor mosaic at the Hippolytus Hall, Madaba.*

Plate 25. Floor mosaic at the Church of St Stephen, Umm ar-Rasas, Jordan.

Plate 26. Floor mosaic at the Church of St Stephen, Umm ar-Rasas, detail of Philadelphia (Amman) and Madaba.

Plate 27. Detail of the mosaic inscription recording the construction of the Dome of the Rock at Jerusalem.

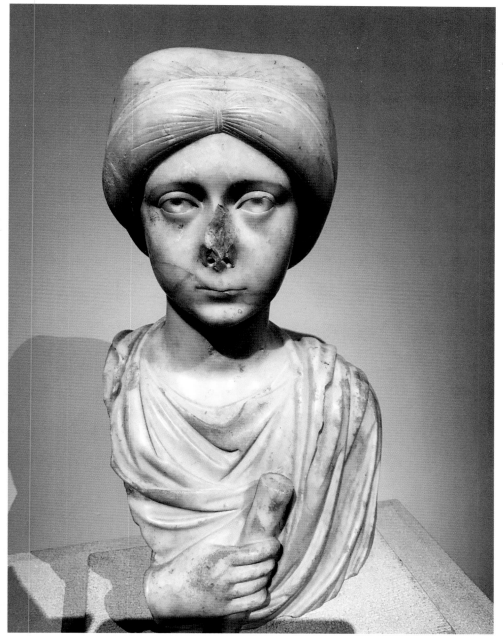

Plate 28. Portrait bust of a young woman wearing a bonnet, Metropolitan Museum of Art, New York.

Plate 29. Mosaic of Justinian processing during the liturgy's Great Entrance, San Vitale, Ravenna.

Plate 30. Transfiguration mosaic, Monastery of St Catherine, Mt Sinai.

Procopius proceeds to note that no doctor or anyone nursing the diseased, nor those disposing of their bodies, contracted bubonic plague from touching the sick or dead. Doctors sought the cause of the disease within the buboes, which they dissected during autopsies, while others were perplexed when those they diagnosed recovered fully. No method of treatment and no intensity of nursing worked consistently. 'The disease lasted four months, while it ran its course in Constantinople, and was at its peak for three', when between 5,000 and 10,000 died each day. Then it passed, leaving families deprived of slaves, slaves deprived of their masters, and many households 'emptied of people altogether'.[29]

Justinian's reign from AD 541

An invasion by Chosroes in 540 ended the eight-year 'eternal peace' between Rome and Persia and saw assaults on several cities in Mesopotamia and Syria, culminating in the sack of Antioch. Breaking through the walls, Persian forces were met by large groups of partisans, the circus factions whose main joy was to support the charioteers. They were no match for Chosroes's troops, and many Antiochenes were killed or captured and led back into Persia, where a new town was established for them, 'Antioch of Chosroes'. Chosroes then directed his sights on Lazica and captured the fortress of Petra, where Roman forces had mustered, killing their general. Having returned from Italy and wintered at Constantinople, Belisarius was sent to the Persian front, arriving in Mesopotamia in the early summer of 541. Aware that Chosroes was operating to his north, Belisarius marched his army from Dara towards Nisibis, engaging Persian forces and capturing a fortress, before disease among his soldiers and the threat of Chosroes's return spurred his retreat. Belisarius returned to Constantinople, evidently with Procopius, who was present when the bubonic plague arrived. In spring 542, Chosroes launched a further invasion of Roman territory, which Belisarius was sent out to address, travelling without an army and using the postal horses to arrive with great speed. However, events were overtaken by the arrival of plague, which ravaged Syria and Palestine and affected both the Roman and Persian armies. Belisarius was again recalled to Constantinople, where he appears to have suffered humiliation at Theodora's instigation.[30]

Justinian had contracted the plague and Theodora feared Belisarius.

He was swiftly sent back to Italy, but with too few troops effectively to contain a new Gothic king, Totila. The Goths had launched a bold assault on Roman lands in southern Italy, setting a siege of Naples, which fell in spring 543. The city's gates were removed and large parts of its walls demolished. Chosroes, meanwhile, had once again invaded Mesopotamia and laid siege to Edessa. The city did not fall and Chosroes agreed a truce and withdrew his troops in return for a large payment in gold. The truce was to last for five years, but excluded Lazica. Additional troops demanded by Belisarius were now sent to Italy, but only a small force under John, a nephew of Vitalian, and not before Totila had surrounded Rome, which he besieged for more than a year before it fell in December 546. It was set to suffer the same fate as Naples, but the task of destroying its massive walls was beyond Totila's means and he turned to confront a Roman army under John, which had marched south. Belisarius was able to drive off the Gothic garrison left at Rome and recapture the city, improvising a restoration of parts of the walls that had been destroyed. He resisted Totila's attempt to retake Rome, but was once again recalled to Constantinople.[31]

In North Africa, a rebellion had once again rocked Carthage. The governor Areobindus, Justinian's nephew by marriage – Areobindus was married to Praejecta, the daughter of Justinian's sister Vigilantia – was murdered by his own men. The rebel leader, Guntarith, a Roman commander of Vandal descent, secured the support of the Berbers and Numidians in his plot, but he was in turn killed by Artabanes, an Armenian prince who had earlier fought with the Persians against the Romans. Artabanes was entrusted with Carthage until a new commander, John Troglyta, arrived from the Persian front. Troglyta knew the territory well: he had served under Belisarius during the original campaign of conquest, when his own brother had been killed, and had stayed on to fight under Solomon and Germanus against both native and rebel Roman forces.

John Troglyta's restoration of control in North Africa is portrayed by the panegyrist Corippus, a native of the hinterland of Carthage, as a sacred battle between Christians and heathens. According to Corippus, the Berbers were finally undone because they attacked on a holy day, perhaps simply a Sunday, when 'The Roman soldiers, occupied with their customary rites, will fear no battle'. But John had anticipated the attack and, like Theodosius I at Frigidus, he spent the night before in prayer and the spilling of tears. As the sun rose, the Christian soldiers trooped out with their standards to a tent in the centre of the camp, a mobile chapel, where a priest draped the altar and conducted the regular service. The congregants wept and together wailed, 'Forgive our sins and the sins of our fathers, we beseech You, Christ'. John was with them on his knees, more tears 'pouring from his eyes like a river' as he intoned a long prayer for victory. Once the priest had performed the Eucharist, it was shown that 'the gifts were acceptable to the Lord of heaven, and at once sanctified and cleansed' the army. Victory was secured in this manner, and those who would die did so purified by their tears and the sanctified elements.[32]

At Constantinople, meanwhile, Theodora had died, in June 548. Returning from Italy, Belisarius was received with honour and made *magister militum per orientem*. He did not depart for Antioch, however, but remained at court, even as the five-year Persian peace lapsed. His former secretary, Procopius, also resident in Constantinople, now completed the first seven books of his *Wars* and sat down to write a rather different account of life at court, the *Secret History*. Procopius passed quite a different judgement from that of Corippus on the war to liberate North Africa, a land he had seen with his own eyes, and knew now to be 'so devastated that despite its vast size' it was virtually uninhabited, the Romans and Berbers equally ruined. Procopius attributed this to Justinian's avarice. Rather than institute sound government, he sent tax assessors and 'appropriated for himself the best fields and prohibited the Arians from celebrating their own mysteries. He was always late in paying his soldiers and generally treated them in a heavy-handed way. This caused many revolts and resulted in widespread destruction.' In his opinion, the same happened in Italy on an even grander scale.[33]

Procopius composed his judgement on Justinian's failed wars while the emperor still lived, but he was freer to do so because Theodora had died. 'No tyrant had inspired such fear for as long as humans have existed, given that it was impossible for dissidents to escape detection.' He vilified the empress's ostentatious public moralising, which he viewed as a mask for her profound vanity and avarice, spite and personal immorality. Still, Theodora's death left Justinian more vulnerable at court, and almost immediately there was a coup engineered by Artabanes, recently returned from Carthage, and his fellow Armenian Arsaces, who sought to involve Germanus and his elder son, Justin. The relatives remained loyal, reporting the matter to the count of the Excubitors, who in turn informed the emperor, and the plot was foiled.[34]

In 550, Justinian chose to renew the treaty with Persia, sending Chosroes more than five times the gold he had sent in 545. Lazica remained exempt, but in 551 a Roman army recaptured Petra and held it against Persian attacks. This war dragged on until 556, preventing the redeployment of troops. In 550, however, Justinian had determined to resolve the situation in Italy, sending his cousin Germanus and his two sons, Justin and Justinian, to raise an army in the Balkans and march west. In preparation, the aged Germanus had married a far younger Gothic woman, Matasuntha, sister to Athalaric, daughter of Amalasuntha, and granddaughter of Theoderic. The marriage was consummated and a son, Germanus's third, was born, Germanus Postumus. Apparently, it was intended that Germanus would recover Italy by force and persuasion, then rule on Justinian's behalf with Gothic support. However, Germanus died, quite suddenly, at Serdica and the plan changed. Justinian sent Narses at the head of a large army and charged him with the defeat of Totila. One of the emperor's closest confidants and holder of high palatine office through decades, Narses had earlier been entrusted with bringing aid to Belisarius in Italy, and had now been made *magister militum* and given control of the whole theatre of war. As a eunuch who was no younger than the seventy-year-old emperor, Narses posed no personal threat to Justinian and proved to be a very capable general and tactician.[35]

Unlike Belisarius, Narses was given the resources needed to complete his assignment, and he marched his army across the Balkans, collecting Germanus's forces at Salona before entering Italy. The Roman army, which may have numbered 25,000, took a full year to arrive, but shortly afterwards a decisive battle with Totila was fought, in summer 552, at Taginae, somewhere between Ravenna and Rome. Narses's clever use of archers, placed in the flanks of a crescent battle line, devastated the Gothic cavalry. Totila himself was injured and died shortly afterwards. The war was renewed by Totila's successor Teias, who mustered the depleted Gothic army, which was once again defeated and their king killed in battle. So ended the Gothic war, and with it the eighth and final book of Procopius's *Wars*. It was not, however, the end of campaigning in the west, as a force was sent on to Spain, where a Roman stronghold was established at Carthago Nova (Cartagena). Roman control of the southern coast resulted in the creation of a province, Spania, which sat to the south of the Visigothic kingdom and allowed some degree of control over the straits of Gibraltar. For decades interactions included both conflict and cooperation across a fluid frontier. The Balearic Islands remained Roman after 624, when all else had been lost to the Visigoths.[36]

Before Narses's decisive campaign, Justinian had systematically skimped on human and material resources to support his campaigns of reconquest. He had many motivations, including the desire to diminish the capacity of his most capable generals to challenge his authority and detain them on foreign soil rather than allow them to reside with their retinues in or close to Constantinople. Sustaining foreign wars and being on a war footing may have suited an emperor who faced many domestic challenges, and the timing of the first war fought on a budget, against the Vandals, which commenced so soon after the Nika riot, is suggestive. The fact that Belisarius had achieved that first victory so rapidly and with few resources provided the emperor with a distorted template for success, but also profound concerns about the ambitions of his general. If local resources and available manpower frequently proved insufficient to achieve victories abroad, Justinian sought to compensate by securing the prayers of monks, legislating on the matter. Within Novel 133, the emperor explains that if monks 'offer prayers to God on behalf of the state with clean hands and pure spirits, evidently the armies will fare well and the cities will prosper'. Monasteries were to become prayer factories, devoted to the interests of the Roman state. The lives of ascetics were increasingly regulated by imperial legislation and the institutional church. Those wishing to found monasteries were now required to consult and secure the permission of local bishops. Much of this was, however, to ensure that monks meddled less often and less effectively in the dogmatic disputes that still fractured eastern Roman society.[37]

In an earlier novel, his sixth issued in 535, the emperor had sought to ensure the purity of the clergy, comparing their divinely granted authority to his own.

> There are two greatest gifts that God, in his love for man, has granted from on high: the priesthood and the imperial dignity. The first serves divine things, while the latter directs and administers human affairs; both, however, proceed from the same origin and adorn the life of mankind. Hence, nothing should be such a source of care to the emperors as the dignity of the priests, since it is for their [imperial] welfare that they constantly implore God. For if the priesthood is in every way free from blame and possesses access to God, and if the

emperors administer equitably and judiciously the state entrusted to their care, general harmony will result and whatever is beneficial will be bestowed upon the human race.[38]

Harmony within the church was Justinian's stated but unfulfilled desire, and he made several earnest efforts to achieve compromise between those who supported and those who opposed the Council of Chalcedon (of 451). However, when he failed to bring the sides together he lapsed into persecution. Perhaps conceived as a combination of carrot and stick, Justinian's religious policy was clever, but not coherent or consistent. Justinian's own support for Chalcedon was ostentatiously displayed, even as Theodora supported Monophysites. The emperor's vacillation between softer and sterner interventions, and the impression that support could be found at court for both sides, hardened opinions and created for the first time a distinctive Chalcedonian identity, centred on Constantinople, which set itself firmly against the Monophysite Church to the east.[39]

The so-called 'Three Chapters controversy' is a clear example of Justinian's approach. In 543–4, he issued a novel that condemned the writings of three theologians whose works had been approved at the Council of Chalcedon. The emperor intended to establish common ground with the Monophysites, who viewed those writers – Theodoret of Cyrrhus, Ibas of Edessa and Theodore of Mopsuestia – as Nestorians. Instead, however, he alienated many in North Africa and Italy, including Pope Vigilius (537–55), who refused to recognise the edict. Summoned to Constantinople, Vigilius remained there for eight years, witnessing the election of Eutychius as patriarch of Constantinople in 552, and in 553 the calling of a fifth ecumenical council, the second held at Constantinople, where Justinian's condemnation was upheld. Vigilius acquiesced and was permitted to return to Italy, but died on the way home.

Patriarch Eutychius also presided over the rebuilding of Hagia Sophia, following earthquakes that cracked the dome. It fell while it was under repair in 558, crushing the ciborium and altar beneath. The great earthquake of December 557 was the third that year, and its aftershocks, which lasted forty days, destroyed part of the city's walls and many buildings, trapping citizens under rubble. The spear held by the statue of Constantine in his forum fell to the ground and drove itself into the earth. When Hagia Sophia was rebuilt, a ceremony of rededication was held on 24 December 562, the festival of encaenia. There was an all-night vigil at a nearby church, before the emperor and patriarch processed to the cathedral. The crowd sang Psalm 23 (24): 'Lift up your heads, you gates; and be lifted up, you ancient doors; that the King of glory may come in.' And so the emperor entered through the doors of his temple in glory, to hear more of his achievement, the creation of a place for God on earth. The regular encaenia celebrations were elaborated with a canticle that was composed and sung on that day alone. Its fifth and sixth cantos describe the church as heaven on earth, a dwelling place for God among men.

> And this wonderful precinct shall become known above all others as the most sacred residence of God, the one which manifestly exhibits a quality worthy of God, since it surpasses the whole of mankind's knowledge of architectural technology ... This sacred church of Christ evidently outstrips in glory even the firmament above, for it does not offer a lamp of merely light that can be sensed, but the shrine of it bears aloft the divine illumination of the Sun of Truth and it is splendidly illumined throughout during day and night by the rays of the Word of the Spirit, through which the eyes of the mind are enlightened, by him [who said] 'Let there be light'.[40]

The light of the cathedral was the light of wisdom, which can be seen today through the many windows at the base of the rebuilt dome (Plate 16), and then also by the many candles and lanterns that furnished the interior. Light was the first of God's creations, and revealed Christ to be the Son of God by his Transfiguration, when 'his face did shine as the sun and his raiment was white as light'.[41]

Some days later, Paul the Silentiary delivered an oration in three parts, its central part a celebration of the church, which was flanked by celebrations in verse of the city, and of the patrons, Justinian and Eutychius. Here the emperor is Light, and he is Wisdom, the embodiment of his own church and of Christ's Church on earth. Each part of the newly rebuilt Hagia Sophia was splendid, as were its furnishings. The altar table was wrought of gold, 'the all gold slab of the holy table raised on columns of gold, standing on gold foundations and bright with the glitter of precious stones'. When a service was to be performed, it was covered in a precious altar cloth of purple silk, embroidered with an image of Christ in gold thread, flanked by 'Paul, replete with divine wisdom, and the mighty doorkeeper of the gates of heaven', Peter.

And on the hem of the veil shot with gold, art has figured the countless deeds of the emperors, guardians of the city: here you see hospitals for the sick, there sacred temples, and elsewhere are displayed the miracles of heavenly Christ, a work suffused with beauty. And upon other veils you see the monarchs joined together, here by the hand of Mary, the Mother of God, there by that of Christ, and all is adorned with the sheen of golden thread.[42]

The canticle delivered at the rededication of Hagia Sophia on Christmas Eve 562 ended with an invocation:

O Saviour, born of the Virgin, preserve this house until the consummation of the world! ... Heed the cries of the servants of your house and grant peace to your people by banishing heresies and crushing the strength of barbarians! Keep the faithful priest and emperor safe and adorned with all piety! And save our souls, for you are God, the life and resurrection of all.[43]

The concerns were real in the last years of Justinian's reign, for which the fullest source is Agathias's *Histories*, supplemented by important fragments of a work by Menander the Guardsman. By then it was clear that the church would never be in harmony and that barbarian strength was far from crushed. Additionally, religious groups that suffered regular persecution or degradation had refused to be cowed. In 529, the Samaritans in Palestine had revolted, following an order to destroy their synagogues. Their property and inheritance rights had been curtailed by legislation, and they were forbidden from holding public office, although wealthier Samaritans must still fulfill the obligations of decurions without enjoying the honours. In 556, the Samaritans rose again. They rioted in Caesarea Maritima, uniting with the Jewish community to attack churches and kill the city's prefect. The ringleaders were caught and killed.

The most enduring legacy of Justinian's last years was the emergence of new threats to the north, which threatened the empire's Balkan lands and focused attention again on Thrace and the approaches to Constantinople (Map 3). In the 550s, arrangements at the Danube frontier broke down. In 529, an agreement had been reached with Mundo, the Gepid general who as a consequence became *magister militum* of Illyricum. This followed the death in 526 of Theoderic the Goth, to whom Mundo had owed allegiance. The expulsion of Goths from Dalmatia prior to the Roman invasion of Italy had established sufficient stability to allow Justinian to restore the walls of Singidunum (modern Belgrade), and to create the city of Justiniana Prima at his own birthplace (the modern location is disputed) and with its own archbishop, who was initially free of any interference from Gothic Rome and the papacy. The death of Mundo at this time, and the Gepid acquisition of Sirmium, initially led to conflict between Romans and Gepids, before Justinian determined to secure the peace by regular tribute payments first to the Gepids and later also to the Lombards. Justinian also gave lands around Singidunum to the Heruls, who contributed federate troops to Roman campaigns in North Africa and Italy. Conflict between Gepids and Lombards threatened the region's stability and offered opportunities to others, notably the Avars and Slavs, but the balance of power was sustained until 548. When it collapsed, it earned the attention of Procopius, whose gaze rarely fell on the Balkans and lands beyond the Danube. Justinian, wishing to recover Sirmium, sided with the Lombards, sending an army with 1,500 Herul federates against the Gepids, who had twice as many Herul allies. The Romans won a pitched battle, which had the unfortunate consequence of uniting the Lombards, Gepids and all the Heruls against them. For several years 'barbarians', including Slavs and others known as Kutrigurs (who later merged with the Avars), were able to enter the Balkans unchecked, often departing with Roman captives and booty and with Gepid assistance.[44]

In 552, the Lombards inflicted a crushing defeat on the Gepids and accepted Roman tribute to defend the frontier, which they did effectively for some years. However, in 559, Avars and Slavs invaded Thrace. They found that the Anastasian long walls had been damaged by earthquakes and were able to press on to the walls of Constantinople itself. Belisarius was brought out of retirement to defend the city, which he did successfully with a band of veterans. The invaders were bought off and returned across the Danube, accompanied by Justin, Justinian's nephew. They were to return to the Balkans in 561, however, and regularly thereafter. It was in this climate that Justinian determined to reach a lasting peace with Persia. Peter the Patrician, the master of offices, was despatched to Dara to negotiate terms with Yesdegusnaph, a high official of Chosroes, known to Menander the Guardsman as the Zikh. Menander provides exceptional detail about the diplomatic process, contrasting Peter's verbosity with the Zikh's concision, recording verbatim the translated text of Chosroes's letter accepting the terms of the treaty, and reporting the terms clause by clause, which were to last for fifty years. Clearly, the Persians had the upper hand, for while it was agreed that the Romans would not 'send an army against the Persians' in any region, it was only established that the Persians' allies would not attack the Romans. Additionally, the Romans were to pay an annual tribute of 30,000 solidi, the first seven years paid in advance, with the next three payments also to be paid in a lump sum after those seven years. The first clause of the treaty related to the abiding issue of Persian defence of the Caspian Gates. This justified the tribute payments, since the Persians would 'not allow Huns or Alans or other barbarians access to the Roman empire'. However, other clauses related to the status of Dara. The Persians would drop their enduring objection to its construction on the condition that no other town was similarly forti-fied. Dara was now recognised as one of just three or four locations where customs posts were established and where goods could be exchanged, and taxed, and intelligence gathered. The others were Nisibis, on the Persian side of the Mesopotamian frontier, nearby Callinicum (modern Raqqa), and Artaxata (Dvin) in Armenia (Map 6).[45]

Menander described the aged Justinian as much diminished by this settlement, 'his bold and warlike spirit had become feeble', but he did not condemn the value of diplomacy and asserted that Justinian would have crushed the barbarians, 'if not by war then with wisdom' had he not died shortly afterwards. In 558, plague had returned to Constantinople in spring and had lasted for six months. In December 560, the Julian harbour in Constantinople was badly damaged by fire, and in both 562 and 563 a drought affected the city, provoking fighting at the cisterns and fountains and the closure of public baths. In November 562, a plot to murder the aged emperor was uncovered. Belisarius was implicated and his retinue disbanded. He died the following year. Then, in 565, 'after raising the doctrine of corruptibility and incorruptibility and issuing an edict to all places that was contrary to piety, with God acting in time', Justinian died, on 14 November 565. The emperor, aged eighty-five, had lapsed into heresy himself, advocating Aphthartodocetism, the notion that Christ's body was incorruptible. His final substantive act was the deposition and exile of patriarch Eutychius, who had resisted him.[46]

Justin II and Tiberius II Constantine

Justin's short reign is well represented by panegyric, notably a long narrative poem by Corippus, recently arrived in Constantinople from Carthage, and Agathias's *Cycle* of epigrams, now integrated into the *Greek Anthology*. History is less abundant. Agathias wrote his *Histories* under Justin, but this work covers only six years, from the end of Procopius's *Wars* until 558. Menander Protector, 'the Guardsman', and Theophylact Simocatta, 'Snub-nosed cat', together offer a historical narrative for the end of the sixth century, when chronicles are almost silent. Marcellinus Comes ends in 548, Malalas with the death of Justinian, and the *Easter Chronicle*, which continues until 628 before ending mid-sentence, reports almost nothing for the reigns of Justin II and Tiberius II Constantine except tax years and the names of consuls.

Justin II was the son of Vigilantia, Justinian's sister. Justin had served as *cura palatii*, major-domo of the Great Palace under Justinian. This office, henceforth, was reserved for a member of the imperial family. The elevated rank of *curopalates*, soon granted to foreign dignitaries as well as to the emperor's relatives, emerged at this time. Justin came to power despite Justinian's failure to nominate a successor, still less to crown a co-emperor. Corippus states pointedly that Justin was shown to be Caesar by his actions, by which he meant his mastery of court politics. In particular, Justin had secured the appointment of a close friend, Tiberius, to command the Excubitors, and relied also on the political skills of his wife, Sophia, a niece to Theodora. Two patriarchs, the deposed Eutychius and his replacement John Scholasticus, claim to have predicted the accession of Justin and Sophia. Certainly, they reigned together and Sophia appeared alongside her husband on imperial coins, an honour even Theodora had not enjoyed. To secure her husband's position, Sophia appears to have dispensed with a second Justin, her husband's greatest rival, the son of Germanus, Justinian's cousin, who was murdered at Alexandria. Two prominent senators were also soon killed. Besides the spilling of rivals' blood, Justin's reign began with some familiar gestures, such as cancelling debts and tax arrears. The new emperor repaid Justinian's debts from his personal funds, suggesting that the imperial treasury was short of cash needed to fund existing commitments. In any event, in a reversal of established policy, Justin refused to pay 'barbarians' their tribute, casting himself as a 'strong man', although he never personally went to war.[47]

A colossal statue of Justin as consul commemorated his restoration of the ordinary consulship in 566, another conscious deviation from the policy of Justinian, who had ceased to appoint consuls in 541. A poem inscribed on the statue or its base offered a confident prediction.

The bold Persian will set up one statue within Susa to our ruler for his victory, wearing the spoils; the long-haired army of Avars another beyond the Danube, cutting a lock from their sun-dried head. This one the ruling city set up here as a result of his consulship, to commemorate flourishing justice. Now stand firm, blessed Byzantine Rome, and repay the divine might of Justin.[48]

The poem reflects the overwhelming interest in Roman relations with the Persians and the Avars, which is captured in the surviving fragments of the history of Menander the Guardsman, among the best sources for the reigns of Justin and Tiberius.

Menander begins in Justinian's last years, when there had been increasing unrest among the peoples settled at and beyond the empire's Danube frontier. When the Lombards and Avars combined against the Gepids, the Gepids turned to Constantinople for support, ceding lands, including the city of Sirmium. The Gepids had dominated the lands to the north and east of Sirmium and Singidunum for decades, although they had been threatened and defeated by the Lombards. When the Gepids were crushed by the Avars and the Lombards moved on to Italy, occupying the northern regions of Venetia and Liguria, Justin was left with a hard border to defend against the Avars and Slavs. The aged eunuch Narses, who had retained control of Rome with the support of the papacy, died around this time. The Avars were ruled by the chagan Bayan, whom Menander portrays as an archetypal Scythian barbarian. The Persians were still ruled by Chosroes, who is cast as an oriental despot, wily and prone to cruelty, but also flexible and accommodating when pressed.

In relations with the Persians, Justin had initially observed the status quo, and in 568 he paid three years of the agreed tribute, amounting to some 90,000 solidi. However, when the next payment became due, the sum for a single year, Justin refused. In the intervening years he had received an embassy from the chagan of the Turks, who proposed an alliance and joint assault against Persia. A Roman delegation accompanied his representatives as they returned to the chagan in 569, and it also discussed trade relations and another mutual enemy, the chagan of the Avars. Diplomatic exchanges were frequent, with five missions exchanged over the following decade. The Persarmenians were at that time in revolt against Chosroes over his policy of building fire temples in Armenia. Chosroes's actions were interpreted as persecution of Christians and in violation of the terms of the fifty-year peace, which protected the rights of Persian Christians. In the war that ensued, Chosroes quickly gained the upper hand. Justin set his sights on the capture of Nisibis and entrusted the campaign to his cousin Marcian, who launched a successful raid with an advance force of 300 men, but then failed to follow up for several months, allowing the Persians to prepare a successful defence. As Marcian besieged Nisibis, Chosroes sent his own invasion force into Roman Syria, with a detachment diverted towards the besieged city. Marcian's army fled and Chosroes directed the Romans' own siege equipment against Dara, which fell in November 573. John of Ephesus relates how the water supply to Dara, to which Justinian had devoted great attention, was diverted, even as a huge tower was built to overtop the walls.[49]

According to two well-informed writers, Evagrius and Theophylact Simocatta, the loss of Dara resulted in the loss of Justin's sanity. Another writer, John of Ephesus, who was well connected at court, reports that Justin's mind was 'agitated and darkened, and his body given over to secret and open tortures and cruel agonies, so that he even uttered the cries of various animals, barked like a dog and bleated like a goat, and then he would mew like a cat and crow like a cock'. This psychotic break may have been associated either with bipolar disorder or hyperparathyroidism, the latter of which would also account for his physical symptoms, including severely painful feet and kidney or bladder stones. Sepsis, the result of surgery to remove those stones, would be the eventual cause of his death. Following events at Dara, however, the emperor lived on for five more years, largely incapacitated, although with moments of striking lucidity, such as when he delivered a speech commending the appointment of his trusted friend Tiberius, the *magister militum*, as Caesar in December 574. Sophia remained closely involved in matters of state at Constantinople.[50]

A one-year truce was negotiated in spring 574, and the following year a further three-year peace was agreed with the Persians. This required even greater tribute payments by the Romans to Chosroes, now totalling 45,000 solidi. Matters were resolved at the frontier, but a settlement of affairs in Armenia was excluded from the treaty. Chosroes promptly launched an invasion of Armenia. His encounter with the Roman general Justinian, son of the late general Germanus and brother of the murdered Justin, resulted in an unexpected Roman victory. In his humiliation, Chosroes issued a decree banning Persian rulers from future campaigns unless these were against fellow rulers rather than their subordinate generals. As peace negotiations resumed, Justinian was recalled and the eastern front was entrusted to a new general, Maurice.[51]

A remarkable speech is attributed to the general Justinian by Theophylact Simocatta, which – if it was actually delivered in 575 and not conceived by the author during the reign of Heraclius when such rhetoric was more fully developed – includes the first offer of battlefield martyrdom to Christian soldiers who died fighting an infidel enemy. Theophylact has Justinian commend death to his troops, rather than surrender, since 'the battlefield demands lovers of danger'. Sacrificing the body would free the soul. 'Death, this sweet thing that we assay daily, is a kind of sleep', but far shorter than 'the day that is to come', eternal life in Christ. Justinian concluded this harangue with the words, 'Today angels are recruiting you and are recording the souls of the dead, providing for them not a corresponding recompense, but one that infinitely exceeds in the weight of the gift'.[52]

Maurice launched an invasion of the Armenian kingdom of Arzanene, transferring 10,000 captives to Cyprus. Although he did not attack either Dara or Nisibis, which remained in Persian hands, Maurice's successful campaigns isolated those strongholds. In a story that paralleled Justin's descent into madness at the loss of Dara, Chosroes is reported to have witnessed this from his summer palace and died shortly afterwards, in February or March 579. By then, Justin had also died, in October 578. Tiberius II Constantine now ruled the Romans and Hormisdas (Hormizd) the Persians, and as they established themselves there appears to have been no substantial military campaigning. In 580, Maurice's army spent the whole summer ravaging within Persian territory, before over-wintering in Cappadocia. In spring 581, Maurice set out to attack Ctesiphon itself. However, the Persians had anticipated his arrival, cutting a bridge over the Tigris and sending an army to devastate Roman lands behind his advance, which fell on Edessa. Maurice was forced to withdraw, his grand campaign a failure. Before his prestige was fully restored, in summer of the following year, Maurice was summoned back to Constantinople. Tiberius was in poor health and wished to reach a constitutional settlement.

Tiberius was married to Ino, with whom he had three children, including two daughters who survived their father. Ino took the name Anastasia upon her husband's elevation as Augustus. Until that point, Sophia had successfully barred Ino from the Great Palace, restricting her and her family to the Palace of Hormisdas. According to John of Ephesus, Sophia intended to force Tiberius to divorce Ino and marry her or her daughter. Her frustration accounts for the suggestion by Gregory of Tours, a western source, that Sophia plotted to install Justinian, the recalled general, on the throne. However, Tiberius and Ino Anastasia prevailed. On 5 August 582, Tiberius betrothed Maurice to his daughter Constantina, naming him Caesar. At the same ceremony, he also raised another man to that rank, Germanus, who was married to Tiberius's other daughter, Charito. Germanus disappears from history after Tiberius's death, when Maurice became emperor, possibly to reappear in 602. Germanus was probably the son of Justinian the general, who also disappears from history at this time. Both were descended from Germanus, who was a trusted cousin of the emperor Justinian, and would have been serious rivals to Maurice. Sophia managed to maintain some influence at court, living at the palace as Augusta until at least 601. At the same time, both Ino Anastasia (d. 593) and her daughter Constantina, who married Maurice in 583, were also Augustae.[53]

Maurice

Maurice took the throne aged forty-three and reigned for twenty years. As emperor, he no longer travelled widely, but remained mostly in the city of Constantinople, where he staged spectacles and lavish ceremonies, including his inauguration and, only months later, his wedding, presided over by the new patriarch, John Nesteutes, 'the Faster'. The city was entertained for seven days, as musicians and conjurors paraded through the streets, pantomimes and races were staged, and banquets were held for men of rank. A year later, Maurice spent similar sums in marking his first consulate. He paraded through the city in a chariot pulled by men, 'not horses, mules or elephants'. Still, there was an elephant to hand, which Maurice sent to the chagan of the Avars at his request, and received back when he had gazed upon it. Maurice and Constantina had nine children, their first, Theodosius, being the first son born to a reigning monarch since Theodosius II. This was the reason for his given name, according to John of Ephesus, who dated the birth to 4 August 583. 'Forthwith there were offered to the infant gifts in abundance by all the nobles and ladies of rank and senators and others besides, competing to outdo each other in the expense of their offerings.' Pope Gregory I in distant Rome became the boy's godfather. Three other sons were given imperial names – Tiberius, Justin and Justinian – and the other two family names. Paul was named after his father's father, and Peter after Maurice's brother. His three daughters were Anastasia, Maria (perhaps also called Cleopatra) and Theoctista, which was also the name of one of Maurice's two sisters (the other was Gordia). Maurice's closest adviser, Domitian, was also a relative, his cousin or nephew. Domitian, a correspondent of Pope Gregory the Great, was appointed as guardian to Maurice's children in 597, but died shortly before the emperor.[54]

For the first year of Maurice's reign, affairs in the east were handled by John 'Mystacon', whose epithet indicates that he sported an impressively hairy upper lip. Transferred from Armenia, at the end of summer 582, John 'Moustache' was defeated by the Persian 'Black Falcon', Kardarigan. John was removed, but continued to hold high command, fighting against the Avars and in Armenia before he became a patrician in 591. The war in the east was entrusted to Maurice's new brother-in-law, Philippicus, whose activities are recorded floridly by the 'Snub-nosed cat', Simocatta. In 585, Philippicus is said to have 'planted his spear proudly in the region of Arzanene and captured a glittering and most distinguished booty', for which he is compared to Scipio Africanus. However, he then fell seriously ill and retired to Martyropolis, which drew the attention of a Persian army, before returning to winter in Constantinople. In spring 586, Philippicus marched east again, pausing at Amida to reject a Persian offer of peace, before advancing further into Mesopotamia. Scouts identified the Persian army advancing on a Sunday, seeking to surprise the Roman troops on their day of rest and worship. To prepare his troops for battle, 'Philippicus displayed the image of God Incarnate ... shaped by divine wisdom, not fashioned by a weaver's hand or embellished by a painter's pigment'. This *acheiropoietos* image of Christ, an icon 'not made by human hand', was probably that known as the 'Camuliana', which had recently been transferred to Constantinople. It was paraded before the troops to inspire them. Next, Philippicus rode into the middle of his troops and, as part of his harangue, threw himself down, 'pouring out an unquenchable flood of tears'. The *acheiropoietos* image was then entrusted to the bishop of Amida, who was installed in a nearby fortress, and 'who propitiated the Divinity on that day, and with an abundance of tears made supplications that the Romans might gain the victory'. In the midst of battle, a mighty voice was heard urging the Romans to attack the Persian cavalry, which fled.[55]

The Persian flight was attributed to divine intervention, but the failure to pursue and press the Roman victory was blamed on Philippicus's subordinates, who had replaced him in the front rank after his tearful exhortation. Simocatta's account of preparations for battle on a Sunday echoed those in Corippus's account of John Troglyta's climactic encounter with the Moors and, before that, Theodosius at Frigidus. Evagrius adds the fascinating detail that Philippicus had also written to Gregory, patriarch of Antioch, asking for the relics of Symeon the Stylite to be sent to the front 'for the protection of the eastern armies'. The *Strategikon*, a military manual written during Maurice's reign, and therefore later attributed to the emperor himself, provided detailed information on the Persians, their temperament and tactics, weapons and armour. It also shows how sacred elements were now integral to battle preparations. A mobile tent chapel was placed at the centre of the camp, in place of the *aedes*, the temple and treasury where the military standards were stored and venerated. It was here that the units prayed together on holy days, but also on 'the actual day of battle before anyone goes out the gate'. Maurice further required that the standards be blessed a day or two before battle; that the 'Trisagion' – 'Holy God, Holy Mighty One, Holy Immortal One, have mercy on us' – be sung by each unit early in the morning and late at night, before and after all duties; and that, as each unit marched out of camp, it should cry in unison 'God is with us' thrice. Military services would become increasingly complex, as later military manuals reveal, incorporating a range of icons and relics.[56]

An important figure emerged in the campaign of 586, and he would come to the fore in 587, when Philippicus was once again ill. Heraclius, inevitably referred to by Simocatta as 'father of the emperor Heraclius', is portrayed as a paragon, capturing an unnamed Persian fortress. When in 588 the sickly Philippicus was replaced as *magister militum per orientem*, however, it was not by Heraclius senior, but by Priscus, whose illustrious military career would last through the next two decades. A story would later circulate that Priscus was offered the throne after the death of Phocas but declined, paving the way for Heraclius junior's accession. Priscus had been sent to enforce a new policy 'about the duration of the soldiers' active service, and about their precise equipment, and about what they received from the treasury'. This was part of a reorganisation of military finances that Maurice considered essential and which he continued to press in the face of sustained resistance. He appointed Constantine Lardys as his top financial official in Constantinople. The troops had already heard that their pay was to be reduced by one quarter, and they did not consider that this was compensated by the offer of better conditions of service and provision of more equipment. Following the Easter celebrations, they rebelled, turning their anger against Priscus, who escaped to Edessa, where he was besieged by 5,000 of his own men until he withdrew to Constantinople. The army raised up Germanus, a local commander, as their new general, acclaiming him emperor and disparaging Maurice. When Philippicus was despatched to replace Priscus, the rebels rejected him and he withdrew to Tarsus. When the Persians sought to benefit from the conflict, Germanus repelled them. Finally, Aristobulus, a civilian bureaucrat, arrived as an envoy from Maurice. 'Partly by bribery and partly by persuasion he mitigated the savagery of the mutiny.' The troops turned their attention to the Persians once again and won a notable victory at Martyropolis, sending booty back to Maurice.[57]

The mutiny ended at Easter 589, when Gregory, the gout-stricken bishop of Antioch, reconciled Philippicus with the troops. Germanus and his subordinates were granted amnesty. Philippicus immediately marched on Martyropolis, which had been betrayed to the Persians by a junior officer. However, he failed to capture it and was recalled, replaced by Comentiolus, who captured a nearby Persian fortress, increasing pressure on Martyropolis. Marching his troops towards Nisibis, Comentiolus won a significant victory, despite losing his horse and suffering a wound. Evagrius credited Comentiolus, in contrast to Simocatta, who heaped praise on to Heraclius, 'father of the emperor Heraclius ... who was conspicuous through his glorious achievements with the spear'. Booty and prisoners were sent back to Constantinople, where chariot races were staged and the circus factions 'danced in triumph'.[58]

The struggle with Persia was brought to an end not by Roman arms, but by a civil war that erupted between Hormisdas, the Persian ruler, his son Chosroes, based in Caucasia, and the general Bahram Chobin. The western Turks had long been in contact with the Romans and had exchanged embassies that discussed an alliance against Persia. In 588, the

Turks had launched an invasion of Persian lands from the north-east, across the river Oxus, seizing several cities in what is today Afghanistan. Bahram was sent to counter the invasion and secure the region of Chorasan. Following a decisive victory he returned to confront the Roman threat, sending only part of his booty to Hormisdas. Maurice had determined to press his advantage in 589. In addition to the forces deployed in Mesopotamia under Comentiolus, he sent armies into Lazica led by Romanus, and into Persarmenia led by John 'Moustache', who besieged Dvin. Bahram was directed against Romanus, but lost a key battle and was humiliated by Hormisdas and dismissed from his command. Bahram refused to stand down and his troops stood with him. The rebellion was swiftly joined by troops defeated by Comentiolus. The rebels killed those of their comrades who had remained loyal to Hormisdas. Two armies sent against Bahram were defeated before Hormisdas was deposed and replaced by his son, Chosroes. Bahram, however, still would not stand down and he marched against Chosroes, who fled into Roman territory.[59]

Chosroes sent a letter to Maurice, delivered by ambassadors whose words are reported at length by Simocatta. His offer, recorded more briefly in the Armenian history of Sebeos, was to cede to him broad regions of Syria and Armenia, in return for Roman recognition and support. Sebeos singles out Nisibis and Dvin and their territories, which Maurice would gain from Chosroes if a Roman army fought with him against Bahram. Maurice agreed, apparently against the advice of members of his council and senate, and directed the Syrian army under Narses to join that commanded by John 'Moustache' in Armenia. The Romans fought alongside the retinues of loyal Armenian potentates, who mustered under their commander-in-chief Mushel Mamikonian. Sebeos numbers them 'at about 15,000, the battalions of each of the nobles in hundreds, in thousands, in battalions, according to their banners. All of these were fully armed elite warriors burning with courage like fire'. Rejecting Bahram's promise to recognise an independent Armenian kingdom, Mamikonian and the Armenians remained loyal to Maurice. When battle was joined, the Persians were driven back. As they fled, the Armenian 'armed nobility, galloping in pursuit, pierced from below the armour of the riders who were on the backs of elephants'.[60]

With the support of the Roman Syrian army, led by the general Narses, and an efficacious prayer to St Sergius, Chosroes won a second victory. Bahram's defeats allowed Chosroes to press his claim on the Persian throne, and he was acclaimed as he crossed the Mesopotamian frontier early in 591. As he entered Dara, Chosroes handed the city back to the Roman commander, and in return Maurice 'exalted, calling Chosroes his son'. As he consolidated his power, Chosroes delivered the lands he had promised to Maurice and sent as a gift a golden gem-studded cross, which Theodora had dedicated to St Sergius in the reign of Justinian, and which Chosroes's namesake had taken as booty in an earlier conflict. The peace between the Romans and Persians endured until Maurice's murder, when the personal relationship between the two men justified Chosroes's retaliation. As well as appeasing the Romans, Chosroes chose for himself a Christian wife. In 592, he married a Monophysite named Shirin, who displayed a devotion to St Sergius. Following the birth of their first son, Chosroes sent offerings to the shrine at Resafa.[61]

Roman attention could now be turned decisively to the northern frontier to confront both the Avars and the Slavs settled beyond the Danube. Maurice continued to send an annual tribute payment worth 80,000 solidi in cash and goods, including worked silver and embroidered cloth, to the Avars. The Slavs were nominally subject to the Avar chagan, but did not share in the tribute and were unbound by the treaties agreed between Romans and Avars. Increasingly, groups of Slavs acted independently and as they crossed the Danube they passed out of the Avar orbit. The further south they travelled the more, it seemed, they commanded their own destinies. Maurice developed a policy to contain the Avars, with treaties, tribute payments, and occasional demonstrations of force, while moving more aggressively against the Slavs. To achieve this, Roman armies operated regularly north of the Danube for the first time in centuries (Map 3). Chagan Bayan had died in around 583 and his son, whose name we do not know (but it was perhaps also Bayan), wished to assert his presence. It was to him that Maurice sent an elephant, and also a golden couch, even as he refused to increase the tribute demanded. Bayan's son set about capturing Roman fortresses, first Singidunum then Viminacium (Kostolac), and both fell quickly despite their restored walls. The Avars proceeded as far as Anchialus (Pomorie) on the Black Sea coast of Thrace, which sat halfway between the Danube and the Long Walls, on the near side of the mountains. John of Ephesus, who was at the time detained in Constantinople, suggests that Roman manpower was so short that clerics were forced to wear military uniforms and stand on the walls. The Avars withdrew when they learned of a threat from the Turks, evidently part of

Roman diplomacy. A new treaty was agreed in 585, which included the larger tribute payment of 100,000 solidi annually. However, this lasted less than a year.[62]

In 586, the Avars assaulted fortified outposts along the Danube before pressing inland. It was probably in this year that an Avar invasion arrived at Thessalonica and set a siege. This was a joint offensive with the Slavs, which failed to breach the strong walls. According to John, the city's archbishop, whose account is preserved in the *Miracles of St Demetrius*, the invading army comprised 100,000, 'or somewhat fewer'. Thessalonica, as the seat of the praetorian prefect of Illyricum, had received a reinforced circuit of walls in the middle of the fifth century. The upper parts of the walls were brick, but the lower parts used marble spolia, largely seating, beams and framing from the disused stadium. According to an inscription over a tower in the eastern walls, construction took place under the care of a certain Hormisdas, suggesting the 440s, when the city received immigrants fleeing the Hunnic attack on Sirmium. This may have happened again in 582–3, and the immigrants from Sirmium are mentioned later in the *Miracles of St Demetrius*. Returning in 587, the Avars sacked Mesembria (Nesebar) on the Black Sea. Confronted by a Roman army led by Comentiolus, they moved west swiftly, attacking but not seeking to capture major, well-defended cities like Philippopolis (Plovdiv) and Adrianople (Edirne). It appears that these were swift cavalry operations intended to sow fear and acquire slaves and booty. Familiar generals were transferred from the Persian front to the Balkans. When Comentiolus was sent east, John 'Moustache' appeared to liberate Adrianople. At his departure, Priscus was given command of the Balkan theatre of war. The pattern remained much the same: the chagan invaded across the Danube, the Roman general sent out a force under a subordinate who won or lost a battle. Treaties were signed then broken at will.[63]

Simocatta's account of the Roman–Avar war of the 590s is detailed but frequently confused and this confusion is carried over into the summation by Theophanes Confessor. There are moments of great clarity, however, which reveal much about the Roman determination to end the threat to the Balkans, defending the approach to Constantinople by taking the fight to their opponents. In the spring of 593, Maurice determined to strike against the Slavs, who were omitted from his treaties with the Avars. He did so by sending Priscus with an army across the Danube. The target was Ardagast, leader of the Slavs who had invaded a decade earlier. He was run down and said to have disappeared in a river, employing a skill attributed generally to Slavs in the *Strategikon*. 'Often when they are in their own country and are caught by surprise or in a tight spot, they dive to the bottom of a body of water', using reeds to breathe. The booty gathered was sent back to Constantinople with a detachment, which returned with Maurice's order that Priscus should over-winter in Slavic territory and press his advantage. Again, this echoed general advice in the *Strategikon* that 'it is preferable to launch our attacks against them in winter when they cannot easily hide among bare trees, when the tracks of fugitives can be seen in the snow ... when it is easier to cross over frozen rivers'. But Priscus's troops scorned the imperial command, forcing their commander to withdraw under threat of mutiny. Peace with the chagan was restored by transferring to him another part of the Slavic booty. For this failure, Priscus was replaced by Peter, the emperor's brother.[64]

In 594, the Slavs struck back against the Romans under Peter, inflicting sufficient damage to his men and reputation that he was replaced, by Priscus, who again in 595 led his army north of the Danube. Simocatta records a set-piece meeting between the general and the chagan at an island close to Singidunum, which resulted in another treaty. After some time dealing with their western neighbours, the Avars ravaged Dalmatia and, in 598, besieged Tomis, which was defended by Priscus. A relief force under Comentiolus was despatched, which the Avars moved to engage and defeat. Comentiolus fled while the chagan marched unopposed into Thrace. Maurice marched out in person leading a scratch force formed from his bodyguard and the circus factions to defend the Long Wall. With disease spreading through the Avar ranks, the chagan agreed to another treaty, which increased the annual tribute payment to 120,000 solidi. Maurice returned to Constantinople humbled and sources began to report ominous portents, including 'creatures of human form, a man and a woman' that appeared in the Nile. At Constantinople, 'two monsters were born ... namely a four-footed child and another with two heads'.[65]

The denouement of the Roman–Avar war came after a final successful Roman campaign. In 599, Comentiolus defended the Danube fortresses while Priscus achieved significant victories north of the river. The chagan lost sons in the campaign, and his Slav allies were routed. Maurice sought a lasting peace by returning Avar prisoners of war and over the next two years the Roman position at the Danube was consolidated. The chagan was again preoccupied to his west until 601, when he agreed an alliance with the Lombards and the Franks. In that year, Maurice again sent Peter to the Danube and his troops conducted military operations north of the river. When seeking to press his advantage and continue operations through the winter, however, Peter faced a mutiny. At the same time, in February 602, Maurice was stoned by hungry crowds as he processed barefoot through Constantinople towards Blachernae. His ostentatious humility turned to humiliation when the circus factions paraded a looka-like on an ass and mocked him with an obscure and obscene song:

> He has found a gentle heifer, and, like the young cock, has leapt on her, and made children like hard seeds, and no-one dares to speak, but he has muzzled everyone. Oh my Lord, terrible and powerful, strike him on the skull to make him less arrogant. And I shall vow to you this great ox in thanksgiving.

With his authority in the capital waning, Maurice was vulnerable. Domitian, his closest confidant, and also his eyes and ears, had died in January 601. He was buried with great ceremony in the imperial mauso-leum at the Holy Apostles, a dignity that Maurice himself had earned but was not to be his fate when he was executed in November 602.[66]

The age of Justinian began with great projects and successes swiftly deliv-ered. Hagia Sophia stands today as testament to this period of energy, creativity and achievement. Likewise, a grand legal compilation, the *Corpus Iuris Civilis*, was completed in just a few years and endured for more than a millennium as the basis for European law. A lawgiver who considered himself unbound by law, Justinian survived the Nika riot, which burned much of Constantinople in 532, by ordering the deaths of 30,000 of his own people. Shortly afterwards, North Africa was recovered from the Vandals in a lightning campaign, establishing the reputation of Belisarius. However, it was never firmly held against the Berbers. Justinian launched a protracted, exhaustive, but only superficially successful effort to recover Italy for the empire, occupying Belisarius far from Constan-tinople and obliging locals to pay dearly for their 'liberation'. Justinian signed an eternal peace with Persia that was anything but, sending armies frequently against those of Chosroes, eventually paying him a huge annual tribute. All of Justinian's great initiatives commenced in the first fifteen years of his long reign. The second, longer part of his reign saw the first appearance in history of the bubonic plague in 541, which Justinian himself contracted, and was marked by the early death of his empress, Theodora, in 548, seventeen years before his own demise.

When Justinian died the poet Corippus composed a narrative on the accession of his successor. This set out the virtues of a great emperor, who must be a wise lawgiver but humble, a philanthropist but frugal, a generous host but sober, Orthodox but tolerant, the most powerful ruler on earth but the meekest before God. Justin II did not heed his advice. His reign began with the murder of a rival and he aroused hostility by his abandonment of Justinian's policy of appeasement, launching an unsuccessful war in the east that ceded land and cities to the Persians. Tiberius Constantine wielded power for much of Justin's reign and for a few years afterwards, relying increasingly on the military prowess of Maurice, who succeeded him and reigned for twenty years. Maurice's achievements were considerable. He secured a lasting peace with Persia and sustained it through personal diplomacy. Strategic principalities, exarchates, became established at this time in Ravenna and Carthage. The Roman position in Italy was secured by paying the Franks to attack the Lombards, with the situation closely monitored by several men called, after 584, exarchs, governor generals, notably Smaragdus (c.585–9, 602–8), Romanus (590–6) and Callinicus (596–9). Direct action was needed in North Africa, where Gennadius, exarch of Carthage from c.591–8, defeated the Moors. His replacement as exarch was Heraclius, father of the emperor, who learned of Phocas's usurpation in Carthage. Most substantially, Maurice secured a defensible frontier at the Danube through long years of campaigning by his trusted generals. However, this was underpinned by a necessary but unpopular reorganisation of military finances and provisioning, which led directly to both his downfall and those of the same generals at the hands of mutineers from the Balkan army.

The first Slavic penetrations through the Balkans into Greece took place at this time, but they are absent from historical sources. There are, however, clear indications of violence in the archaeological record, as newcomers and natives clashed over control of resources. Cities occasionally came under threat, and Athens may serve as a case study. Of the thousands of coins discovered in the decades of excavation at the Athenian agora almost all were lost and discovered singly. Of the few hoards discovered in Athens, at least five can be dated to the early 580s by their latest coin, struck for Tiberius II Constantine. All hoards were discovered in a layer of ash and burned debris, and many of their coins showed signs of fire damage, surely associated with a Slavic assault on the city. One parcel of twenty-four copper coins, principally half-follis pieces struck at Thessalonica, seem to have been kept in a small shed near the Tholos, a round temple at the south-west of the agora. A far larger collection of around 400 copper coins was discovered in one room of a flour mill just outside the late Roman city wall. These coins were principally nummia, tiny coins barely larger in diameter than a pencil, which shoppers would have used to buy flour or bread. There were, however, too many to have been lost in cracks in floorboards or carelessly overlooked by the miller. They comprised the miller's takings, perhaps kept in a dish or drawer at the northern end of the room, which fell and was buried during a devastating fire of *c*.580. The largest copper hoard, of 594 coins, was lost shortly after its final coin was struck in 578. Discovered at the Dipylon, the western gate of the city's outer walls, it comprised 34 folles (400-nummia pieces), 74 half folles, 18 pentanummia (five-nummia pieces), and 472 nummia. Of the folles, twenty-five were struck recently, that is since 574/5, and of those twenty were struck at Thessalonica. This may have been the cash kept on hand by a shopkeeper or money-changer. Clearly, the periodic eruption of violence did not end city life in Athens in the 580s. However, the city and many like it remained under threat so long as the Danube frontier was open.[67]

The city of Sirmium, a key to Rome's northern defences, disappeared at this time (Map 3). In 579, the Avar chagan built two bridges across the river Danube and set siege to the city. It fell in 582 and seems never to have been recovered by the Romans. Sirmium had extensive late Roman fortifications, which were repaired under Justinian. The heyday of the city was in the third and fourth centuries, when a Tetrarchic palace complex, featuring a hippodrome, was constructed, alongside imperial granaries, baths and urban villas. The number of coins circulating in Sirmium dropped significantly in the later fifth century, after the city was sacked by the Huns in 441 and later was held by the Gepids. The palace complex had been abandoned by this time and the marble head of a fourth-century statue of Sirmium's *tyche* has been found in its ruins, as have the shattered remains of at least three porphyry busts of Tetrarchs. However, there was a consistent trickle of coins after that until a copper half follis was buried in a layer of burned debris, evidence for destruction within the city's walls. That coin, among the last to circulate in Sirmium, was struck in Thessalonica in 578 and lost shortly afterwards. The signs of destruction and abandonment appear to stand in contrast to written accounts, which suggest that the Avars were generous in their victory. However, these also record a devastating fire in 583, which forced the Avars to flee. Some were later buried with their horses within abandoned houses. The city appears afterwards to have served as a forward military base for Roman armies, but it was not rebuilt and no further coins have been found that date to before 829.[68]

To the east, Antioch was now a shadow of its former self (Map 5). The city was frequently struck by waves of plague and earthquakes through the later sixth century. Its resilience would be most sorely tested in 588, when another devastating earthquake struck. Evagrius estimated the dead at 60,000, a quarter of the number said to have died in the 526 tragedy, although the earlier number may be exaggerated, as may the 588 death toll, which is both very high and an indication of the decline in the city's population. When Symeon the Younger Stylite had ascended his pillar in 541, the city was newly rebuilt and had a secure future. At Symeon's death, a decade into the reign of Maurice, a stone monastery was built at the site of his pillar, almost unique as a new construction. Within decades the city would fall into Persian hands, as waves of invasions pressed through Syria and deep into Asia Minor, reaching the very walls of Constantinople.[69]

As the age of Justinian closed, the divisions in the eastern Church were deeper than ever, and this gave writers reasons for the successes of Rome's enemies. In 577, when the Chalcedonian Eutychius, deposed by Justinian in his last substantive act as emperor, returned to the patriarchal throne in Constantinople, he was received at Blachernae in a grand ceremony by both Justin and Tiberius. He died in 582, and his life was celebrated in the *Vita Eutychii*. A contrasting portrait is provided by John of Ephesus, who was at this time imprisoned in Constantinople, where he wrote his great Miaphysite ecclesiastical history. John understood the deaths of John Scholasticus and Eutychius as God's punishment for their suppression of Monophysites. More than this, he wrote, since the Chalcedonians had 'rooted up the churches of the Orthodox [i.e. Monophysites] during their persecution, so ... the altars of their churches throughout Thrace, and up to the very walls of the city, were razed to the ground by the barbarians', the Avars and Slavs. At the same time, those who fought the Avars, Slavs and Persians did so increasingly as Christian soldiers. Altars were raised in their camps, services conducted and fasts commended on the eve of battle, while commanders prayed and spilled tears of contrition.

Soldiers were offered spiritual rewards for mundane sacrifices, including the status of battlefield martyrs. The powerful idea of violent martyrdom was adopted into a new religion that emerged in Arabia, inspiring its warriors to amazing feats that ended the world of antiquity and established a new order across the eastern Mediterranean, Egypt and North Africa.[70]

9

THE HERACLIANS,
AD 602–*c*.700

Maurice's demise was widely foretold. A monk had run from the Forum of Constantine to the palace gate wielding a sword, 'proclaiming to all that the emperor would die, murdered by a dagger'. A rumour circulated that he would be killed by a man whose name began with the letter Phi (Φ). Maurice suspected Philippicus, his general, as perhaps was intended. However, soon he received news that a centurion named Phocas had been elevated on a shield and was marching towards Thrace. The circus factions were ordered to guard the walls and Maurice looked to eliminate threats at court. First he turned on his son, Theodosius, already an emperor, who was beaten with rods. Germanus, a patrician who was once Caesar and was now father-in-law to Theodosius, was accused of plotting and sought exile in Hagia Sophia. When the Excubitors were sent to remove Germanus from the church, the city erupted in violence. When Germanus tried to seize the throne, he was stopped by the circus factions, who also set fire to the palace of the treasurer, Constantine Lardys.

With the city in flames and no supporters remaining, late at night on 22 November 602, Maurice 'took off his imperial robes, put on civilian dress, boarded a barge with his wife and children' and sailed away. The family made it less than fifty miles before they were detained and brought back. On 23 November, Phocas's inauguration as emperor took place at the Church of St John at the Hedomon Palace complex, seven miles outside the walls of Constantinople, also site of the imperial parade grounds. Cyriacus, in his eighth year as Constantinople's patriarch, presided, showing how firmly matters had turned against Maurice. On 25 November, Phocas processed into the city through the Golden Gate and along the Mese as far as the Great Palace 'with no one at all opposing him, but all acclaiming him'. On 27 November, Maurice's family was slaughtered. 'The five male children of the emperor were put to death first' as the father watched. Then Maurice was beheaded, his body thrown into the Bosphorus, and his head sent to the Hebdomon to be shown to the troops. The statues of Alexandria stepped down from their pedestals to announce what had befallen to a drunken calligrapher nine days before official news of Phocas's accession reached the Augustal prefect of Egypt.

Phocas

Phocas's purge of Maurice's family and followers seems to have included his first son, Theodosius, and his trusted adviser Constantine Lardys, who had fled together towards Persia. Maurice's brother Peter was murdered and Comentiolus's 'body was eaten by dogs'. Constantina and Philippicus were permitted to retire to family monasteries. All of this is reported by the *Easter Chronicle*, which picks up in 602, just as Theophylact Simocatta ends his history with the observation that Maurice's death allowed Chosroes to reignite Roman–Persian hostilities, ostensibly to avenge the death of his 'father'. Chosroes saw an advantage in circulating a rumour that Theodosius, Maurice's son, was alive and in his custody. We cannot know certainly that this was not true. Phocas's usurpation coincided with the end in Persia of the rebellion that Bahram, now dead, had begun more than a decade earlier. The great Turkic chaganate that Bahram had confronted was also less threatening. In the last decade of the sixth century, Turkic potentates had been drawn ever further into the orbit of Tang China. The chaganate dissolved into civil war, splitting into eastern and western empires under diminished rulers.[1]

Chosroes was unbound and set about preparations for war, encouraged by the refusal of the Roman general Narses, who had fought with him against Bahram, to recognise Phocas. Narses seized Edessa and invited Chosroes to aid him in resisting a force sent out by Phocas. The

Persian army under Chosroes inflicted a swift defeat on the Romans besieging Edessa, killing their general and setting a siege at Dara. Chosroes then defeated a second Roman army under Leontius, commander of the eastern armies, 'setting his elephants in a fort-like formation … he captured many of the Romans and beheaded them'. Chosroes retired in triumph, entrusting the siege of Dara to a general. It fell in the summer of 604, inspiring Chosroes to contemplate a more expansive campaign, first conquering Mesopotamia and Armenia, and then driving through Syria and deep into Asia Minor.[2]

Maurice had circulated his trusted generals, relieving but rarely punishing them for defeats. In contrast, Phocas brought Leontius to Constantinople in iron chains. Like Maurice, Phocas sought to reinforce his regime by promoting family members and his military comrades. Domentziolus, Phocas's nephew, replaced Leontius as *magister militum*. Another Domentziolus, the emperor's brother, became master of offices. Priscus, Maurice's only general to have survived the purge, became commander of the Excubitors. Later he was married to Phocas's daughter Domentzia and made urban prefect in Constantinople. Phocas's position was far from secure. He ruled through fear and failed to inspire loyalty. The situation in the city was febrile. In 603, the circus factions rioted and burned the Mese 'from the Palace of Lausus and the praetorium of the urban prefect as far as the treasury opposite the Forum of Constantine'. As punishment, the controller of the Greens, John Crucis, was executed by burning in the same location. Whether this riot, or later unrest, was associated with a palace coup is unclear. However, a palace coup was launched, possibly in 605, by the praetorian prefect Theodore and a certain Elpidius, which Theophanes later attributes to hopes that Maurice's son Theodosius was alive. Theophanes states that Theodore was killed by flogging, while the *Easter Chronicle* has him beheaded. Both sources agree that Elpidius was dismembered and burned. Many other high officials who were implicated in the plot were killed, as were members of Maurice's family who had hitherto survived. The empress Constantina and her daughters, the patrician Germanus and his daughter, were all executed.

Theodore of Sykeon, a Galatian holy man whom we met in an earlier chapter, appears to have prayed for Phocas on the condition that he stop murdering his rivals. Priscus, 'who could not suffer the unjust murders and other evils … wrote to the patrician Heraclius, who was governor general of Egypt, that he should send his son Heraclius' to depose Phocas, 'for he had heard that a revolt was being planned in Africa'. At Carthage and Alexandria, Heraclius the exarch had struck coins depicting himself and his son as consuls, an honour then preserved for emperors. They cultivated support across North Africa, including in distant Egypt. When Nicetas, son of the exarch's brother Gregory, led an army towards Alexandria, he encountered Phocas's loyalists and defeated them outside the city. Nicetas was installed as Augustal prefect and held back the grain ships, causing hunger and hardship in Constantinople. Phocas sent his general Bonosus to recover some ground. He quashed resistance in Syria, including at Antioch, and set sail for Egypt from Caesarea Maritima. However, after some success Bonosus failed in two attempts to recover Alexandria and returned to Constantinople. Benefiting from the unrest, the Persians occupied Armenia and pressed into Cappadocia.[3]

In late spring 610, Heraclius, son of the exarch, set sail for Constantinople, burning with personal ambition, relics and icons of the Virgin Mary lashed to the masts of the ships that carried him and his Berber federates from Carthage. Heraclius was sailing to civil war, and the Virgin protected him, according to George of Pisidia, because 'she alone knows how to conquer nature, first by birth and then by battle'. Heraclius's fleet sailed unopposed into the Hellespont and captured Abydus, meeting little resistance as it proceeded towards Constantinople (Map 11). After a single encounter at sea, Heraclius anchored near the Hebdomon Palace on 3 October. Having arrived back in the city, Bonosus was entrusted with its defence, but he fled his position on the walls on 4 October. He was captured, killed and his body later burned. On 5 October, Phocas was seized and escorted to the Hebdomon, where 'his right arm was removed from his shoulder, as well as his head, and his right hand was impaled on a sword'. Phocas's body parts, and those of his closest followers, were paraded along the Mese and burned at the Forum of the Ox. Heraclius was made emperor on the same day and promptly married Fabia, who took the name Eudocia.[4]

The circus factions took to the streets to settle scores. Phocas's treasurer, Leontius the Syrian, was dragged alive towards the hippodrome where he was beaten to death. The hippodrome 'race-starter and the sergeant of the city prefect', whose role was to keep the peace, were also both killed. Among Heraclius's first imperial actions was to stage races, and Leontius's head was brought into the hippodrome to be burned alongside an image of Phocas. During Phocas's reign this image had been paraded like a holy icon into the hippodrome with lighted candles 'by foolish men wearing white robes. At the same time the Blue flag was also burnt'. Heraclius would favour the Greens.[5]

Heraclius

Heraclius's long reign is quite well documented, although, in the words of his modern biographer,

> The first years of the new emperor were mixed. Primary sources are fragmentary, limited and diverse with respect to their character and provenance. There is no detailed history, no archive of documents, no trove of correspondence, nor any diary for the historian to probe. Instead there are glimmerings of information in saints' lives, chronicles, inscriptions on coinage, papyri, and religious tracts, which require filtering, assembly, interpretation, and synthesis.[6]

This is true for all periods we have addressed so far. Moreover, a recent analysis of the seventh-century written record has established quite how rich it is. Invaluable regional perspectives are offered in the Armenian history attributed to Sebeos and the reconstructed *Chronicle* of Theophilus of Edessa, parts of which are preserved in several later works, including Theophanes Confessor's *Chronicle*. John of Nikiu's history, written in Coptic but preserved only partially in a later Ethiopic translation of an Arabic version, provides precious information on the early Arab invasions, although his account of the Roman–Persian wars that preceded them is lost. The picture that emerges from these writings must be supplemented with material evidence. For example, coins offer us not only inscriptions, but also images and mint marks, which convey additional information, such as where temporary mints were located during expeditionary campaigns. Examining their metallic composition and their distribution helps us answer different questions about fiscal policy, military and economic activity.

For his first decade in power, Heraclius relied on the support of his family and a few trusted potentates, including his cousin Nicetas, who held Egypt and North Africa, his brother Theodore, who controlled the palace as *curopalates,* and the powerful patrician Bonus, who as *magister militum praesentalis* commanded imperial forces in the city and acted as regent when Heraclius was away campaigning. Heraclius also forged a strong relationship with the patriarch of Constantinople, Sergius, who had been appointed in April 610 and died a little more than two years before Heraclius himself. Their partnership survived the patriarch's public censure of the emperor's incestuous second marriage. It is likely that Sergius was less critical privately, as he performed both of Heraclius's wedding ceremonies, baptised his children, and served on regency councils in his absence from Constantinople.

Heraclius had a daughter and a son, Epiphania Eudocia and Heraclius Constantine, with his first wife Fabia Eudocia, before she died in August 612. Epiphania Eudocia was crowned Augusta two months after her mother's death. Heraclius Constantine was also crowned as an infant, being his father's co-emperor from 613. Heraclius had an illegitimate son, John 'Athalaric', whom he sent as a hostage to the Avars, and who would far later serve as the figurehead for a failed coup in 637. Heraclius's second wife, Anastasia Martina, was also his niece, daughter of his sister Maria. Heraclius would have at least ten children with Martina, including four who died as infants or children. Heraclius's first two sons with Martina were known to have disabilities: Fabius had problems with his spine; Theodosius was born deaf and never learned to speak. Disapproval of the incestuous union was vehement, including from within the imperial family, and criticism included observations that their dead and disabled children were divine rebukes. This is hardly surprising, for in Roman tradition, Claudius's marriage to his niece, Agrippina, had led to the intolerable tyranny of Nero. Incest was sacrilege and in Roman law an incestuous marriage was no marriage at all. Those born of incest had a mother but no legal father, and they could not hold public office. This fate awaited Martina's children after Heraclius's death.[7]

The exact date at which Heraclius married Martina is unknown. Theophanes dates the marriage to 613–14, but it probably took place a decade later, long after their relationship began and after their first children were born. Martina stayed close to her husband, even travelling with him on campaign while pregnant or accompanied by young children. It is hard to escape the conclusion that this was to protect them from hostility at court and from the senate at Constantinople. It is likely, therefore, that Heraclius chose to spend a great deal of time away from Constantinople later in his reign precisely because he felt safer when he was in the field or travelling with a skeleton staff of trusted advisers. Rather late in his reign, Heraclius integrated one of his sons by Martina, known as Heraclonas, into the imperial college, and he sought to protect the interests of his other surviving children with Martina by investing them with high ranks.[8]

Heraclius's earliest struggles against Phocas's relatives and supporters allowed Slavs and Avars time and opportunity to invade the Balkans, and the Persians to launch a series of invasions under Shahrbaraz ('Boar of the Empire') and Shahin, generals of Chosroes II. The Persians swiftly captured Syria, including Antioch, and pressed into Anatolia, occupying Cappadocian Caesarea. Heraclius, wishing to strike back, found himself in a struggle with Priscus for control of the army. According to Nicephorus, who wrote a short history of these events a century later, Priscus (whom he calls 'Crispus'), was even offered the crown before Heraclius. In 612, Priscus was given command of an expeditionary force sent to drive the Persians back from Caesarea, which he placed under siege. When Heraclius appeared at his camp leading his own army, Priscus was shocked that a reigning emperor had taken the field. Priscus feigned illness, refusing to attend an imperial audience. For this humiliation, and for the fact that Priscus failed to press his advantage over the Persians, Heraclius had him tonsured. Philippicus, in contrast, was brought out of monastic retirement. According to the *Life of Anastasius the Persian*, to which we shall turn shortly, Philippicus led a diversionary army into 'Persia', probably Persarmenia. Heraclius will have known Philippicus as his father's commander two decades earlier, and perhaps even accompanied him on campaign as a youth. It was a blow, therefore, that Philippicus died on this expedition. Heraclius lacked experienced generals he could trust, which in large part explains his determination to lead his own armies.[9]

In 613 Heraclius led his army into Syria, seeking to drive the Persians from Antioch. He stopped at Sykeon on the way and met the aforementioned holy man Theodore, who died shortly afterwards. One Persian army under Shahin advanced into Cilicia, capturing Tarsus and Melitene, while another under Shahrbaraz took Caesarea Maritima and Jerusalem, which fell in 614. Sebeos reveals that the city surrendered quickly to the Persians, but a Christian revolt some months later led to severe reprisals. Antiochus Strategius, a monk at nearby Mar Saba, wrote a long eyewitness account in Greek of this tragedy, which is preserved in later Georgian and Arabic versions. Strategius wrote of being taken from his monastery by the Persian army and led to the walls of Jerusalem, where he watched angels who were defending the city recalled to heaven, 'and thereby we knew that our sins exceeded God's grace'. He wrote of burning churches, the capture of the patriarch and the murder of priests. He witnessed the seizure of the True Cross and accompanied it on its journey to Ctesiphon, where it was entrusted to the care of a Christian community transplanted from Antioch. Heraclius returned to Constantinople in failure and remained there for most of the next eight years.[10]

In 615, the Persian army pressed right through Asia Minor as far as Ephesus and turned north to envelop Chalcedon, opposite Constantinople. Shahin, the Persian general, received Heraclius himself, who brought gifts and sought to negotiate from his ship. In an extraordinary gesture, Heraclius offered to relinquish the throne and allow Chosroes, who had refused to acknowledge any emperor since Maurice, to install an emperor of his choice – presumably the man he claimed was Theodosius, the son of Maurice – or accept Heraclius as his client. An embassy despatched to Ctesiphon by the Senate of Constantinople carried a letter of supplication. The usually concise *Easter Chronicle* records it verbatim. The compiler of the chronicle was probably a member of Sergius's staff at this time. Heraclius, the letter stated, had not wanted the throne, but wanted only to avenge the murder of Maurice, who had treated Chosroes as a son. Now, Chosroes should treat Heraclius 'as a true son, one who is eager to perform the service of your serenity in all things'. Sebeos, an Armenian source, appears to preserve a haughty response from Chosroes, but at this point his history is confused, conflating events from 615 and 626. Still, it is clear that after more than a decade of war, the Persian capture of key cities was regarded as irreversible and their dominance recognised. This was underlined when Persian forces entered Egypt in 616. Nikiu, later home to the chronicler John, and Pelusium both fell quickly. Alexandria capitulated after a lengthy siege in 619. These losses were felt immediately in Constantinople where the bread dole, which relied on grain from Egypt, was cancelled. It was at this time that, according to Nicephorus, Heraclius proposed moving the imperial court to Carthage, but was dissuaded by Sergius and some powerful senators.[11]

Heraclius relied on his cousin Nicetas to defend Egypt and North Africa while he remained in Constantinople, confronting invasions by Avars and Slavs and dealing with domestic concerns. We know only a few details of this period, for example that Salona, the greatest city of Dalmatia, was razed, and that other well-defended Balkan cities, including

Naissus and Singidunum, were captured. The Avar chagan plundered Thrace at will, his horsemen passing through the damaged Long Walls and ravaging as far as the Golden Gate of Constantinople. Isidore of Seville claimed that at this time the Slavs seized Greece, perhaps referring to sieges of Thessalonica in 615–16. Thessalonica did not fall, but further west King Sisebut of the Visigoths captured cities in Spain, negotiating their retention in 617. In Italy, the exarch of Ravenna was murdered, and his replacement rebelled, negotiated with the Lombards, and had himself crowned emperor before he was killed in 619. Heraclius was obliged to accept all of these losses and to make peace with the Avars, naming the chagan guardian of his son, in order to concentrate on the eastern front. The absence of familiar mint marks during this period suggest that no coins were struck at Antioch after its capture, nor were the imperial mints functioning at Cyzicus (after 616) and Nicomedia (after 619) (see Map 11). Coins struck at Seleucia in Isauria suggest that a temporary mint was established there in 616, to service defences against the Persians, who had occupied neighbouring Cilicia.[12]

In summer 621, the Persians returned to Anatolia and captured Ancyra, in Galatia. Heraclius's immediate response was to bring the remains of Theodore of Sykeon from Galatia to Constantinople, to protect them and to harness their power to defend the city. In winter of 621, Heraclius withdrew to his winter palace and seems to have read Maurice's *Strategikon* rather carefully. He set out from Constantinople in April 622, 'to the country of the *themata*, where he collected his armies and added new contingents to them. He began to train them and instruct them in military deeds', staging mock battles and teaching them 'the battle cry, battle songs and shouts, and how to be alert'.

> Taking in his hands the likeness of the Man-God, the one that was not painted by human hand … the emperor placed his trust in this image painted by God and began his endeavours after giving a pledge to the army that he would struggle with them unto death and would be united with them as with his own children, for he wished his authority to be derived not from fear but from love. [13]

Through his own studies, careful preparation of his troops, and by human and divine inspiration, Heraclius won his first significant victories against the Persians. He left the army to over-winter in Armenia while he returned to Constantinople. There he discovered that the peace with the Avars had broken down and that tens of thousands of Romans had been taken captive. Heraclius took a large entourage out to Heraclea Lyncestis, to meet the chagan and impress him with horse races and ceremonial. However, it was only by offering 200,000 solidi and numerous important hostages to the chagan, among them his illegitimate son John 'Athalaric' and his nephew Stephen, that Heraclius secured another ephemeral treaty. The Romans could barely afford this tribute: Heraclius had seized church silver to finance his Persian campaign. The loss and destruction of cities had undermined the imperial tax base. State expenditure on official salaries had been cut in half and a law issued recalibrating the official rate of exchange between gold and silver. Taxes, when they could be collected, were still demanded in gold, but state payments were now made in a new silver coin, the hexagram. The desperate situation was reflected in the inscription on this coin, which read 'May God help the Romans'.[14]

In March 624, Heraclius marched into Armenia and this time he did not return to Constantinople for four years. Expeditionary warfare was conducted over vast distances, and the Romans achieved remarkable victories. According to Theophanes, upon entering Persia, Heraclius exhorted his men to 'fight to avenge the insult done to God ... Let us be inspired with faith that defeats murder'. He concluded his harangue with a telling observation directed at each soldier: 'Danger is not without recompense; nay it leads to the eternal life. Let us stand bravely and the Lord our God will assist us and destroy the enemy.' At the conclusion of a victorious campaigning season, the emperor 'ordered that his army should purify itself for three days'.[15]

In the following year, according to Theophanes,

> Heraclius gathered his troops and gave them courage by assuaging them with these words of exhortation: 'Be not disturbed, O brethren, by the multitude of [the Persian army]. For when God wills it, one man will rout a thousand. So let us sacrifice ourselves to God for the salvation of our brothers, may we win the crown of martyrdom so that we may be praised in future and receive our recompense from God.'[16]

In Heraclius's harangues we find a remarkable imperial guarantee of martyrdom to Christian soldiers fighting against the infidel: martyrdom would shortly afterwards form the kernel of jihad, the spiritual reward (*ajr*) promised to Muslims killed in battle against Christians.[17]

Heraclius's successes allowed the Romans to establish some order in north-western Anatolia, indicated by reopening the imperial mints at Nicomedia and Cyzicus. However, that region was to suffer its greatest shock in spring 626, when Chosroes sent Shahin against Heraclius, while a large force under Shahrbaraz marched towards Constantinople. Heraclius followed suit, splitting his forces, with a large contingent under his brother Theodore sent to confront Shahin, another marching with him towards Lazica, and a third force despatched towards Constantinople. Theodore's army won a famous victory over Shahin, who died shortly afterwards. Chosroes now looked to Shahrbaraz, whose forces had encircled Chalcedon. Shortly afterwards, on 29 June 626, an army of Avars and Slavs led by the chagan himself arrived at the walls of Constantinople. In response, Patriarch Sergius paraded a Marian icon around the city's walls and the imperial family led public prayers. Fires were set, destroying churches and many other buildings beyond the walls of the two cities. Materials from the destruction were employed to construct siege engines. The Avars and Persians sought to coordinate their assaults. When a Roman delegation was sent out to the chagan it encountered Persian legates who had brought him gifts. The chagan sent ships to transport some Persians across the Bosphorus, so that they might participate in his attack on Constantinople. However, the Roman forces under Bonus established an effective naval defence, which hindered communications and won two decisive naval encounters.

According to Theodore Syncellus, who delivered a homily on the siege shortly after its conclusion, fighting at Constantinople's land walls was fiercest on the ninth day of the Avar siege, immediately before the city's divine deliverance. As Slavs and Avars set sail along the Golden Horn, they were met by Roman ships and the Mother of God herself, 'in front of her holy church of Blachernae'. According to the homilist, 'The fact that the Virgin herself won this fight, and won this victory, was shown clearly' in the number of drowned enemies. In George of Pisidia's *Bellum Avaricum*, another account of the siege, Mary and her tears were both inspiration and salvation, 'For the more you spread the flow of the eyes, the more you prevent the flow of blood' of Christians. The *Easter Chronicle*, which also offers a detailed account of the siege, reports that even the chagan had seen Mary, 'a woman in stately dress rushing around the wall all alone'.

Witnessing the destruction of his fleet, the chagan retreated, setting fire to his siege engines and the environs of Blachernae, including the Church of Saints Cosmas and Damian. The fire was so fierce and its smoke so thick it was seen at Chalcedon, where the Persians imagined Constantinople had fallen and been put to the torch. When he became aware of the Avar withdrawal, Shahrbaraz retreated.[18]

With Shahin dead, Chosroes determined to remove Shahrbaraz, whom he suspected of acting in his own interests, in order to install new generals of unquestioned loyalty. He miscalculated, allowing Shahrbaraz to make a pact with Heraclius and stand down the forces under his command, which withdrew to northern Syria. There they experienced a bitterly cold winter. A massive volcanic eruption somewhere in the northern hemisphere threw up a dust cloud to rival that of 536–40. Theophilus of Edessa observed that in Syria, 'There was an eclipse of the sun that lasted from October until June' 627. At that time, Heraclius was engaged in active diplomacy and appears to have promised his greatest prize, his empress daughter Epiphania Eudocia, to Tong Yabghu, chagan of the western Turks. The two warrior rulers met in person near Tiflis (modern Tblisi, Georgia), where together they set a siege.[19]

By forging strategic alliances with two powerful infidels, Shahrbaraz and Tong Yabghu, the devout Heraclius was able to invade Persia itself and conclude his holy war. He did so from the north in winter 627, taking Chosroes by surprise and drawing the Persians into battle. On 12 December, the Romans won the decisive battle of Nineveh on a mist-shrouded plain near the Great Zab river. A story was swiftly circulated that Heraclius had met the Persian general 'Razates' (Roch Vehan) in single combat, 'and by God's might and the help of the Theotokos, threw him down'. The emperor's tawny horse, 'the Gazelle', was wounded in the thigh and struck in the face, but fought on. When news of the Roman victory reached Chosroes, he was at the royal palace at Dastagard. He fled as Heraclius marched south, burning and looting smaller palaces on the way. Heraclius celebrated the feast of Epiphany at Dastagard, where the Romans found so many silken garments and precious carpets that the emperor ordered them burned, since they were too heavy to carry away. There were also 300 Roman military standards, captured by the Persians at different times, and exotic animals in abundance, including lions and tigers in the hunting grounds. Fleeing towards Ctesiphon, Chosroes was put in chains and then killed.[20]

During his protracted campaigns, Heraclius sent regular despatches back from the front to be read in Constantinople. These provided material for George of Pisidia's epic poems and also a good deal of military detail. However, more than anything they reflected the sacred atmosphere that suffused the campaigns. One despatch in its original form, preserved in the *Easter Chronicle*, celebrated the fall of Chosroes. It was read out in Hagia Sophia on 15 May 628.[21]

> May you, the whole world, raise a cry to God, may you serve the Lord in merriment, enter His presence in exultation and know that the Lord Himself is God ... For the arrogant Chosroes, who fought with God, is fallen. He has fallen and been overthrown, cast down to the underworld, and his memory has been obliterated from the earth. He who was puffed up and spoke unjustly in his arrogance and contempt against our Lord Jesus Christ the True God and his undefiled mother, our blessed Lady, the Mother of God, the ever-virginal Mary, he, the impious one, has perished with a crash ... God and Our Lady the Mother of God acted with us and our Christ-loving expeditionary forces in a way beyond human comprehension.[22]

Chosroes was, in fact, murdered by his own son, who reigned briefly as Kavad II. Kavad also murdered his brothers and half-brothers, ensuring the succession of his young son, Ardashir, when Kavad died only months later, apparently of bubonic plague, in September 628. Shahrbaraz remained the most powerful Persian leader, however, and it was with him that Heraclius negotiated the Persian evacuation of Egypt and Syria. It was agreed that Heraclius would support Shahrbaraz in attaining the Persian throne, and that Shahrbaraz's daughter would marry Theodosius, Heraclius's disabled son. Shahrbaraz's son, Nicetas, was made a Roman patrician. The True Cross and other precious relics of the Passion, the holy lance and holy sponge, were returned to the Romans.[23]

Chosroes's fall resonated most of all in those Christian communities overrun by Persian armies, where stories circulated that were intended to sustain their morale and integrity, to maintain boundaries and ensure separation from Zoroastrian, and very soon Muslim, encroachment. The conversion to Christianity of a Persian soldier, Magundat, offers a tale of martyrdom distinct from that which the emperor had promised his soldiers on the battlefield. According to the martyr's passion, which was written within two years of his death on 22 January 628, Magundat was the son of a Zoroastrian priest, a magus 'who instructed him from childhood in the magic arts'. Before 614, Magundat was recruited into Chosroes's army in Ctesiphon. However, when Jerusalem fell to the Persians, and the 'precious wood of the life-giving and most sacred cross of Christ was seized and carried into Persia', Magundat was moved to investigate Christianity. Unable to resist the power of the True Cross, 'shooting forth its rays of love and power everywhere', Magundat deserted the Persian army and fled to Heliopolis (Baalbek). There he became an apprentice silversmith in the workshop of a fellow Persian, a Christian, with whom he attended church, prayed, marvelled at icons, and became enamoured of the stories of martyrs, 'the atrocious torments that had been inflicted on them by tyrants, and their superhuman endurance'. He travelled to Jerusalem to be baptised and took the monastic habit at the 'monastery of abbot Anastasius of blessed memory', for which reason he adopted his new, Greek name. That was in 620–21, and for the next seven years, the Persian studied Greek so passionately that later he would refuse to speak Persian. He memorised scripture and 'read on his own in his cell the combats and struggles of the victorious martyrs ... because he wanted to be equal to the saints in their combat for the praise and glory of Christ'.[24]

Anastasius fulfilled this wish in 627–8 when, following a vision, he left his monastery to travel to Caesarea Maritima, and there challenged the Persians in command of the city to punish him. For a period, Anastasius was first forced to work in chains, so that his suffering strengthened the resolve of local Christians. At this point, the abbot of his monastery took a direct interest, sending two brothers to 'comfort and minister to him', to ensure that his resolve did not crack, and to report back on and spread the word of his ordeal. One of the brothers accompanied Anastasius when he was transferred into Persia, to be imprisoned in the village of Bethsaloe, six miles from Dastagard. He was, again and frequently, offered the opportunity to abjure Christ and gain his freedom. The offers were made by officials of increasingly elevated rank, ultimately in a letter from Chosroes himself. Anasatasius refused all offers, however, suffering tortures until finally he was beheaded. No reason is given for the determination that Anastasius should be martyred; it was not because he was a Christian. There was no Persian law, as there later was for Muslims, that apostasy was punishable by death. It has been suggested that Chosroes was at the time persecuting Christians, but we cannot discern that in the text itself, where we meet many other Christians, including the monks from Anastasius's monastery, and even the man in charge of his prison in Bethsaloe. Certainly, many other Christians were martyred alongside Anastasius – 'about seventy', according to the *Life*, although none is named. Probably Anastasius was punished because he was a deserter who refused to return to the army.

The motif of the live-giving cross in Anastasius's story is telling. Although the recovery of the True Cross was never a reason for Heraclius's campaigns, its recovery became a dominant theme of his victory and there was no bigger story at the time that Anastasius's life was written. George of Pisidia and Sophronius, soon to become patriarch of Jerusalem, both wrote in praise of the restoration. According to Antiochus Strategius, when Heraclius secured the return of the great relic, 'he took it to Jerusalem on the occasion of his going there with the empress Martina ... And when he had entered Jerusalem, on 21 March, he re-established in its own place the glorious and precious tree of the Cross, sealed as before in a chest, just as it had been carried away'. Heraclius's visit to Jerusalem was part of an extended period in the region, when he was based partly at Hierapolis. It is clear that he intervened in religious affairs, meeting with Anastasius, the Monophysite patriarch of Antioch, and recommended a formulation that promoted a compromise with Chalcedonianism. Pressure was also put on the Armenian archbishop, Catholicos Ezr, to accept Chalcedon. Cyrus, a Chalcedonian, was appointed patriarch of Alexandria, to replace the Monophysite Benjamin, who was exiled.[25]

Treasures seized in Persia were displayed at Heraclius's triumph, where he paraded four Persian elephants. Gold and silver were used for rebuilding and renewal, most notably at Jerusalem and Constantinople. In return for treasures taken to finance his campaigns, Hagia Sophia received a generous annual subsidy. The Magnaura, a grand reception hall at the Great Palace of Constantinople, was rebuilt, as were the adjacent baths that 'time had held prisoner like the barbarians previously held cities'. The silver David Plates, to which we shall turn in the very last pages of this book, were made at this time, surely in Constantinople as an imperial commission (see Plate 18, and Figure 30, p. 351). The future looked bright as Heraclius married his son and co-emperor, Constantine, to Gregoria, the daughter of his late cousin Nicetas, confirming his ties to North Africa. On 7 November 630, while the emperor was in Syria, an imperial grandson was born and named Heraclius. Around the same time,

Epiphania Eudocia was sent off to marry Tong Yabghu at his city of 'One Thousand Springs', cementing a powerful alliance between the Romans and the western Turks. Later tradition also suggests that Heraclius made diplomatic approaches to new powers in the northern Balkans, Croats and Serbs, seeking to undermine Avar hegemony. Certainly, there was no immediate prospect of a Persian counter-offensive. Ardashir's short reign had been ended by Shahrbaraz, who marched on Ctesiphon and occupied the throne for forty days in 630, before he was killed by a loyalist of Ardashir, Farrukh Hormisdas. Boran, daughter of Chosroes II, was placed on the throne, with Hormisdas as her chief minister. Warring factions would soon replace Boran with Shapur, son of Shahrbaraz, and then Shapur with Azarmidokht, Boran's sister, before Azarmidokht was removed and Boran reinstated. Finally, in June 632, an eight-year-old grandson of Chosroes, who proved acceptable to both factions, was placed on the throne. He would reign until 652 as Yazdgerd III.[26]

After three decades, the affairs of the Roman state had stabilised. However, if political fortunes could be plotted on charts like stocks, 630 marked a considerably lower high than had been achieved in 395 and 540, and the market in good fortune would soon crash. Empress Eudocia, en route to central Eurasia, was promptly recalled when news reached Constantinople that Tong Yabghu had been murdered by his uncle. The powerful western Turkic chaganate would collapse, following the conquest of the eastern Turkic empire by Tang China. It has been suggested that the end of the nomadic chaganates was hastened by freezing temperatures, a result of the rapid climate cooling event of 626–7, which had killed countless horses and flocks of sheep. It is clear, however, that the western Turks were sufficiently powerful and organised in 629 to exploit Persian weakness, launching a grand invasion in that year. It is possible that only the death of Tong Yabghu saved Persia from conquest. As the Turkic khanates fragmented, new nomadic confederations began to form, including those that came to be called the Khazars and Bulgars. Under pressure from these new powers, the Avar confederation would fracture and new 'barbarian' armies would move west from inner Asia. New Rome would confront these challenges in turn, and Heraclius had already forged an alliance with the Bulgar khan, 'Kuvrat, the nephew of Organas and lord of the Onogundurs, [who] rose up against the chagan of the Avars'. In fact, Kuvrat had lived in Heraclius's palace as a boy, when he was brought to Constantinople for baptism, an honoured prince and hostage securing the loyalty of Organas. Kuvrat was given the rank of patrician, an honorific that was now conferred regularly upon foreign kings and princes. It was also awarded to Nicetas, son of Shahrbaraz, who was thus recognised as a peer to another ally and *patrikios*, Jabala, king of the Ghassanid Arabs, whose lands in the northern Arabian peninsula acted as a buffer against Bedouin raids into southern Palestine.[27]

Heraclius was quite familiar with Arabs and their military prowess, since they had fought regularly on both sides of his war with Persia. Some Bedouin had taken the opportunity offered by the Roman–Persian war, which occupied the Ghassanids, to launch raids into southern Palestine, but there was no indication that they desired more than slaves and portable treasure. Through the quarter century of war with New Rome, Persia had claimed authority over the whole of Arabia, but it is clear that Chosroes lacked the desire and resources to govern the peninsula effectively. In this power vacuum several charismatic local figures emerged claiming authority derived from God, some of whose names have been preserved, including Musaylima, Tulayha, Aswad, Sajah, Laqit, Ibn Sayad and Muhammad. The last of these was a leader-prophet in the central west Arabian city of Mecca, an advocate of austere monotheism. His religious teachings were collected into a concise volume, the Quran, within which one finds a reflection on recent Roman defeats by the Persians and an assurance that Allah would give victory to the monotheists. Forced to leave Mecca for Medina, in 622 Muhammad recognised that peaceful advocacy was inadequate to his sacred task, so he established a community (*umma*) devoted to 'fighting in God's path' (*jihad fi sabil Allah*). Through a combination of successful raids on neighbouring tribes and marriage diplomacy, Muhammad achieved what his rivals could not, to unite the fractious nomadic tribes of western Arabia and direct their energies towards a singular cause, launching an attack on Roman territory in 630. It is possible that Muhammad had attracted support also from the tribes of eastern Arabia, separated from Mecca and Medina by vast deserts, but it was the achievement of his immediate successor, Abu Bakr, to end a rebellion by these tribes (*ridda*) and lead the first campaign of conquest (*futuh*) to the north in 633–4.[28]

In February 634, a Roman army engaged Abu Bakr's Arabs in southern Palestine, twelve miles east of Gaza, and lost its commander. A force was sent out from Caesarea Maritima to impose order but failed. A further encounter nearby seems to have involved Heraclius's brother Theodore,

who fled when the Arabs fell upon his camp. These episodes were con-
flated in later Muslim sources, which refer to a single decisive battle of
Ajnadayn. Clearly, the Arab army had gained ground and momentum and
it took Bostra as its base for further operations. The raids attracted the
attention of Sophronius, newly elevated as patriarch of Jerusalem, who
wrote of the Arab army as a punishment for Christian sins. By December
of 634, Sophronius was unable to travel to Bethlehem to deliver a homily
on the Nativity, because 'the army of Godless Saracens' had captured the
city. News of success had attracted many more horsemen to ride north
towards Damascus, encouraged by Abu Bakr's successor, Umar. Heraclius, now aged about sixty, travelled again to Syria and mustered Roman
forces near Antioch. Deteriorating health meant he could no longer lead
his troops into battle, but he monitored developments carefully from
well-fortified cities to the north, including Antioch, Edessa and Emesa.
Spies gathered information about the Arabs, their movements and motivations. Reports of their religious zeal and brutality will not have shocked
the battle-hardened commander, who had commended martyrdom to his
own men and benefited from their sacrifice.[29]

Overall command of the Roman field army was given to an Armenian
general, Vahan, who marched south past Emesa, collecting a second large
Roman contingent led by the imperial treasurer (*sakellarios*), Theodore.
The army arrived in July 636 near Jabiya (Gabitha), high ground with
abundant pasture and water beside the Yarmuk river. This was home
to Jabala's Ghassanids, who joined Vahan, bringing the Roman army to
around 20,000. Muslim forces had withdrawn south before the Roman
march, but won an initial clash on 23 July. After some weeks of delay,
the smaller Muslim force feigned retreat, drawing Vahan into advancing
across the difficult terrain cut through by wadis, dry riverbeds. Roman
forces were split, dividing the left flank from the main body of troops and
infantry from cavalry. The Muslims exploited these gaps, falling upon
exposed foot soldiers and encircling the Roman camp. Unable to retreat
in formation, detachments of Romans attempted to fight their way out
or surrender. The Muslims, however, were instructed to take no prisoners
and fleeing Romans were pursued as far as Damascus. The slaughter was
immense and opened Syria and Palestine to the invading Muslim army.[30]

Heraclius withdrew to Constantinople, pausing to recapture and
burn Melitene (Malatya) to deny the Muslim army a base for the capture
of Mesopotamia and further penetration into Anatolia. However, no

concerted Roman counter-offensive was possible and without imperial support, local Christian potentates negotiated with the invaders. The walled cities of Damsacus, Emesa and Jerusalem all opened their gates to Muslim armies after shorter or longer sieges. Caesarea Maritima, with a strong garrison, held out until 641, allowing it to be the first prize won by the new Muslim commander in Syria, Mu'awiya. He slaughtered 7,000 men but did not burn the city. Mu'awiya made Jabiya his headquarters, displacing the Ghassanids, upon whom regional defence had rested. Many Ghassanids withdrew to settle in Anatolia. Muslim forces had by then also captured Ctesiphon, driving the young Persian ruler, Yazdgerd, far to the north. An indecisive siege of Ctesiphon in winter 636–7 was followed by a decisive Muslim victory at Qadisiyya, in January 638. Just as in Christian lands, Persian potentates began to negotiate local agreements with their conquerors, seeking to preserve as much wealth and influence as they could. This led to internecine score-settling, which undermined coordinated regional defence and benefited the Arabs.

Egypt was invaded in 639 and lost by the end of 642. Nicephorus reports these campaigns briefly, John of Nikiu at greater length. Both name 'Amr as the Arab commander and explain that John of Barqa was the first Roman general to engage the Arabs in North Africa, dying in battle. Nicetas, Heraclius's late cousin, appears to have had strong ties to Barqa (modern al-Marj, in Libya, then in Cyrenaica), and his daughter Gregoria was perhaps involved in persuading John to take the field when others were ready to refuse the orders of a distant, failing emperor. Peter, commander of troops in Numidia, refused to march east, looking to his own interests and the likelihood that he would need to defend his homeland soon enough. Heraclius, therefore, sent a contingent of troops from Thrace, which was cut down under their commander Marianus. Theodore, the Augustal prefect, appears to have personally led the unsuccessful defence of Arsinoe (Fayum). A year-long siege finally saw the fall of the powerful garrison city of Babylon. 'A panic fell on the cities of Egypt and everyone fled towards Alexandria, abandoning all their wealth, possessions and cattle.'[31]

The central motif of the story of Egypt's fall is that Patriarch Cyrus of Alexandria determined, like many others before him, to reach a local agreement with the Muslims, paying them a substantial sum that, according to later accounts, 'Amr recognised as the *jizya*, or poll tax. Cyrus also proposed to Heraclius that he marry a daughter to 'Amr. Heraclius rejected

Cyrus's negotiated terms and instructed Theodore to mount a defence. The payment had already been made, however. Cyrus died shortly before Alexandria was handed over to 'Amr, who entered in September 642. The loss of Egypt altered the whole basis of the Roman economy and imperial system in the east. In the very last years of his life, Heraclius ordered his treasurer Philagrius to undertake a new census to assess the greatly diminished tax base. Expropriations of church plate, hitherto emergency measures, became a regular means to finance government as well as military campaigns.[32]

The shock of defeat in 636 and accumulating losses thereafter threatened Heraclius's hold on power. It undermined his claim to divine protection, reversed his hard-won gains, and exposed him to censure for his sinful, fecund marriage. At Antioch, the emperor had confronted his own brother Theodore, who interpreted the Arab advance in light of Heraclius's marriage to Martina. According to Nicephorus, having declared 'His sin is continually before him', Theodore was taken to Constantinople to be publicly humiliated and imprisoned. Heraclius was suffering from edema, the condition from which he would die. His legs and feet were swollen and his abdomen was so distended that 'his private parts turned around and discharged urine in his face'. In 637, he refused to re-enter Constantinople, remaining at the Hieria Palace (today Fenerbahçe), where he was made aware of a coup that involved various unnamed senators, Theodore's son (also called Theodore), and Heraclius's own bastard son John 'Athalaric'. Both suffered brutal mutilations, their noses and hands cut off. Athalaric was exiled to the nearby Princes Islands in the Sea of Marmara; Theodore, who also lost a leg, to the far more distant Gozo, near Malta. Another ringleader, David Saharuni, managed to escape and set himself up as prince in Armenia, confident that Heraclius could not oust him. The emperor's response was to seek to protect the interests of his second family, raising his son by Martina, the Caesar Heraclius, known as Heraclonas, to the rank of Augustus, placing him immediately below Heraclius Constantine, his mature adult son by Eudocia. The seniority of Heraclius Constantine is captured in the record of the entry all three emperors made into Hagia Sophia on 1 January 639. The two older men wore the *chlamys*, the younger the *toga praetextata*.[33]

The *Book of Ceremonies* preserves an acclamation offered to the imperial family standing together in the Augusteus, a reception room within the palace, from which they were about to depart for the hippodrome to watch the start of a race: 'Heraclius Augustus, conquer! Anastasia Martina, conquer! Constantine Augustus, conquer! Heraclius Augustus, conquer! Augustina Augusta, conquer! David Caesar, conquer! Martin, most noble, conquer!' The acclamation reveals the relative ranks within the imperial family. Notable by their absence are Heraclius's oldest daughter Epiphania, who was possibly dead (although she had very recently been suggested as a bride for 'Amr), and Heraclius Constantine's wife Gregoria and three children, who were very much alive. As events would reveal, however, Constantine was far from isolated at court and had the support of the senate. Now aged almost thirty, Constantine had spent far longer at court than his father since 622 and powerful interests were invested in his succession and that of his son.[34]

On the day of the recorded imperial acclamation, 4 January 639, Patriarch Sergius had been dead for less than a month. Among his last substantive acts was to ensure that Heraclius issued his *ekthesis*, a statement of faith Sergius had written in 636 that sought to reconcile the Chalcedonian and Monophysite communities by forbidding further discussion of the problem of Christ's one or two energies and wills. Henceforth, Christ was to have undiscussed energies, and a single will. Supporters of the doctrine, including Sergius's successor Pyrrhus, were called Monotheletes, 'single willers'. Naturally, the matter was still more widely discussed. While the *ekthesis* had been agreed in advance by Pope Honorius in Rome and Patriarch Cyrus of Alexandria, it was rejected by Chalcedonians, especially Sophronius of Jerusalem, and was vociferously attacked in Carthage by Maximus the Confessor, who by his efforts to crush Monotheletism became regarded as the most important theologian of his day. Almost inevitably, another imperial attempt to impose unity in the Christian community had created still greater disunity.[35]

Constantine III, Heraclius II and Constans II

Heraclius died early in 641, probably on 11 February. He was buried in the mausoleum built by Justinian at the Church of the Holy Apostles, where he had placed his first wife Fabia Eudocia many years earlier in a sarcophagus of green Thessalian marble. Heraclius insisted that his own sarcophagus, fashioned from white Docimian marble, be left open and his body be tended by eunuchs for three days. He had heard the story that Zeno was buried alive and screamed for that long. Within four months, Heraclius was joined by his oldest son, Heraclius Constantine, whose tomb was of Proconnesian marble and large enough to accommodate, some years later, his wife Gregoria. Three more sarcophagi would be placed in the mausoleum before the end of the century: for Fausta, wife of Constans II, in green Thessalian marble; for Constantine IV, son of Constans, in Sagarian marble; and for Eudocia, first wife of Justinian II, in variegated rose-coloured Docimian marble. In death, Anastasia Martina and her family were denied imperial honours.[36]

In his will, Heraclius had specified that his sons, Constantine and Heraclonas, both already crowned as Augustus, should rule as co-emperors – Constantine III and Heraclius II – and that they must continue to honour Martina as senior Augusta. Patriarch Pyrrhus and the treasurer Philagrius played key roles in the transition of power. Heraclius appears to have anticipated that Constantine would seek to drive Martina away – she retired to her own palace immediately after his acclamation – so he had entrusted Pyrrhus with a large sum of money to ensure she was 'not lacking in funds'. Philagrius made Constantine aware of this and 'Pyrrhus surrendered the money against his will'. However, when it became clear that Constantine, convalescing at Chalcedon, was terminally ill, the money was sent with 'Valentinus, an adjutant of Philagrius, whom he despatched to the army' to organise opposition to Martina and her children. Valentinus was a capable and ambitious general from a noble Armenian house. When Constantine died, Heraclius II (Heraclonas) 'shared the administration of the empire with his mother' and had Philagrius tonsured and exiled to Septem (Ceuta), the most distant point of North Africa. A rumour circulated that Martina had poisoned Constantine. With a Roman army encamped with Valentinus at Chalcedon, Martina and Heraclonas accepted that they must elevate the son of Constantine.

Nicephorus, who provides these details, offers the interesting observation that Heraclius, as was customary, had been buried with his crown. Heraclonas recovered this, because it was valued at seventy pounds of gold. There were no hereditary Roman 'crown jewels', but Heraclonas placed his father's crown on the head of his nephew, Heraclius's grandson, named Constantine, who is remembered as Constans II. Pyrrhus escaped to Carthage, where he debated Christ's energies with Maximus the Confessor. Paul became the new patriarch, Valentinus became commander of the Excubitors, and the Caesar David was made Augustus with the name Tiberius. At this point, the *Short History* of Nicephorus breaks off for twenty-seven years. John of Nikiu, writing half a century later in Egypt, recalls a struggle that involved the Bulgar khan Kuvrat, who sided with Martina, and the shadowy David 'the Matarguem', perhaps David Saharuni, tempted from Armenia by the promise of marriage to Martina. Martina's supporters were bested and Theophanes, whose entries are cursory in these years, records that the mutilation and exile of the imperial family was by order of the senate. Martina lost her tongue and Heraclonas lost his nose, which supposedly rendered him ineligible to rule. However, Justinian 'Slit-nosed' (*rhinotmetos*) would later demonstrate that this disfigurement was not an absolute barrier to recovering the throne. Heraclonas was allowed no recovery: shortly afterwards he was killed, along with Martina and his brothers David and Martin. Only Theodosius, Heraclius's deaf and now oldest surviving son, was left untouched. The powerful interests that Heraclius had feared for much of his life ensured that in death only his first marriage was considered legitimate. The children by his second, incestuous marriage had no legal father and no right to hold public office. Those same powerful interests, however, also constrained Heraclius's legitimate grandson and successor.[37]

The early years of Constans's reign were marked by an absence of military success. In September 642, Roman troops were evacuated from Alexandria to Cyprus as part of an agreement reached between Patriarch Cyrus and the Arab leader 'Amr, who continued his advance into the Pentapolis and then Tripolis. At the same time, Mu'awiya opened a second front, launching a massive invasion into Armenia in 643. Theodore Rshtuni, successor to David Saharuni, mounted a successful defence, earning recognition from Constantinople as ruler of Armenia with the elevated rank of *curopalates*. Another Armenian named Theodore, called by Sebeos 'the Greek general', was sent east from Constantinople, to ensure closer cooperation. The young Constans remained at Constantinople in these years, initially under the tutelage of Valentinus. Although Constantinopolitan sources are silent on the matter, John of Nikiu and Sebeos suggest that Valentinus had been made co-emperor and given command of the army. His daughter was betrothed to Constans, and he was determined to quash resistance and dominate Constantinople with his Armenian retinue, as Zeno had with his Isaurians. However, in 644, after two years of military dictatorship that is unreported in extant written sources, Valentinus was overthrown and burned to death.[38]

The following years are equally obscure. Nicephorus preserves nothing and Theophanes reports only brief excerpts from his Syrian source, Theophilus of Edessa. From these we learn only tall tales, such as that the Arab caliph Umar had 'started to build the temple at Jerusalem', the original Aqsa mosque,

> but the structure would not stand and kept falling down. When he enquired after the cause of this the Jews said 'if you do not remove the cross above the church on the Mount of Olives, the structure will not stand'. On this account the cross was removed from there, and thus the building was compacted. For this reason, Christ's enemies took down many crosses.

Shortly afterwards, Umar was murdered by a Persian apostate. Uthman, the new caliph, continued to rely on Mu'awiya in Syria and appointed Abi Sarh as governor in Egypt. In 645, Gregory, the exarch in Carthage, rebelled and was acclaimed emperor, benefiting from widespread doctrinal and fiscal grievances. Gregory maintained an active interest in theological disputes and had presided over a debate between Maximus the Confessor and the deposed patriarch Pyrrhus. At the same time there was a crisis in Italy, where the exarch was killed in battle against the Lombards.[39]

In 647, Mu'awiya pressed deep into Cappadocia, besieging Caesarea. No attempt was made to hold territory, but enormous booty was taken back to Damascus, which Mu'awiya had made his new base. If the Anatolian plateau proved no hindrance to the Arab horsemen, the Mediterranean Sea may have seemed a barrier to further advances. It was not. As the Arabs arrived at the eastern Mediterranean littoral they saw the need to construct a fleet. They pressed the shipyards of Alexandria and Palestine into service and, in 649, launched their first naval campaign, directed against Cyprus, where the capital Constantia was taken by storm. Further success followed and the strategic island of Arwad, off the Syrian coast, was captured in 651 after a great siege. When Mu'awiya pressed into Isauria, the Romans proposed a three-year truce, which was agreed at Damascus. The Arabs took the opportunity it offered to settle matters to their north. By 652, when Yazdgerd III was murdered, most of Persia was already under Muslim control. In 653, Theodore Rshtuni negotiated a treaty with Mu'awiya, giving his son as a hostage. Sebeos portrayed the agreement as a pact with the devil, but the Armenians received what they had long desired – virtual autonomy in political, military and religious affairs. The Arabs were far less comfortable in the mountains than in the lowlands that they now dominated and Mu'awiya's attention was increasingly drawn west.[40]

When Constans directed his energy towards Armenia, marching in person to the region, Mu'awiya planned a joint land and sea assault on Constantinople itself, pressing every shipbuilder and arms factory into service. In 654, the Roman fleet under the personal command of the emperor sailed to intercept the Muslim armada, but was routed by Mu'awiya. Constans barely escaped alive. Cos was taken, Crete pillaged and Rhodes was sacked. A fiction circulated later that Mu'awiya had demolished the Colossus of Rhodes, '1360 years after it was set up. And a Jewish merchant of Edessa bought it and loaded 900 camels with its bronze'. Mu'awiya pressed on overland, marching as far as Chalcedon, where he watched his fleet arrive. According to Sebeos, 'they had stowed on board mangonels, and machines to spew fire, and machines to hurl stones' and towers to surmount Constantinople's sea walls. But as the ships were lined up to sail towards the city, 'the Lord looked down from heaven with the violence of a fierce wind, and there arose a storm, a great tempest' which wrecked the fleet. 'The towers collapsed, the machines were destroyed, the ships broke up, and the host of sailors drowned.' As the storm raged on for six days, Mu'awiya withdrew. At the same time, Roman forces defeated a second Arab army in Cappadocia.[41]

In the years before these unexpected setbacks for the Arabs, they had appeared invincible, even as centrifugal forces had been growing across the Roman periphery. Armenia and North Africa had rebelled, and imperial control in Italy was weak. Through this difficult decade, the Monothelete controversy had raged on. The *ekthesis* enjoyed strong support in Syria and Palestine, where Christian communities cohered under Muslim rule. It was rejected in Carthage, however, and after the death of Pope Honorius, also in Rome. Pope Severinus had refused to acknowledge the *ekthesis*, dying without imperial recognition. Next came John IV, who reigned for less than two years and regarded Monotheletism as a heresy. Pope Theodore I organised the Lateran Council of 649 to condemn Monotheletism, but died some months before it was convened by his successor, Pope Martin. Martin had fallen under the sway of Maximus the Confessor, who had fled North Africa to Rome. In 648, Constans had sought to defuse the issue, retracting the *ekthesis* with an edict, called his *typos*, that prohibited further discussion of Christ's wills and energies. However, clerics were not so easily silenced and at the Lateran Council they condemned both *ekthesis* and *typos*. The original acts of the council were published in Greek, demonstrating Maximus's influence and the fact that the condemnation was aimed squarely at Constantinople. As Pope Martin wrote to Constans, 'the safety of the state is contingent upon correct belief, and only if you rightly believe in Him will the Lord grant success to your arms'.

In 649, Constans ordered Olympius, the new exarch in Ravenna, to arrest Pope Martin. But when he arrived in Rome, Olympius witnessed Martin's popularity and made peace with him, proceeding on to Sicily where he drove back a Muslim attack. The loss of Sicily would have been a devastating blow for the empire. Since the capture of Egypt, and with Asia Minor increasingly threatened, Sicily was now the empire's principal source of grain and revenue. At the peak of his popularity, Olympius was proclaimed emperor, but he promptly died. Meanwhile, to the north, the Lombards completed their capture of Liguria and Venetia. Olympius's successor, Calliopas, did Constans's bidding, arrested Martin and conveyed him to Constantinople. There he was put on trial, condemned and exiled to Cherson, north of the Black Sea. A new pope, Eugene, was elected. Maximus was also arrested and tried, somewhat later in 655, when it was alleged that he 'single-handedly betrayed Egypt, Alexandria, Pentapolis, Tripolis, and Africa' to the Muslims, starting two decades earlier when he urged Peter not to march east. Maximus was condemned and imprisoned at Bizya (Vize) in Thrace, close enough to Constantinople that he was recalled for a second trial and condemned again in 662. His hands were removed so he could write no more letters before he was sent into distant exile in Lazica, where he died shortly after arrival. Evidently, there was a concerted effort to make Martin and Maximus, leaders of the Chalcedonian 'heretics' of Italy and North Africa, scapegoats for Roman failings. Martin wrote a letter to a friend in Jerusalem assuring him that he had never sent letters to the Muslims or a statement telling them what to believe, although he did correspond with bishops in the region now under Arab control.[42]

In 655, Theodore Rshtuni fell ill and his arrangement with Mu'awiya collapsed. Constans seized his chance and marched to Armenia even as the Arabs came under assault further to the east, from the distant Parthian region of Media. After decades of successful warfare, the Arab armies experienced defeat and frustration. Resentment had grown among their leaders against Caliph Uthman and his Qurayshi regime in Mecca. In 656, during the *hajj* pilgrimage, Uthman was murdered. Subsequently, between 656 and 661, Mu'awiya, who had been proclaimed caliph in Syria, fought Ali, the son-in-law of Muhammad and ruler in Medina, for supremacy among the Arab peoples. The civil war, the first *fitna*, allowed the Romans some time to regroup, and their ravaged towns and cities some time to recover and refortify. Furthermore, Mu'awiya even agreed to pay the Romans a nominal annual tribute. With this breathing space, Constans turned his attention to the Balkan provinces, launching a punitive campaign against the Slavs who had settled in the peninsula. Constans was successful, and transported several thousand Slav prisoners to devastated regions of Asia Minor, enrolling many in the imperial army. In 659, Constans raised his younger two sons, Heraclius and Tiberius, to the rank of Augustus. Shortly afterwards, he left them with their oldest brother, the ten-year-old Constantine, in Constantinople, and marched to Armenia, to receive oaths of loyalty from the princes, chief among them Hamazasp Mamikonian. Constans also looked beyond Armenia to Transcaucasian Iberia (Georgia), traditionally a client of Persia. The Iberian prince Juansher, who had been raised up by Yazdgerd, was now recognised as a Roman client with the rank of 'first patrician'. Constans gave him the title 'King of all the eastern peoples', along with a precious fragment of the True Cross. This was the high point of Constans's reign. Confident that his eastern flank was secured, Constans looked west.[43]

In the year after his return from Transcaucasia, Constans marched an army across the Balkans and into Greece. He would never return to the capital. Setting sail from Athens, he landed at Tarentum in autumn 662. The emperor wintered at Naples and in July 663 he visited Rome, the first Roman emperor to set foot there since 476. He brought a gift for the pope and his church, but left with far more, seizing gold and silver treasures to fund his future campaigns, which he intended to direct from Syracuse on Sicily. There is no reason to believe that Constans intended to move the imperial court permanently to Syracuse, although there are good reasons why he preferred to be away from Constantinople. He had recently put to death his own brother, Theodore, perhaps after a failed coup. Sicily was distant from Arab raids into Anatolia, which had recommenced in 662–3 after Mu'awiya's victory over Ali. Holding the rich island, with its grain and taxes, was vital to the empire. It was also an ideal base from which to direct the energies of the western empire towards the defence of North

Africa. Upon arrival at Syracuse, Constans ordered the construction of a new fleet, built in part with forced labour, and levied extremely high land taxes and a poll tax, emulating his enemy. His unpopularity was such that no sources report anything he achieved in this period, only his impositions and failures. Meanwhile, raids pressing deep into Asia Minor became an annual event, designed to gather booty and prestige, disrupt economic activity and destroy the defensive capabilities of the Roman army. In 664, an Arab army marched as far as Smyrna then over-wintered in Anatolia for the first time. In 665, Constans's arrangement in Caucasia fell apart and Arab suzerainty was recognised once again. In 667, taking advantage of a Roman rebellion in Melitene, a large Arab force led by Mu'awiya's son, Yazid, marched as far as Chalcedon, where they took captives. Yazid then besieged Constantinople in spring 668, but was driven back when disease spread through his army. Returning via Amorium, Yazid left a large garrison within the city, which was later surprised and slaughtered by a Roman detachment that had climbed into the city one night and 'slaughtered all the Arabs, all 5000 of them, and not one of them was left'. The Romans enjoyed this victory, but it was an ephemeral morale boost. Arab attacks continued by land and sea.[44]

Even as they plundered Roman territory, Mu'awiya projected a new, benevolent image towards his Christian subjects, visiting holy sites in Palestine. Presenting himself as an alternative to a distant Christian ruler bent on pursuing another existential war, Mu'awiya received a Roman delegation, along with representatives from Caucasia and Transcaucasia, who together proposed to murder Constans and reach an accommodation. Juansher, who had travelled to Damascus to recommend the plan and swear loyalty to Mu'awiya, received generous gifts, including fifty-two horses, an elephant and a parrot. Later that same year, back at his Iberian palace, Juansher was murdered by a confidant. Two months earlier, on 15 July 669, Constans had been murdered while he washed his hair, struck on his soapy head with a heavy bucket. Mu'awiya had eliminated two Christian rulers in one season, exploiting the fears and hopes of their courtiers. Constans's three sons, already Augusti, were recognised in Constantinople. However, the Roman forces who had been under Constans's command declared for an Armenian patrician, Mzez.[45]

None of this is mentioned by Nicephorus, whose *Short History* picks up again only with Constans's death. Nicephorus's omission of Constans's long reign, and the cursory coverage in Theophanes, suggests that no

Constantinopolitan sources for this period survived in the later eighth century. Nicephorus, a high-ranking bureaucrat before his accelerated promotion to patriarch, had access to imperial and patriarchal archives, but seems to have found nothing useful there. Similarly, there is no evidence for dossiers of documents in the tenth-century *De Administrando Imperio*, a secret imperial compilation, where information on the war with Mu'awiya is drawn from Theophanes. If Constans's reign and achievements were excised from the archives, and therefore the historical record, this was surely because of his insistent promotion of Monotheletism and his brutal treatment of Martin and Maximus. Those actions ensured that Constans's coverage in western sources is universally hostile. It is easy to imagine that Constans portrayed his existential war with Islam in the same manner as his grandfather Heraclius, sending regular despatches from the field to be read at Constantinople. We might also presume that Constans had his own George of Pisidia, to celebrate victories won under the sign of the cross against a ruthless enemy that sought to remove it from churches and undermine its power. Constans placed the first Greek inscription on his coins, 'By this conquer', depicting himself holding the cross. Indeed, material and numismatic evidence suggests that he was a very active ruler, intent on taking the war to the Arabs and defending Roman territory. Control of the coastal cities of Syria-Palestine was in the balance during his reign, and the mountainous lands north of Antioch were full of brigands, including runaway slaves, known as Mardaites. The triumph of Islam was certainly not inevitable there or further north in the Caucasus. Constans exerted influence in Armenia and beyond, cultivating distant potentates in the manner of Heraclius. Constans campaigned as extensively as his grandfather, marching through Anatolia to Syria, Mesopotamia and Armenia, then leading his troops through the Balkans to Italy and Sicily, looking to the defence of North Africa and the recovery of Egypt. Many of his coins have been excavated at sites across the empire, evidence for troop payments and provisioning and useful for dating the construction of fortifications, for example at Ancyra (Ankara) and on Cyprus. Heraclius, we might conclude, was fortunate to have won his greatest victories before he embraced Monotheletism, which gave those who came after less incentive to destroy the record of his achievements.[46]

Constantine IV and Justinian II,
Leontius and Tiberius III (Apsimar)

Mu'awiya's goal was now the capture of Constantinople. He curtailed the drive into North Africa and directed a fleet that had just taken the island of Djerba towards the Aegean. By 670–71 his fleet had captured Cyzicus, controlling the approaches to Constantinople. All of this took place while a Roman fleet, led by Constantine IV, made for Syracuse to oust Mzez and take command of Constans's forces. A story circulated that, to bypass the Arab blockade of the Hellespont, Constantine had emulated Xerxes, the ancient Persian king, by carrying his ships overland. In 672 a Muslim squadron was established in Smyrna, which launched summer and winter raids around Constantinople. Constantine IV, however, had now returned with his father's new fleet allied to his own, and he was prepared to defend the capital. In 674, Mu'awiya sent his ships against Constantinople. Under the greatest pressure, defences held, and a vigorous Roman counter-offensive drove away the Arabs. The decisive factor was the infamous 'Greek Fire', a primitive form of napalm consisting probably of crude oil mixed with various sticky resins, which was heated and propelled by a pump on to enemy vessels where it was ignited. This major Roman victory highlighted the unprecedented, dominant role the fleet had come to play in military affairs.[47]

Nicephorus and Theophanes both use the same lost source at this point, which suggests that there was a years-long naval blockade of Constantinople. However, it is clear that this is a fiction and that 674 was the nadir for the Muslim fleet during a longer period marked by more and then less successful raids. The Romans, meanwhile, had launched a counter-offensive in northern Syria, which catalysed local Christian resistance and directed the attention of the Mardaites south from the Amanus Mountains into Syria and 'as far as Jerusalem'. Mu'awiya sued for a lasting peace, agreeing to pay a token tribute to the Romans of 3,000 solidi, fifty freed prisoners and fifty horses. This was to be paid each year for thirty years. Mu'awiya died shortly afterwards in 680. The first Umayyad caliph had arranged the succession of his son Yazid I, but this was resisted by many who were opposed to hereditary succession, and others who had witnessed Mu'awiya's recent humiliation. Opposition at Mecca and Medina cohered around al-Zubayr, who advocated for election of the caliph by *shura*, selection from among the Quraysh by a conclave of leaders. Yazid and his son Mu'awiya II, the second and third Umayyad caliphs, were both killed early in this civil war, the second *fitna*, which lasted until 692. The fourth, Marwan, ruled for a year, before his son Abd al-Malik succeeded. There was little respite in Constantinople, however, since the empire's Balkan provinces already faced new threats.[48]

The Slavic threat to Thessalonica had been revived in the 670s, with perhaps two sieges of the city driven back. In 680 a Turkic tribe called the Bulgars (*Boulgaroi*), who had settled north of the Danube, crossed the river. Led by a certain Asparuch, they settled in the vicinity of Varna, whence they 'subjugated the so-called Seven Tribes of the neighbouring Slavic peoples'. Legend recounts that Asparuch was one of five sons of Kuvrat, who had led the Bulgars away from lands north of the Black Sea and had resisted both the Avars and the Khazars. Nicephorus mistakenly dates Kuvrat's rise to the 'reign of Constantine', meaning Constans, 'who died in the west'. However, this was the same Kuvrat who had supported Heraclius and received the rank of patrician. Nicephorus attributed the Bulgar invasion to chance and notes that they acted swiftly to subjugate Slavs who previously had acknowledged the emperor's authority. Through the two centuries of their recorded existence the Slavs had not formed a coherent polity. The Bulgars, in marked contrast, established a powerful empire able to resist, for more than three centuries, Byzantine attempts to recover the northern Balkans, between the Haemus (Balkan) Mountains and the Danube. Asparuch emulated Roman policy by transferring Slavs to strategic areas where they might act as border guards. To neutralise their threat to Thrace, Constantine agreed to pay tribute and recognised their occupation. Nicephorus had every reason to regret the Bulgarian occupation of ancient Moesia, which culminated in 811 (that is, during his own patriarchate) in the death of the emperor, also called Nicephorus, at the hands of the Bulgarian khan Krum.[49]

In November 680, the third council of Constantinople was convened, which by its conclusion ten months later had condemned Monotheletism. Since the greatest number of Monophysites, residing in the east, were now living under Muslim rule, the value of the compromise around the single human-divine will was no longer evident to the heir of Heraclius and Constans. Constantine IV, who presided over many of the council's sessions, was hailed as a new Marcian and a new Justinian, the destroyer of heretics. Rapprochement with Rome facilitated a negotiated settlement with the Lombards. Constantine's situation had never looked better. He took the opportunity to disinherit his younger brothers, Heraclius and Tiberius, and promote his young son, Justinian. When a Roman army marched on Chalcedon demanding that the brothers be reinstated and Constans's settlement respected, Constantine slit his brothers' noses, so that they would be considered ineligible to rule. The leaders of the rebellion were captured and executed, solidifying Constantine's authority. In 684, he led his army into Cilicia and threatened an invasion of northern Syria, forcing Marwan and Abd al-Malik to retreat and pay tribute. Within a year, however, Constantine was dead at the age of thirty-five. No source offers a cause of death.[50]

Constantine was succeeded by his sixteen-year-old son, Justinian II. His reign receives fuller coverage than those of his father and grandfather, but it is coloured by the perspective of a major lost source employed by both Nicephorus and Theophanes, the so-called *History to 720*.[51] Soon after his accession, Justinian sent an expeditionary force under the general Leontius to Armenia and Transcaucasia where it operated free of constraints and raised a great sum in taxes. Roman movements in those regions were not in violation of the treaty with the Arabs and Abd al-Malik could not respond effectively so far north. Therefore, in retaliation 'he occupied Circesium and subjugated Antioch', suggesting that the great city was not yet fully annexed. The caliph's situation became still worse when, in 685, the Khazars launched a devastating invasion of Transcaucasia, posing a similar threat to the Muslim position as the western Turks had once posed to the Persians. A new treaty was agreed between Justinian and Abd al-Malik, with terms even more favourable to the Romans: 1,000 solidi, one freed captive and one horse *per day* were to be paid. Additionally, Abd al-Malik ceded half the tax revenues of Cyprus, Armenia and Iberia to the Romans. In return, Justinian agreed to withdraw rebel fighters, the Mardaites, which our source refers to in Homeric language as a 'brazen wall'. Some were resettled in distant lands and enrolled in the local military, part of a broader policy of population transfers. In 687 Justinian II broke his treaty with the Bulgars and led an expedition against them and the Slavs. The fact that the emperor had to fight his way from Constantinople to Thessalonica reveals how completely imperial authority in the Balkans had broken down. Emulating his father, Justinian transferred large numbers of Slav prisoners to Asia Minor where they were drafted into the provincial armies. However, as we shall see shortly, events proved that he could not command their loyalty and many of these conscripts rebelled during a campaign against the Arabs, which Justinian initiated.[52]

Justinian also concerned himself with unfinished ecclesiastical business, convening in 691–2 a Church Council in a domed hall (*troulos*) within the Great Palace. The council came to be called, after its location, the Council in Trullo, and far later also the Quinisext (*penthekte*) Council, after its remit to complete the work of the fifth (553) and sixth (680) ecumenical councils. Among the many matters discussed for which rulings were issued, it was determined that Christ should be depicted in his human form, not metaphorically.

> Now, in order that perfection be represented before the eyes of all people, even in paintings, we ordain that from now on Christ our God, the Lamb who took upon Himself the sins of the world, be set up, even in images according to his human character instead of the ancient Lamb. Through this figure we realise the height of the humiliation of God the Word and are led to remember His life in the flesh, His suffering and His saving death, and the redemption ensuing from it for the world.

Justinian immediately placed an image of Christ in human form on his coins, the first time this had happened, under the inscription 'Jesus Christ, King of Kings' (Figure 23), with the emperor's portrait relegated to the reverse. At the same time, Abd-al Malik removed all figural images from his coins, replacing them with calligraphy. The removal of crosses and icons from Christian locations under Muslim rule intensified. Here we can detect changes to Islamic doctrine on figural images and the introduction of an official policy of aniconism, an absence of portraits.[53]

To renew war with Abd al-Malik, Justinian deliberately broke the thirty-year treaty established in 677–8 and renewed in 686. His given reason was a provocation by the caliph, who paid the annual tribute in coins minted especially with an inscription that denied the divinity of Christ. Justinian appears to have demanded the caliph use Roman coins displaying the image of Christ. As another provocation, Justinian ordered the Cypriots not to pay the Arabs their designated share of Cyprus's taxes and initiated a partial evacuation of the island, intending to settle Cypriots in the vicinity of Cyzicus. The Arabs were aware that Justinian was inciting them to renew conflict, and when he 'indicated his preference for

Figure 23. Golden coin (solidus) of Justinian II, showing Christ as 'King of Kings'

battle, they affixed the written peace treaty to a tall standard and ordered it to be carried forward'. When battle had been engaged, a contingent of Slavs from among those Justinian 'had armed and named "the Chosen People"' defected. In defeat, Justinian repaid the Slavs by killing their wives and children. When the Romans were defeated once again in the following year, Armenia fell back under Arab authority, and Slavs were employed in attacks on Roman lands.[54]

Justinian is judged by the hostile *History to 720*, followed by both Nicephorus and Theophanes, to have been 'foolish in the administration of affairs', notably in abandoning peace with the Arabs. However, his policy of population transfers was a continuation of an imperial initiative to revitalise the regions that had borne the brunt of Arab assaults. The devastated lands of Asia Minor had been among the most productive of the empire, and to bring them back under cultivation was essential for the revival of the greatly diminished state. At the same time, essential troops were drafted into regional armies and the navy. There is material evidence, in the form of lead seals struck in 694–5 by an official named George, for the provisioning of the Slavic prisoners of war relocated to Asia Minor and Anatolia. These seals demonstrate that Slavs were settled not just in Bithynia, but also in more distant Cappadocia, Caria and Lycia.[55]

Justinian appears to have suppressed dissent in the army by imprisoning his enemies, including Leontius, formerly a renowned general, 'who had spent three years in gaol as the result of an accusation'. When Leontius was released, he led a successful coup, releasing other prisoners, 'a numerous band of brave men, most of them soldiers, who had been confined six or eight years'. Rumours were spread that Justinian had ordered the death of the patriarch and other citizens, who joined Leontius. 'The whole multitude raised the shout "May Justinian's bones be dug up!" ... They brought Justinian out into the hippodrome through the Sphendone and, after cutting off his nose and tongue, banished him to Cherson'. Leontius was made emperor. He lasted long enough to lose North Africa. Carthage fell in 697, Leontius shortly afterwards. The fleet sent to defend Carthage retreated to Crete where the marines murdered their admiral and acclaimed his subordinate, Apsimar, as emperor. Apsimar sailed to Constantinople, anchoring at Sycae, across the Golden Horn. After some time, even as plague once again ravaged the city, he found support among those entrusted with keys to the gates at Blachernae. Leontius's nose was slit before he was confined to the Dalmatus Monastery, an ancient and active foundation within Constantinople. Apsimar was acclaimed emperor, taking the name Tiberius III.[56]

Very little information has been preserved about the decade in which Leontius and Apsimar held power, which is treated by our main source as an intermission in the cruel reign of Justinian. Apsimar made no effort to recover North Africa, but his brother Heraclius – surely not his original name – mounted a vigorous defence of Anatolia and even pressed into Arab territory, raiding the region of Samosata. In response, Abd al-Malik rebuilt and garrisoned Mopsuestia in Cilicia. He continued to send raiding armies deep into Asia Minor, but was content to regard Cilicia as the border. The condominium in Cyprus was respected and Cypriots were returned to the island from Cyzicus. The Armenians were inspired to rebel in 703, but were crushed by the army of Muhammad ibn Marwan, Abd al-Malik's brother. Heraclius appears to have won some significant victories in Cilicia in 704. In the following year, however, his brother was removed from power and Justinian restored.[57]

During his exile, Justinian escaped Cherson and made his way to the court of the Khazar chagan at Doros, further north on the Crimea peninsula. If he was without a tongue, he remained persuasive and soon was married to the chagan's sister, who took the name Theodora. Soon, however, Justinian fled south. Apsimar had approached the chagan with gifts and promises. Justinian, in turn, sent gifts and made promises to Tervel, ruler of the Bulgars, grandson of Kuvrat. Together, Justinian and Tervel encamped before the walls of Constantinople, but Apsimar had already fled. Entering the city along the Valens aqueduct, Justinian raised a shout of 'Dig up his bones!' and regained his throne. Immediately, he began to round up and eliminate his enemies. Heraclius and his commanders were brought from Thrace and impaled on the land walls. Apsimar was brought back to the city where he and Leontius were paraded in chains to the hippodrome. The two former emperors were thrown at Justinian's feet while he watched the hippodrome games. As he trampled their necks, the people cried 'You have set your foot on the asp and the basilisk, and you have trodden on the lion and the serpent.' They were beheaded at the Cynegium. Six years later, in 711, Justinian was himself beheaded following another military coup.[58]

Justinian's second reign is portrayed rather like his first, comparatively fully and as a series of betrayals and missteps. Tervel was acclaimed as Justinian's Caesar before he returned to Bulgaria. Some time later, Justinian's Khazar wife was brought back to Constantinople to be crowned empress. She brought with her a young son whom they called Tiberius and immediately made co-emperor, perhaps to remove the stain of Apsimar's reign. Tiberius joined his father on the reverse of golden solidi, both clutching the same tall cross-on-steps. On the obverse was a young Christ. Breaking his treaty with the Bulgars, Justinian suffered a humiliating defeat at Anchialus in 708. In 709, an army under Maslama, Abd al-Malik's son, pressed north from Cilicia into Cappadocia, where it besieged and captured Tyana, decisively defeating a relieving army. Finally, Justinian turned against the Khazars, sending a fleet across the Black Sea to capture Cherson. His armada comprised 'ships of every kind – dromons, triremes, transports, fishing boats, and even barges – from contributions raised by the senators, artisans, ordinary people, and all the officials that lived in the city'. In this way, all suffered the tragedy when, their mission accomplished, the fleet was wrecked on the way home by an October storm.

The newly installed governor of Cherson, Elias, promptly rebelled along with Bardanes, son of an Armenian patrician, who was exiled there. 'When Justinian heard of this, he slaughtered the children of Elias in their mother's lap and forced her to marry her cook, who was an Indian and extremely ugly.' After this he fitted out another fleet and sent it north, with siege equipment, under the patrician Maurus. When the Khazars intervened an agreement was reached and Maurus joined the rebellion. His men acclaimed Bardanes as emperor with the name Philippicus. The story of Philippicus's victory is the denouement of the Justinianic tragedy preserved by Nicephorus and Theophanes. Both report Justinian's death at the hands of Elias, formerly his own bodyguard. Anastasia, widow of Constantine IV reappears, protecting her grandchild Tiberius. The child clutches at the altar of the Mother of God at Blachernae before he is torn away and 'killed with a knife, like a senseless animal'. In this way, the Heraclian dynasty was extinguished as violently as it had been established 101 years earlier.[59]

During the sixth century, equilibrium existed between the two great Near Eastern empires, Rome and Persia, which for long periods allowed both to focus their energies in other directions, externally and internally. Emperors of the sixth century pursued and celebrated the restoration of Roman power. They built extensively throughout the empire and restored older structures, including churches and palaces, city walls and aqueducts. The crowning jewel of the age was Hagia Sophia, the Church of Holy Wisdom, which for the next nine centuries was the spiritual heart of the Christian empire. Romans learned to control and direct the energy of 'barbarians', and frequently defeated them in battle. Roman armies campaigned extensively and exhaustively, advancing through North Africa and far into the west, across Italy as far as southern Spain. They encamped and fought north of the Danube and devised new tactics to deal with now familiar enemies, the Avars and Slavs. Disputes between Christian churches and their powerful champions, bishops and holy men, were frequent, and major doctrinal differences were unresolved. Frequently they were inflamed by imperial interventions aimed at achieving consensus or suppressing dissent. However, in the year 600, the destruction of vibrant, fractious Christian communities across the eastern empire was unthinkable. By the end of the seventh century, it was a fact.

The geopolitical situation changed with Maurice's death, when Persia rose against the Romans and an existential war, charged by religious rhetoric on both sides, erupted. Our sources record little of Phocas's reign other than the emperor's violent outbursts and ignorant blunders. Yet it was in his reign that the Roman–Persian war was renewed. This last great war of antiquity endured for a quarter century and it was increasingly charged with religious rhetoric on both sides, an orchestration that reached a crescendo in the years immediately before the Islamic notion of jihad was formed. It was prosecuted principally by Heraclius, an emperor who commanded his own troops, the first to do so in the east since Theodosius I. Heraclius became emperor aged around thirty-five and his reign might have ended quickly and in the same manner as those of Maurice and Phocas. Instead, he clung to power for three decades and established a dynasty that endured for a century. Heraclius won the existential war with Persia but failed to contain a new force, the Arabs who professed a new faith, Islam. Heraclius's heirs were rather more successful, stopping the Arab advance and, at times, driving it back. The Romans benefited from civil wars fought for leadership of the new Muslim empire, but also by developing their military institutions and naval capacity. More than this, the Heraclians sustained a web of alliances that drew the western Turks, Bulgars and Khazars into the Roman orbit against, first, the Persians and Avars, and then the Muslim Arabs. If the capacity of the diverse Slavs was never fully harnassed, the energy of Christian Arabs, Armenians and Georgians was frequently directed against the common enemy. Most importantly, Constantinople, which was besieged several times by land and sea, never fell to a foreign power. Emperors and pretenders were mutilated and murdered, but the institutions of imperial government and the church survived. New Rome endured.

PART 3

THE END OF
ANTIQUITY

10

THE END OF ANCIENT CIVILISATION

Shortly after 700, looking out from his study in Ravenna, a city that was governed by a Roman exarch, an armchair geographer saw a world of Roman cities linked by Roman roads. In his description of the world, the Ravenna Cosmographer employed the classical division of the earth into three continents, Europe, Asia and Africa, adding Christian motifs wherever possible, such as the names of Noah's three sons, who each settled in one continent. He invoked Christ's help so that he might 'designate in detail the lands of all nations, placed around the great circuit of the impassable ocean'. The Cosmographer described twenty-four routes, each corresponding to an hour of the day. This practice may derive from Roman land surveying, but the Cosmographer justified his organisation as divinely inspired, because God created the luminous bodies in the sky that inscribe the passing of time.[1]

The Cosmographer's twenty-four routes contained more than 5,000 place names from India to Ireland. Throughout his work toponyms are frequently garbled and great efforts have been made to identify places he records and sources he used. His distances are also inaccurate, but for the Cosmographer writing the world was a way to organise knowledge rather than to measure distances. The Cosmographer's world was a Christian Roman ideal, a world of information archived and perhaps memorised. It was uninformed by travel. 'Though I was not born in India nor raised in

Ireland', he reflected, 'though I have not meandered through Mauretania nor scrutinized Scythia nor ambled to the four corners of the earth, nevertheless I have imbibed intellectually the habitations of the whole world and its diverse peoples from the books written in the ages of many emperors which describe the world'. The Cosmographer did not need to travel, because he could base his observations on materials that were available to him in his own library.[2]

Perhaps the armchair geographer, when he ventured beyond his library, saw that Ravenna, and its port at Classe, remained a bustling nexus for international trade where imported goods were regulated and taxed, then redistributed across northern Italy. Ravenna's city walls enclosed 166 hectares, far less than Rome or even Milan, which it had replaced as the administrative capital of northern Italy, although this space was not all built over or densely populated. Ravenna was not a large city with a well-developed infrastructure when it was chosen, but was, like Constantinople, an ancient foundation with notable strategic advantages that grew as a consequence of a concatenation of political decisions. Trade and enterprise at Ravenna followed the development of its imperial and archiepiscopal complexes. Wine imported from the eastern Mediterranean was a key commodity, although far more arrived from southern Italy, and olive oil, fish sauce and fine wares came principally from North Africa. The latter trade was sustained under Gothic rule, when connections with Carthage in particular thrived. After Justinian's reconquest, African red slip ware was replaced by Phocaean red slip ware, manufactured in western Asia minor, which appears in huge amounts. Grain appears to have been imported from the eastern Mediterranean in the sixth century, along with far more amphorae from the Levant and Aegean islands, notably Samos. Amphorae from Samos have been associated with the transportation of olive oil and wine processed on that island for distribution as part of the *annona*, in this case to provide for troops stationed in northern Italy.[3]

Even in the lifetime of the Cosmographer, Ravenna was transformed. In the heart of the city, from the middle of the seventh century and through the eighth, small residential buildings of wood and clay begin to crowd the administrative centre, ecclesiastical buildings and river. A grand house on the Via D'Aglezio, built in the sixth century, was replaced in the seventh by a graveyard and two smaller residences. Ravenna's medical school, active in the sixth century, had ceased to function by the eighth, although some form of higher learning continued. Trade slowed considerably and warehouses at Classe, from the second quarter of the seventh century, were repurposed and smaller houses were built within their crumbling walls. More than fifty seventh-century burials have been identified in areas that had been prime commercial land, alongside far smaller, wooden structures used for storage and modest production, including that of local pottery.

The last recorded Roman exarch at Ravenna, Eutychius, held office until 751, when the city was taken by the Lombards. Their suzerainty did not endure. In 755 Ravenna was handed, by order of Pepin, King of the Franks, to Pope Stephen II. However, Ravenna's leaders disputed papal authority for the rest of the eighth century. Chief among those leaders was the archbishop, who had grown accustomed to distant imperial authority and preferred to deal directly with a new secular power, soon to be an emperor, Charlemagne.[4]

The transformation of urban life

The survival of the exarchate until the middle of the eighth century guaranteed some degree of continuity at Ravenna. Across the empire of New Rome, other cities were transformed more or less dramatically, and many disappeared altogether. The very meaning of the designation 'city' changed. When we read the word *polis* in a Greek text of the seventh century, we must imagine an absence of many defining features of the ancient city, such as the school and library, the thronged stoa and flowing aqueduct, the noisy baths and latrines, and the busy interest of decurions. It is clear that many functions once performed by decurions, notably those relating to taxation, had been taken over by officials appointed and despatched from Constantinople. Civic treasuries were deprived of their full share of income and were incapable of rebuilding from their own resources after natural disasters or invasions. Procopius, in his *Secret History*, attributes this to Justinian's initiative, so that 'whatever revenues the residents collected for their own political or cultural events, he likewise dared to divert away and mix in with the imperial income'.

> From then on there was no concern for doctors and teachers, nor was anyone in a position to make plans for public works. Lamps no longer illuminated public spaces in the cities at night, and there was nothing

left to make the citizens' lives more pleasant. For the theatres, hippodromes and wild beast shows had all mostly been closed down ... Thus
depression and misery overcame both private and public life, as if yet
another calamity had struck from above, and life was without laughter
and joy for everyone.[5]

It is clear, however, that this was a process that started long before
Justinian. It is equally clear that urban elites remained, but now civic
governance was entrusted not to councillors operating within an ancient
tradition of corporate governance, but to grandees with access to those
holding high imperial office, who will themselves once have held high
office. We read now of *ktetores*, *oiketores* and *proteuontes*, general designa-
tions applied to men of senatorial rank who had earned their reputations
and fortunes in imperial service. A novel of Justin II (149), issued in
569, made the notables of each city collectively responsible for taxation,
and also granted them a role in electing provincial governors. However,
notables were hard to bully and coerce, and in 575 a novel (163) of Tiberius II remitted taxes for every fourth year, excepting Mesopotamia and
Osrhoene, then a theatre of war. A principal function of the decurions had
been to secure payment of the *annona*, the tax in kind levied to support
the military. Imperial functionaries known as *kommerkiarioi* appear, from
the middle of the seventh century, to have taken over this task, also storing
and distributing rations, wheat and other provisions, in and from imperial
storehouses (*apothekai*).[6]

Bishops were given increased authority in civic matters, although evidence for their involvement in mundane administration is scarce. Bishops
appear most frequently at times of emergency. One benefit of ecclesiastical
involvement in civic affairs is the church's habit of record-keeping, which
was otherwise largely lost. Civic archives ceased to be kept. However, when
we read of bishops of certain cities participating in Church councils, we
cannot consider this proof of continuity in urban settlement. Much has
been made of the records of such councils, and of the *notitiae* – the lists
of episcopal sees and the patriarchs to which they are subject – which fill
a gap in the extremely meagre documentary record. But we must beware
of drawing firm conclusions from such sources, since neither tells us anything about the condition of the cities or bishoprics in question, nor if
the city and bishopric even coincided. A further consideration is whether
a bishop was still resident in his bishopric, or lived permanently in exile.[7]

Written evidence is insufficient in itself to allow us to assess the fate of cities generally, as well as to assess what happened to any given city. To deepen our understanding and to reach specific and broader conclusions we must turn to the abundant material and numismatic evidence uncovered in recent and ongoing excavations and field surveys.

Asia Minor and Anatolia

Evidence for disruption in city life can be found across Asia Minor and Anatolia. Sardis, like many cities across the Roman world, had acquired a new perimeter wall in the later third or fourth century AD.[8] This was as much a means of defining and projecting an urban identity as providing for the city's defence, following the model established by the Aurelian walls at Rome and the Theodosian walls at Constantinople. The wall at Sardis, although it enclosed 130 hectares, excluded several large buildings, including a major public bath complex that was still functioning. The space contained within the walls was redeveloped frequently with new colonnaded streets, including the so-called 'Marble Road', a twenty-metre-wide paved street that connected the western part of the city to a grand gate, and a second intersecting street, eighteen metres wide and unpaved, that ran to the south-west gate. An inscription indicates that this was paid for with private funds, in other words by decurions. The wealthiest citizens of Sardis constructed mighty urban palaces along these streets. Just south of the Marble Road at the western gate, four buildings were combined to form a sprawling residence of some thirty separate rooms with spaces arranged around a central peristyled courtyard. Many rooms were elegantly decorated, including a grand apsidal dining room with wall paintings and inlaid marble designs. Water features abounded, including six basins, tanks or fountains and two private toilets. Private access to so much water was a rare privilege. This palace was completed no earlier than AD 500. A little over a century later it was gone. Marcus Rautman has characterised the situation succinctly.

The prosperity Sardis enjoyed in the later fourth to sixth centuries may be the result of local circumstances, but the far-reaching changes it saw around the turn of the seventh century are those seen among its neighbors as well. Many public spaces were poorly maintained if not abandoned altogether; houses were subdivided and put to new purposes; imported pottery decreased in variety and quantity; and the supply of currency dwindled ... Private houses underwent complex changes that saw individual spaces selectively maintained, subdivided, stripped, and put to new purposes. Hearths, ovens, vats, drains, and latrines were cut through or set atop marble or mosaic floors, sometimes in grand apsidal rooms.[9]

Several coin hoards at Sardis attest to a Persian assault on the city in 616. At a row of shops adjoining Sardis's gymnasium, more than a thousand copper coins have been discovered. Most date from the reigns of Maurice, Phocas and Heraclius, suggesting active trade that ended quite suddenly in 615/16. After this date, no further coins are found at the shops, and no coins are found anywhere in Sardis that were struck before the reign of Constans II (641–68). Half of the coins date to before the shops were built, proving that far older coins remained in circulation, even those that were struck before Anastasius's currency reform. The greatest number of coins found in the shops, struck after c.500, were from the nearest large imperial mints at Constantinople (c.300 coins), Nicomedia (66), and Cyzicus (40), with rather fewer from the more distant mints at Antioch (17) and Thessalonica (26). There were no coins from Alexandria, and one each from Rome and Carthage.[10]

In a residential building in a western suburb of the city, a citizen of Sardis stashed a collection of twenty-one green-blue glass weights in a locked wooden box buried in the earthen floor of an interior room. Each weight was marked with a circular stamp containing a cruciform monogram, which was first used in the 530s. Two bronze coins of Heraclius, each struck in c.615, were found near the glass weights. In the same room were found a bronze lock plate, surely from the room's door, a pair of tweezers, a bronze spatula with traces of gold leaf, a stone pestle, and sherds of pottery, including from an African red slip plate of a type made between c.580/600 and 650. Three similar glass weights were found in one of the aforementioned shops adjoining Sardis's gymnasium, together with part of a bronze scale and a bronze coin weight. The largest of the glass shop weights is identical to the largest of the hoard of weights. These both weigh 4.34 grams, approximately the weight of a gold solidus at this time. These were all coin weights, officially stamped, kept in a locked box in a locked interior room, and abandoned only in dire circumstances, when the city was attacked. Byzantine coin weights were centrally produced and show great consistency. Glass was a preferred medium for coin weights, as it could not easily be altered and damage would be clear to all. In the sixth century, most glass weights appear to have been manufactured in Egypt, although in the later sixth and early seventh century some were produced in the Levant (Plate 19).[11]

The glass weights are a fraction of the Roman glass finds at Sardis, which together reveal much about changes to patterns of production and distribution in the sixth and seventh centuries. Natron glass from the Levant is found in large quantities at both Sardis and nearby Pergamum, fashioned into a range of items including glass vessels, windowpanes and jewellery. Chunks of raw glass, vitreous slag and deformed objects are clear evidence that glass was worked extensively at Pergamum from the first century AD onwards, although luxury items were imported. The chemical composition of glass found at Pergamum changes before the fifth century, as a regional glass manufacturing tradition developed in western Asia Minor, using minerals from the borate deposits between Pergamum and Sardis in place of natron. Very little glass is then found in Pergamum from the eighth century until the tenth. Later glass is based on plant-ash recipes, although there was extensive recycling of earlier glass.

At Ephesus, archaeologists discovered evidence for at least 1,150 different glass objects in a single villa, 'Hanghaus 1', and analysed 106 glass samples, which demonstrated occupation from the first century AD to the later sixth or early seventh century. Across the city, many more coins of the fourth century have been found than of the sixth, and far fewer for the later sixth than for the early sixth century, suggesting a gradual decline in circulation over time, followed by a rapid decline at the onset of plague and foreign invasions. While many restorations at Ephesus were undertaken with reused materials, and new buildings were smaller and more crowded, the standard of construction remained high until well into the sixth century, when the situation changed drastically. Coin finds prove that the upper agora at Ephesus was abandoned in *c.*614 and that the buildings lining the *embolos* were ruined, while the lower, main agora lay disused. The immediate reason for this was a Persian attack, and the city's failure to recover can be attributed to the continuous Arab assaults into Asia Minor thereafter. Monetary exchange ended quite suddenly in the mid-seventh century. Towards the end of the seventh century, as a defence against the Arabs, a new city wall was erected encompassing the harbour and the northern section of the old city of Ephesus. The southern section was left in ruins, and the *embolos* served merely as the access route to the road south. Eventually, the old city was abandoned altogether for a heavily fortified inland settlement on the hill of Ayasuluk.[12]

The harbour at Ephesus, called Panormus, was connected to the city by a canal. Both harbour and canal required regular dredging, but even so the harbour moved gradually westwards until it silted up completely. The site of the late Roman harbour today is indicated by a small lake, Çanakgöl, to the west of the archaeological site. Sediment cores taken from around the lake show the changing landscape, as land became sea and then the sea, once again, became land. In a twenty-centimetre stratum of coarse brown sand and pebbles, the earliest marine levels, microfaunal analysis has identified antique ostracods, tiny marine crustaceans (seed shrimp) and single-celled organisms called foraminifers. Above this, less coarse grey sand pebbles contained marine gastropods and bivalves; then abundant plant remains, more gastropods, and the spines of sea urchins; and then coarser sand, barnacles and an olive pit, discarded around 200 BC. In other words, for around six thousand years, this was the sea. In the layers above, there is evidence for decreased salinity, as the water of the Roman harbour became brackish and lagoonal. Finally, through siltation, land displaced the water and the sand changed from brown to yellow, intermixed with mica, quartz and limestone.[13]

Ancyra (Ankara) in late antiquity occupied a strategic junction of the main highway from Constantinople to the eastern frontier. For that reason it became a commercial centre of considerable importance, a major supply base and the winter headquarters for troops. But for the same reason, it bore the full brunt of the Persian invasions that swept across Asia Minor in the early seventh century. Greek and Syrian sources both record the fall of Ancyra in 622 to the Persian general Shahrbaraz, who is said to have killed or enslaved all of the inhabitants who had not fled. The excavations of the Roman gymnasium at Ancyra provide graphic confirmation of the destruction: the ruined building is covered with a layer of ashes and debris, within which have been found gold coins of Heraclius. Further confirmation for the date of destruction is provided by excavated copper coins, which are found in large numbers from the third century to the sixth, but drop suddenly in the seventh. Only one coin from the later part of Heraclius's reign has been found, and only two of Constans II, after which just two coins represent the next two centuries. The gymnasium had been destroyed and its site was left virtually unoccupied for over 200 years. The topography of the modern city has prevented excavations on a larger scale, but when Ancyra reappeared in the later seventh century it was as a heavily fortified town on a hilltop. These fortifications consist of an outer wall with square towers at forty-metre intervals, and an inner bastion with closely spaced pentagonal towers. The most likely date for their construction is the reign of Constans II, probably 656–61, using a large amount of material from the ruined lower city.[14]

The ports of Lycia, on the southern coast of modern Turkey, flourished in antiquity, offering the last safe harbour for many ships setting off across the Mediterranean for the Levant and Egypt. Alexandria was due south of Lycia across the open sea, and reliable north-west winds expedited the voyage. For the same reason, in the later seventh and early eighth centuries, as the Arabs made a series of seaborne assaults on Constantinople, the region and its cities, including Xanthus and Myra, were exposed to Arab attacks. In the middle years of the seventh century the acropolis at Xanthus had been given a defensive wall and this became the settlement, high above the river, several miles inland. The city of Myra lay three miles from Andriace, its port on the coast. It was the metropolitan see and civil capital of Lycia, and details of life in the age of Justinian are provided in the *Life of St Nicholas*, which records the arrival of bubonic plague in 542, and the presence of a church dedicated to the miracle-working 'Santa Claus', an earlier St Nicholas, which brought many visitors to the city as they travelled to and from the Holy Land. St Nicholas himself sailed from there directly to Ascalon, one of the closest ports to Jerusalem. We have seen in an earlier chapter that seventy-five silver vessels and church furnishings were discovered at Korydala near Myra, thirty of which had fivefold silver stamps dating them to the period 550–65. Shortly after this, much of Andriace was abandoned, and a new wall built enclosing one-third of its former extent. Inland at Myra, the acropolis was fortified and settlement transferred there, away from the pilgrimage church below, which was given its own fortification.[15]

The very conditions that made most cities poorer and smaller led to the expansion of a few. Amorium, located near the source of the Sangarius river, was 170 kilometres (106 miles) south-west of Ancyra along a major road that passed through Pessinus. It was also connected by roads to Constantinople, via Dorylaeum to its north-west, and via separate roads to its south-east to both Isauria and Syria. A significant minor city in antiquity,

Amorium was a neighbour to Orcistus, which we visited in an earlier chapter, and became a metropolitan see with five suffragan bishoprics. Its walls were constructed at the time of Zeno, reflecting the city's importance on the route between Constantinople and Isauria. Some charred wooden beams discovered inside one of the wall's towers have been dated by dendrochronology to exactly this time. However, nothing in the fifth and sixth centuries indicated the importance that would attach to Amorium in the eighth and ninth centuries, when it was the third city of the empire, after Thessalonica and Constantinople. The leap in status can be attributed wholly to Amorium being so well connected by military roads, which led to its designation as headquarters for the army of the Anatolikon, the empire's largest military district. It was for that reason too that, in August 838, the city was razed to the ground by the caliph al-Mutasim.[16]

Amorium comprised two separate fortified settlements, a smaller upper city and a larger lower city. The walls of the upper city were constructed in the early seventh century using spolia from Roman buildings and, remarkably, many gravestones, suggesting that it was built hurriedly in response to an imminent threat. Excavations have revealed much about life in the lower city, including a substantial development in the sixth century. Older buildings were cleared to make way for a new monumental leisure complex, including a bath house with a polygonal hall, dated by brick stamps to after 518. In the seventh and eighth centuries, the area around the baths became increasingly densely occupied, with the creation of an industrial and commercial sector. The baths, complete with marble and opus sectile floors, appear to have functioned until the ninth century, by now adjacent to a winery. Large stone presses, used for crushing grapes, prove that there was extensive local vine growth and wine production. However, olives were not grown locally and there appears to have been a dearth of olive oil. Very few large transport vessels, such as amphorae, have been excavated, in contrast to many large storage vessels, suggesting that most production was local.

A few bricks from a lower city church, as well as those at the baths, are marked with stamped names. Brick stamps are rare outside Constantinople, and it is clear that these bricks were produced from clay dug from local pits. Roof tiles are abundant at Amorium, suggesting domestic use and local production. There are clear indications of what the population ate, including a range of cereals, notably wheat and barley, and larger animals indicated by the number of bones, mostly sheep and goat, some cows, fewer pigs and far fewer chickens. Moreover, and remarkably, there is evidence that the way of lighting houses changed quite suddenly. The use of oil lamps formed from terracotta in moulds was standard until the middle years of the seventh century across the eastern empire. These disappear at Amorium and are replaced in the eighth century by local wheel-made lamps, which is both a reflection of the scarcity of olive oil locally and of changing patterns of production and trade in domestic items over longer distances. A parallel development is reflected in the disappearance of certain types of ceramics, notably fine red slip ware, which is replaced by coarser local pottery. Kilns have been excavated in the upper city at Amorium. Two folding iron stools and a partial set of black and white gaming pieces were found together in a layer of burned brick and roof tiles, evidence for a last game of draughts or *petteia*.[17]

Euchaita also benefited from the military build-up in Anatolia, although far less than Amorium. Euchaita was given the status of city, an archbishop and a circuit of walls by Anastasius, according to an inscription discovered in the village of Avkat (now Beyözü), 'inspired by the martyr' St Theodore the Recruit, whose tomb had become a site of pilgrimage. Recent surveys have identified the line of the wall, the location of the main city gate and that of the martyrium of St Theodore. Euchaita lay fifty-five kilometres to the east of Amasia (modern Amasya), a strategic point where roads from Constantinople and Ancyra met, leading then towards Armenia. Near the Black Sea coast, Euchaita was sufficiently isolated to have served as a place of exile. However, by the middle of the seventh century, it had attained a new significance in the empire's strategic geography. It was sacked by the Persians in *c.*612, when the Church of St Theodore was destroyed then rebuilt. It was sacked again in *c.*640. Probably in the 650s, the city's acropolis, Kale Tepe, was fortified with walls, towers and a ditch. Whereas the Anastasian walls had no great defensive value, the acropolis now served as a powerful citadel.[18]

The Balkans and Greece

Cities to the north and west of Constantinople, from the Danube in the north and as far south as the Peloponnese, suffered like those to the east from major invasions, in this case by armies of Avars, Slavs and others.

We can detect indications of their enduring impact in the archaeological record throughout the Balkan peninsula. Let us recall the devastation of Sirmium, on the banks of the Danube in modern Serbia, which was never restored after it fell to an Avar siege in 582. Similar signs of destruction can be found at cities across the Balkan peninsula, including at Capidava, Histria and Tomis on the lower Danube and Black Sea coast; Stobi and Heraclea Lyncestis, which are inland; Butrint, on the Adriatic coast overlooking Corfu, in modern Albania; and Athens and Corinth, in Greece. Thessalonica, a major administrative and cultural centre, experienced the same phenomena somewhat differently.

Capidava, on a bend in the lower Danube in modern Romania, is around seventy-five kilometres (about forty-six miles) north-west of the Black Sea port of Tomis (modern Constanţa). The Black Sea coastal road that passed through Tomis met the inland road from Capidava at Histria. Capidava was established as a military fortress by Trajan long after the ancient cities of Histria and Tomis were founded. All three cities, and twelve others, are recorded in Hierocles's *Synecdemus*, a sixth-century table of administrative districts (*eparchiai*) and their constituent cities (*poleis*), in the eparchy of Scythia. More than 200 coins of the sixth and seventh centuries have been discovered in excavations at Capidava, all of them copper, and the majority struck at Constantinople. Of other mints represented, more were struck at Nicomedia than at Thessalonica, in contrast to Tomis where coins from Thessalonica are second only to those from the capital. Coins are found at Capidava for almost every year between 538 and 580. Of these, 90 per cent are full or half folles, rather than smaller nummia pieces, which is in keeping with a military installation where salaries were paid in those higher value coins and regular provisions were supplied by the *annona*, reducing the need for smaller transactional currency. Many more nummi have been found at Tomis. More than a hundred lead seals, struck by imperial officials in the sixth and seventh centuries, have been discovered at Tomis. Of these, many concern the regulation of trade and the *annona*.[19]

Evidence suggests that provisioning of troops stationed at Capidava was undertaken principally from the Black Sea, hence through Histria and Tomis, cities with well-established long-distance trade connections. Almost a third of the amphorae discovered at Capidava are Pontic, bringing wares to and from the Black Sea ports. However, eight amphorae brought expensive wine from Gaza, and a handful of coins struck at

Antioch has been found at Capidava, as has a Syrian pilgrim flask, all dating from the second half of the sixth century. No coins from Alexandria have been found, and only a single example from Carthage, although there are several examples of North African ceramics, and even a flask from the pilgrim site at Abu Mena in Egypt (compare with Figure 2, p. 26).

The only gold coin discovered at Capidava was an antique forgery, a gold-plated copy of a solidus of Justinian. This would have been made from a cast of a real coin and weighed the same. A glass coin weight, an *exagium solidi* exactly the weight of a solidus, was found in one of the city's granaries, proving that genuine gold coins did circulate. A hoard of fifty-one copper coins buried *c.*580, including many heavy folles of Justinian that had remained in circulation, was found in a layer of debris and ash from a collapsed building. This medium-sized, three-room structure, 'building C1', was in use only through the sixth century, and was destroyed by fire in its last years. This is but one of many hoards from the lower Danube dating to a few years around 580, all of them containing specimens of Justinian's heavy folles, which were withdrawn from circulation shortly afterwards and melted down to strike many more smaller coins to meet urgent military needs. The Capidava hoard was found with sherds of pottery, amphorae and terracotta lamps, all charred by a fierce fire, and with lumps of charcoal, perhaps the drawer or box in which they were deposited. Large areas of the fortress appear to have been destroyed at this time, and when a new perimeter wall was constructed it enclosed a far smaller area. A defensive ditch that ran outside the wall passed through the middle of what had been 'building C1'. The last Roman coin found at Capidava is a follis of Heraclius, struck in 612/13. There is no evidence for life inside the walls of Tomis after this date, until the city re-emerges as Constantia centuries later. The end of urban life at Histria is marked by a similar coin, struck for Heraclius in 613/14. An episcopal basilica built there early in the reign of Justinian, the third on the same site, ceased to function around that time when a small hoard of eleven copper coins was buried outside its northern wall.[20]

Urban life at both Stobi and Heraclea Lyncestis ended at around the same time. We have looked at their churches and splendid mosaics in an earlier chapter (see Figure 6 and Plate 8). By the times these mosaics were rendered at Stobi, the theatre had been abandoned, its materials used for construction elsewhere. Stobi came under threat as early as the later fourth century and was sacked by the Goths in 450. The Theodosian palace was not rebuilt; instead several small houses were constructed over the ruins that in places used the old walls, between which dry-stone walls were erected. Thick layers of rubbish began to accumulate around these houses, mostly ashes, animal bones and pottery sherds. One habitation, the 'House of the Fuller', appears to have collapsed during an earthquake in 518, trapping one inhabitant, a child, whose skeleton has been excavated along with household objects and possessions, including a marble plaque pitted to serve as a gaming board, and kitchen wares, including a bowl that contained forty-six snails. A very similar situation pertained at Heraclea Lyncestis (modern Bitola), whose large episcopal basilica with its magnificent mosaics was abandoned. However, we know the site was later visited by Heraclius, who entertained the Avar chagan with horse races.[21]

Buthrotum, today Butrint, which was connected to Heraclea Lyncestis by the Egnatian Way, has been extensively excavated by a British team of archaeologists, who have concluded that 'excavations strongly sustain the image of dramatic change during the seventh century, showing a major contraction of settlement that lasted until the tenth ... The last people to use coins to make day-to-day financial purchases in Butrint may well have been able to remember the great baptistery and basilica being erected'. The basilica and baptistery were both erected in the first half of the sixth century. The baptistery's mosaic floor features a cantharus urn with vines and peacocks, known also at Stobi and Heraclea. At the same time, the so-called Triconch Palace, which had been rebuilt and expanded twice in the early fifth century, ceased to function as a luxurious home. It was subdivided by dry-stone walls, within which huge dumps of mussel shells have been found, suggesting that part of the site was now used for large-scale processing of shellfish. There is also evidence for blacksmithing, and, interspersed with dumps of rubbish, there are graves.[22]

Butrint was a coastal settlement with access to the Mediterranean and to seaborne trade. For this reason it had a better chance of retaining a semblance of prosperity while other sites declined. And for a while, at least, it did. Levantine glass imports of the fifth to seventh centuries, confirmed by compositional analysis, have been discovered, and amphorae and North African slip pottery have been discovered in seventh-century levels, although far less than earlier. These finds eventually disappear. The landward western fortifications at Butrint, which were most vulnerable to assault, were strengthened from the sixth century onwards. During the eighth century, this fortified complex constituted the city, and even its towers were used as habitations. Around AD 800 a fire engulfed one of the towers, preserving beneath its collapsed tile roof in a layer of ash various white ware jugs from Constantinople, nine smashed amphorae from Italy and the Aegean, and a crate of glass goblets and unworked glass. No rebuilding took place, but rather later the towers were repurposed for shellfish processing.[23]

If we look at the numismatic profiles of prosperous ancient cities like Butrint, produced and refined by decades of meticulous excavation, we see that the use of coinage as a medium for exchange was disrupted during the sixth century and collapsed in the seventh century. Balkan cities, as centres of civic life and government and as foci for daily trade and monetary exchange, ceased to exist, and the populations dispersed across the countryside until they needed to huddle together in smaller fortified compounds. In Butrint the bastion was within the western defences, at Athens atop the Acropolis, and in Corinth on the similarly imposing Acrocorinth. The *polis* of Corinth, in the fifth centuries BC and AD, comprised both the urban core and its rural hinterland. Surveys of Corinthia have turned up an abundance of imported pottery, with the greatest amount of material dating from the fourth to sixth centuries AD, including North African red slip ware and many types of amphorae, all indicators of wealth and vigorous communications between Corinth and the wider world, between the city and its surrounding villas and villages, and between those villages and surrounding farmsteads. The fifth- and sixth-century landscape was vibrant, with evidence for rebuilding after attacks or natural disasters.[24]

At the end of the sixth century and the beginning of the seventh we see a sudden spike in the numbers of coins that are hidden in hoards and not recovered across the whole region. The greatest number of hoards discovered in the Balkans and Greece date from the period 578–86, with a second slightly lower peak between 613 and 616, both periods of Avar and Slav incursions. We have addressed the situation in Athens in the 580s and 590s in an earlier chapter. The Athenians, however, did not abandon their city, since they could withdraw to their acropolis. There is evidence for rebuilding in Athens during the seventh century and even planning for future construction. In the Stoa of Attalus, which had been subdivided into rooms, hundreds of roof tiles that had fallen during the 630s were stacked neatly. In the Baths of Isthmia, a group of rooms was built into the north-western corner with facilities for grinding grain and baking. A basilica was built at the site of the Library of Hadrian around the time Constans II visited Athens. Elsewhere in Greece, newer, more easily defended settlements emerged apart from older cities; for example in western Macedonia, Diocletianoupolis was replaced by Kastoria, now situated on a promontory jutting into a lake surrounded by mountains. Still more impressive was Monemvasia in the Peloponnese, atop a towering island linked to the mainland by a narrow causeway. Old cities gained new walls and new names that reflected their defences. In Thrace, Abdera became Polystylon, its fortifications 'having many columns' repurposed from the old city, and Plotinopolis was renamed Didymoteichon, 'having twin walls'. All of these settlements had a formidable bastion or citadel that made full use of landscape and natural defences.[25]

Inscriptions attest that Thessalonica's land walls were rebuilt in parts in the sixth and seventh centuries, following a series of devastating earthquakes, and sea walls were added in the 630s, when the Arabs had constructed their first fleets and naval warfare became far more common. At around this time Thessalonica's imperial mints ceased functioning. In the sixth century the imperial mint at Constantinople struck perhaps twenty times the number of solidi as that at Thessalonica. Still, Thessalonica's production amounted to between five and six million gold coins in Justinian's long reign, and even more per year under Justin II, before falling sharply after 578. Around one solidus in ten discovered in excavations at sites in Illyricum and Greece was struck at Thessalonica, proving that circulation was principally regional. Silver coins were minted rarely, for ceremonial occasions, at Thessalonica, but copper coins were minted there annually at a separate 'public mint' (*moneta publica*), and distributed regionally until the 570s, after which numbers found at sites including Athens and Corinth drop dramatically, suggesting disruption in circulation, and production, as a result of Avar and Slav invasions. However, there is also a marked peak in total coin production empire-wide in the first two decades of the seventh century, the result of war on multiple fronts and the need to pay troops.[26]

The significance of Thessalonica's ramparts to the inhabitants of the city is captured in a seventh-century mosaic in the basilica church of St Demetrius depicting the eponymous saint, his head surrounded by a glowing nimbus, flanked by the eparch and bishop of Thessalonica. Behind the heads of both men are crenellations from the city's walls, which together with the divinely protected saint defended the city against all attacks (Plate 20). There were many attacks, as is clear from the seventh-century

Miracles of St Demetrius, from Avars and Slavs. On one occasion, as set out in miracle thirteen of a collection by John, who was Thessalonica's archbishop in the first decades of the seventh century, Demetrius appeared on his walls armed with a lance, which he used to drive back a Slav who was climbing a ladder. This was probably in 586. Archbishop John witnessed another siege of his city in 615–16, at exactly the time he was writing the *Miracles*. A further, anonymous collection of miracle stories involving Demetrius was attached to that of John, reporting on Demetrius's activities through the seventh century, including at another great siege in 'the fifth indiction', probably 676. By this time many Slavic groups had moved out of the Avar orbit and settled with imperial permission and according to agreed conditions in the regions around Thessalonica, including in the Strymon passes, and further to the south in the Pindus and Taygetus mountains.[27]

Generally, Demetrius appears unarmed in his miracles, and it was only through his efforts as reported in them and subsequently that he became known as a military saint. Demetrius's transformation, therefore, mirrors that of Thessalonica, from a cosmopolitan port city with an imperial palace complex, complete with hippodrome, triumphal arch and rotunda-mausoleum, to a fortified outpost of civilisation, frequently besieged and defined by its powerful ramparts. It was connected to Constantinople and the Aegean islands by sea. Its harbour continued to function, receiving grain imports that were stored in large granaries and exporting salt from the city's pans.[28]

Italy, the Mediterranean islands and North Africa

An expansion in trade accompanied the Justinianic reconquest of North Africa, but it was curtailed by the calamities of the following period. Quantitative analysis shows that there was a great deal more African red slip ware (ARS) along the North African coast and across the Mediterranean in the sixth century, rising to a peak in the middle years of that century, followed by a steep decline through the following century, with an abrupt end to production shortly before AD 700. Perhaps most telling is the evidence of the Crypta Balbi ceramic assemblages excavated at Rome, which appear to reflect the goods imported for a wealthy monastery in the later seventh and early eighth centuries. More than 100,000 sherds have been excavated from a dump that was created in around 690, of which almost half are pieces of amphorae. Of ceramic types that have been identified, more than 80 per cent originated outside Italy. Of seventh-century amphora types, the most common were from North Africa, notably the smaller *spatheia* type. There were also sherds of ARS and other imported fine wares. In a second dump dated to *c.*720, almost all the ceramics originated in Italy itself. The most distant point of origin appears to have been Sicily. Glass was also found in the two Crypta Balbi dumps. In the seventh-century level there were more than 10,000 glass fragments, around half of which have been identified. Around 3,000 were pieces of glass vessels, and only slightly fewer were from windowpanes. Fifty of the latter have plaster attached, indicating that they had been removed from buildings. Analysed fragments and lumps of raw glass show that there was substantial use of imported natron glass, which was blown locally. Analysis of glass found in the eighth-century dump indicates continued functioning of local workshops, but use of glass with a somewhat different chemical composition – high copper, lead or antimony content – suggesting greater use of recycled glass, including mosaic tesserae.[29]

The sixth-century walls of Leptis Magna, Libya, enclosed around 28 hectares, including the port, but left much of the earlier Roman city outside. The port city continued to manufacture globular amphorae and to trade in olive oil, the region's greatest export, on a greatly diminished scale towards the end of the seventh century. The Libyan valleys to the south of Leptis showed a marked decline in settlement and productivity after *c.*500. Some coastal cities like Hadrumetum were given new fortifications by their Muslim conquerors to guard against attacks launched from the Mediterranean islands to the north. For similar reasons, new Roman fortifications appear on the southern coast of Sicily between *c.*675 and *c.*750, including at Selinunte, Agrigento and Licata. At Selinunte in south-western Sicily a square fort with sides of more than forty metres (*c.*45 yards) was constructed to house a garrison monitoring the sea lanes. Far larger was the monumental citadel of Kassar, forty kilometres (twenty-five miles) inland from Agrigento, whose circuit wall was punctuated by eleven polygonal towers and enclosed 90 hectares (222 acres). Kassar appears to have been constructed to offer the agricultural workers of Sicily a refuge in the event of Arab assaults. Field surveys in various parts of the island, including the Platani valley between Agrigento and Kassar, and the hinterlands of Syracuse and Catania, have identified unfortified villages ranging in size from one hectare to more than six hectares (2.5 to 15 acres), with clusters of housing, agricultural storage and processing facilities, and often a church, which appear to have been occupied during the eighth and ninth centuries. The same surveys have identified villages of the same period that shrank or were abandoned.[30]

With the loss of Egypt and later North Africa, Sicily became the empire's principal grain producing and exporting region. From the fourth century, when the ships from Alexandria were diverted away from Rome towards Constantinople, Sicily had sent much of its grain northwards. The wealth this generated had supported the construction of luxurious villas, most famously the Villa del Casale near Piazza Armerina, built in the first part of the fourth century, which we considered in an earlier chapter. During the seventh century, many of that villa's rooms were filled with debris, covering and preserving the mosaics and marble opus sectile. Wall murals and marble revetments fared less well as the structure was altered to accommodate artisan production, agricultural storage and processing, and also burials. A boundary wall was built around the complex some time before use ended in the eighth century.[31]

An imperial mint operated at Syracuse between the later seventh century and 878, when the city fell to the Arabs. Gold and bronze coins were minted to service the administrative and military needs of the island, which were substantial, and its economy remained highly monetised. Constans II chose Syracuse as his headquarters in the west for its proximity both to North Africa and to southern Italy, over which he had recently reasserted control and imposed a burdensome regime of taxation to fund his war preparations. Both Crete and Cyprus were rather closer to Egypt than was Sicily, but for that reason also far more vulnerable to Muslim attack. Crete and Cyprus had a good number of crops and products in common: timber, fruit from orchards, perhaps cotton, certainly mastic, although the Aegean island of Chios already dominated that trade. Both produced wine and olive oil in abundance, but also honey and wax, and purple dye. Surveys and catalogues have recorded eighty churches on Cyprus that functioned between the fifth and seventh centuries, and sixty churches in Crete, including at Gortyn.[32]

Gortyn is remarkable for the state of preservation of its antique water supply system. The Roman city was supplied with water from the second century AD through aqueducts linked to springs at Zaros on the slopes of Mount Ida, fifteen kilometres (nine miles) to its north-east. A siphon bridge allowed the aqueduct to cross the Mitropolitanus River. Water towers fed water into clay pipes that ran under the city's streets, supplying the great *nymphaeum* and large public baths until the end of the fifth century, when the pipes were damaged and fell out of use. It is tempting to attribute this to earthquake damage. As we shall see, Crete was affected substantially by seismic activity between the fourth and seventh centuries, when the western half of the island was raised between one and nine metres higher above sea level as a result of tectonic movement ('coseismic uplift'). The fortunes of Gortyn are bookended by massive earthquakes in 365 and 670. Although there are signs of occupation into the eighth century, the city never fully recovered before the Arab conquest in 828.[33]

The last centuries at Gortyn saw major substantial infrastructure projects. A new axial street was built, running west from the city's ancient *embolos*, and the water supply system was restored in a new form. More than fifty rectangular cistern fountains were built of stone and brick at ground level, fed directly by the aqueducts, which discharged water through tapped lead pipes. Four larger tanks, without taps, served as water storage cisterns, including one situated between the baths and the basilica church of St Titus. At the same time the siphon bridge was rebuilt and two new decompression chambers were added to the system, to the north-east and east of the city, to reduce water pressure and redirect water through new pipes, formed from *spatheia* amphorae, to the cistern fountains. This efficient system served what excavators have called the 'Byzantine quarter', at the heart of which stood a complex of houses and public buildings, built between the sites of the temple of Pythian Apollo and the praetorium. Four acclamatory inscriptions on columns right outside the praetorium have been interpreted as suggesting that Heraclius was behind the city's restoration, including the city's walls, which encompassed around 80 hectares (*c.*200 acres). By the end of the seventh century much of this lay ruined or abandoned, and part of the praetorium had become a monastery. The episcopal church of Gortyn was certainly in use in these last years, although apparently remained in a semi-ruined state after the 670 earthquake.[34]

Cyprus was strategically vital in the seventh-century war between the Romans and the Arabs, and the southern and eastern coasts of the island became important staging points for attacks on Arab holdings in Egypt, Syria and Palestine. The numismatic record suggests that Cyprus saw a massive influx of coin from Heraclius to Constans II, surely payments to troops. Hoards appear as evidence for instability, including at Larnaca, and these attest to one or other of the two Arab naval raids on the region in 649 and 654. These mark the period prior to the political and economic agreement known variously as the condominium, the period of neutrality, or the treaty period, where Cypriots were required to send tribute in two directions, to Constantinople and Damascus. The annual total was 7,000 solidi, and there is no suggestion that the Cypriots failed to pay. Al-Walid, the sixth Umayyad caliph, was later said to have spent the gold and silver sent on eighteen Cypriot ships, as well as seven years of taxes from Syria, on the construction of the Great Mosque of Damascus. The agreement required that Cypriots remained productive, but the economy was increasingly less monetised, with the number of coins in circulation falling by a factor of ten after 700.[35]

Survey work across Cyprus has been extensive and intensive, producing voluminous information on fluctuating artefact densities across the landscape, and leading some more radical archaeologists to question the usefulness of identifying 'sites' at all. Still, the distribution of artefacts does indicate how settlement patterns changed, and that Cyprus was clearly an island packed with smaller and larger farmsteads, villages, towns and cities into the seventh century, but with far fewer after that. The northern coast of Cyprus is greatly extended by the Karpas peninsula, which was never heavily urbanised. Partly for that reason, patterns of settlement and agricultural production, principally animal husbandry, do not appear to have changed substantially after the seventh century. Survey and excavation at Koutsopetria, to the east of Kition (modern Larnaca) on Cyprus's southern coast, have together identified a harbour town of some wealth and size occupied through the late Roman period, furnished with a basilica church and supplied with roof tiles from the island's interior, and Cypriot red slip ware produced in the island's west, found alongside both African and Phocaean red slip ware, imports from North Africa and Asia Minor, and locally produced amphorae used to export locally grown and processed olives and oil. There are few signs of occupation after 700.[36]

To the west of Larnaca, at Kiti, a remarkable church attests to the wealth of the region in the age of Justinian, following the arrival of citizens from Kition who were, according to local legend, fleeing Arab raids. It is more likely that the harbour at Kition had silted up and that earthquake damage was so extensive as to recommend relocation rather earlier than the seventh century. The Church of the Panagia Angeloktiste, 'Church of the Virgin, founded by Angels', as it stands today is of the eleventh and twelfth centuries, but it incorporates a far earlier apse, of the fifth century, with an exquisite mosaic, added in the later sixth century. The mosaic depicts Mary holding the Christ Child, standing on a jewelled podium flanked by the archangels Michael and Gabriel, who have wings of peacock feathers and hold golden staffs and blue glass orbs, their insignia of high heavenly office (Plate 21).[37]

Maroni Petrera, to the west of Kiti, had a large basilica church built towards the end of the fifth century and abandoned in the middle of the seventh. Glass fragments from the site are of 'Levantine I' natron glass, probably exported from Apollonia-Arsuf and blown locally. The nearest major city, Amathus, is today an archaeological site in a coastal suburb of the city of Limassol. There are signs across the excavated site of prepara-tions for war and of destruction during the later seventh century. Around this time a long fortification wall was built across the southern face of the city's acropolis. Its western gate appears to have been attacked and burned twice, in a short span, and a basilica on the hill abandoned. Similarly, in the lower city, the Church of St Tychon shows signs of destruction in the seventh century, and the large cathedral church, built in the second part of the fifth century, was destroyed at the time of the Arab raid of 654, accord-ing to a coin in the burned layer. Bodies showing evidence of violence have been excavated from a mass grave at the city's eastern cemetery. Exactly what combination of earthquakes and assaults was responsible for the widespread destruction is unclear, but it all took place in a short period between 650 and 680. It is also apparent that the city was still occupied in the eighth century. The site of the basilica on the acropolis became farmed land, and the cathedral was subdivided to become small dwellings, work-shops and a storeroom. The Church of St Tychon appears to have been rebuilt and lead seals suggest an ecclesiastical presence in the city through the eighth century.[38]

The 'Dor 2006' shipwreck, one of the twenty-six shipwrecks discov-ered in the Dor/Tantura lagoon immediately north of Caesarea Maritima (Map 5), attests to regular trade between Cyprus, Asia Minor and the Levant into the seventh century. Although the ship's main cargo has not been definitively identified, marble slabs have been seen nearby, and the ship's size, at 25 metres (82') long, would have been sufficient for their transportation. High-quality ceramics have been identified, including both Cypriot and Phocaean red slip ware. Four wheel-thrown Cypriot bowls were found, similar to one buried at Salamis some time after AD 643 and to others known from sites on the island abandoned or destroyed in the mid-seventh century. A cooking pot found with the bowls appears to have been manufactured in the Levant, as were storage jars of the 'Gaza' type. These ceramics surely belonged to the ship's galley, and Egyptian bricks from the ship's hearth have also been found. Wine may have been stored on board in an Egyptian amphora and served in jugs and juglets from southern Palestine.[39]

Material evidence proves that by AD 700 far less olive oil and wine were transported, very little over great distances, and that fine red slip ceramics and mass-produced terracotta lamps were no longer manufactured or exported. Glass-blowers had begun to use local materials as well as to recycle fragments and tesserae, no longer importing natron glass from Egypt and the Levant. Marble was extensively recycled and monumental building is rarely identified. There is evidence, however, that regional trade continued into the eighth century, linking Cyprus to the Levant and the southern coast of Asia Minor. The Tantura F shipwreck, also from the lagoon at Dor, has been dated to the period *c.* 650–750. The boat, once 16 metres (52') long, is well preserved and appears to have been transporting fish and fish sauce in ceramic vessels. According to petrographic analysis, several ceramics on board were made of silt from the Nile delta. However, twenty globular amphorae were manufactured at Cyprus or the coast of Asia Minor immediately to the north. Within eight of these were found the bones of small fish, tilapia, between 2 cm and 7 cm (1–3") long, which once swam in the Jordan river system. No fish processing facilities have been identified in this region, but with the loss of the supply of fish sauce and its residue, a boney fish paste (*allec*), from North Africa, one can imagine that demand was now being met regionally.[40]

A transformed environment

The archaeological record shows beyond question that settlement patterns shifted at the end of our period, when cities declined in size, new fortified compounds were constructed, and finds of Roman coins ended for some time, in every part of the empire. The vast majority of Roman coins are discovered in cities, with vigorous circulation well into the sixth century when, quite suddenly, the volume of coins declines or the presence of Roman coins ends. In some cases, coins indicate an abrupt end to life in a city that had existed for many centuries. In others, life continued, and exchange was conducted in other ways. Not all cities suffered similarly, however, and there are significant regional variations in data proving that some cities flourished, at least for a while, as a result of new or enhanced military and administrative functions. Moreover, cities cannot be understood without appreciating their complex interactions with their hinterlands.[41]

The greatly diminished empire now had far fewer people and far less land under cultivation. Pollen analysis, which allows us to reconstruct aspects of the vegetation and cultivation history of past cultures, demonstrates a clear rupture in the seventh century. Until that time and for up to a millennium across Greece and Anatolia, as well as in Cyprus and the southern Levant, pollen records are dominated by 'anthropogenic' plants, which were grown for and by humans, principally tree crops (olives, nuts, fruits, etc.), vines, cereals and grazing grasses. These all decline precipitously across Anatolia for around three centuries after c.600, suggesting a decline in total cereal production of around 80 per cent, and an 85 per cent drop in tree crop cultivation. These declines were even more substantial in south-western Anatolia, where cereal and tree crop pollens indicate cultivation at 7–9 per cent of earlier levels. The decline in grazing was less pronounced, and while the amount of grass is not a direct indication of the number of animals grazing, pollens suggest only one-third of the grasses were now available to them.[42]

In Greece, pollen records from Lerna, to the south of Corinth, indicate a spike in the cultivation of cereals between AD 400 and 600, followed by a rapid decline. Extensive cereal production in Attica ends still earlier. This change in land use must be attributed to invasions by Avars and Slavs, and natural disasters and disease, rather than to climate change. In Asia Minor, the hinterlands of heavily populated coastal cities were farmed most intensively. At Miletus in Lycia, there is evidence for uninterrupted agricultural activity from the fifth to the fourteenth century, although not always at the same intensity. The interior uplands of Anatolia as far as Tarsus were farmed less intensively, and this land was abandoned and rewilded, which could happen quite suddenly. The end of cultivation across Cappadocia can be pinpointed to the 670s, and therefore attributed to the devastation caused by warfare at that time. Varves (annual layers of sediment) identified in cores taken from the bed of Lake Nar show a sudden and decisive shift in the types of pollen being deposited. In little more than a decade, between AD 664 and 678, pollens related to human activity were replaced entirely by pollens from woodland and secondary scrub, which grows very quickly on abandoned agricultural land. Broader archaeological surveys suggest that between c. AD 600 and c.800, the number of settlements in south central and south-west Anatolia fell by at least 70 per cent.[43]

Greater continuity in agricultural activity is evident the closer one gets to Constantinople. However, there are indications that cereal and fruit cultivation gave way to pastoralism in the coastal regions of Bithynia. In an age of frequent raids by land and sea, with attendant destruction or seizure of land and provisions, pastoral farming makes sense. Animals are easier to move and flocks simpler to replenish than orchards, since olive, nut and fruit trees all take many years to mature. Still, enough grain was produced to support the greatly depleted population of Constantinople; and lead seals struck by *kommerkiarioi*, officials involved in moving and storing goods, suggest that some grain was imported from Sicily. More broadly, lead seals attest to radical changes in provincial administration and defence at the end of our period. The army could no longer be centrally financed or effectively provisioned through the *annona*, leading to the establishment of a new system for organising and financing provincial defences, which came later to be called the *themata*, or 'themes'. At the same time, an absence of coins minted in the later seventh century in cities across Asia Minor, Anatolia and the Balkans demonstrates conclusively that the apparatus of state was seriously degraded. Provincial mints were closed and coins no longer made their way beyond the walls of Constantinople. Taxation revenue could no longer be raised in coin and the difficulty in transporting levies in kind meant that these were spent locally.

The new administrative structures also reflected the determination by many Christian Romans to abandon city life and live in smaller, lower density, and more easily defended communities. As we have seen, this accelerated a decline of urban institutions, including curial government, the classical system of education and the literary culture that it supported. The quality and abundance of the material culture enjoyed by Romans over centuries declined dramatically, with the disappearance of imported ceramics and fine wares (Phocaean and North African red slip), glass (Levantine and Egyptian natron glass), and of wines from Crete and olive oils from Tripolitania, goods of distinctive and distinguished quality shipped in identifiable transport vessels (amphorae). Ceramic vessels were never fully replaced by barrels, because wood was scarce in the Mediterranean. However, terracotta oil lamps ceased to light homes, churches and public buildings, and were replaced by candles. The grandest buildings, public and private, basilicas and villas, all disappeared. They were destroyed and not rebuilt, or subdivided and repurposed as dwellings and graveyards, as sites of small-scale industry or processing, or as rubbish tips.

At the Wadi Faynan ore field, today in Jordan, eight thousand years of copper mining and smelting ended quite suddenly in the middle of the seventh century. A sedimentary core taken from a dried reservoir bed shows that environmental pollution levels had peaked in the fifth and sixth centuries, the result of a growing demand for copper, surely for the production of coinage. Pollution levels then fell away rapidly, with a distinct sedimentary deposit free of new pollutants representing the eighth century. Copper mining and smelting ended when the Arabs breached Rome's desert frontier. As the smelters were quenched and polluting aerosols dispersed, clean air swept across an apocalyptic landscape. Truly it had been a 'Gehenna of fire' where generations had sacrificed their health and children to the Roman metal economy.[44]

11

APOCALYPSE AND THE
END OF ANTIQUITY

In the year 472, Mount Vesuvius erupted, devastating much of Naples and the rich agricultural lands of Campania. 'Nocturnal darkness overshadowed the day, and it showered the whole of Europe with fine particles of dust.' The volcano erupted again in 512, with devastating local consequences that were addressed by Theoderic 'the Great' and captured in a letter preserved by Cassiodorus. 'How strange', he observed, 'that one mountain alone should thus terrify the whole world! Other mountains may be seen with silently shimmering summits; this alone announces itself to distant lands by darkened skies and changed air.' These two eruptions occurred shortly before and somewhat after the deposition of the last western emperor. They were events of local and regional significance, whose effects could be mitigated by tax relief for those who cultivated devastated fields. They were also commemorated in liturgical calendars, including at Constantinople, whose citizens 'celebrate the memory of this fearful ash on 6 November'. However, they had no global or enduring environmental impact.[1]

Volcanoes, climate forcing, plague and earthquakes

The eruption of a volcano of which Cassiodorus had no knowledge, and which has still not been firmly identified, was the cause of a far greater dust veil, a cloud of ash and sulphur dioxide, that was seen across the whole northern hemisphere in 536.[2]

> The Sun, first of stars, seems to have lost his wonted light, and appears of a bluish colour. We marvel to see no shadows of our bodies at noon, to feel the mighty vigour of his heat wasted into feebleness, and the phenomena which accompany a transitory eclipse prolonged through a whole year. The Moon too, even when her orb is full, is empty of her natural splendour. Strange has been the course of the year thus far. We have had a winter without storms, a spring without mildness, and a summer without heat. Whence can we look for harvest, since the months that should have been maturing the corn have been chilled by Boreas? How can the blade open if rain, the mother of all fertility, is denied to it? These two influences, prolonged frost and unseasonable drought, must be adverse to all things that grow. The seasons seem to be all jumbled up together, and the fruits, which were wont to be formed by gentle showers, cannot be looked for from the parched earth.[3]

Contemporary eyewitness accounts report on the same phenomenon from as far afield as Ireland, Syria and China, and Procopius, who was then in Italy with Belisarius reported:

> It came about during the year that a dread portent took place. During the whole year the sun gave forth its light without brightness, like the moon, and it seemed extremely like the sun in eclipse, for the beams it emitted were not clear like those it usually makes. From the time that this happened men were neither free from war nor pestilence nor any other thing that brings death.[4]

Signals of this volcanic eruption have been identified in ice cores and in tree rings. Ice-core data from various glaciers in Greenland show greatly elevated levels of acidity and impurities in the period around 536. Tree-ring data from forests across the northern hemisphere – although not in the Mediterranean basin, where such well-established forests are not present – show sharp and immediately diminished growth in the following year, the result of rapid cooling and drought. The same peat bog

in north-western Spain, where elevated levels of lead pollutions are associated with Roman metallurgy, shows a huge spike in ash deposition at
exactly this time. At the same time in parts of the Near East sedimentation
records show instances of severe flash flooding. Together, the data suggest
that the volcanic eruption took place in the northern hemisphere.[5]

The latest research has identified at least one more huge eruption in
540 that has left signals in ice cores drilled in Greenland and Antarctica,
and which was probably a tropical volcano. Climate modelling suggests
that together these massive volcanic eruptions resulted in 'climate forcing',
namely a rapid cooling of the earth by several degrees Celsius, and that
it took up to ten years for the sulphur dioxide aerosols to disperse fully.
Lower summer temperatures, diminished sunlight and drought may have
caused massive crop failures and, consequently, famine.[6]

A further volcanic event of almost equal magnitude is recorded in
the same data sets, although it appears to have received far less attention
from contemporary writers. In 626, as Heraclius's existential war with
the Persians reached its climax, a volcano in the northern hemisphere
erupted. Historians of China have associated the period of rapid cooling
that followed with the collapse of the eastern Turkic khanate, as freezing
temperatures killed countless horses and flocks of sheep. Theophilus of
Edessa observed that in Syria, 'There was an eclipse of the sun that lasted
from October [626] until June [627], that is for nine months. Half of
its disc was eclipsed and the other half not. Only a little of its light was
visible.' Reports from Ireland and Japan confirm that the phenomenon
was observed across the northern hemisphere, and ice-core data from
Greenland show a spike in the deposition of volcanic sulphate impurities
almost as high as that which resulted from the 536–40 events, with a corresponding drastic reduction in tree growth.[7]

These periods of 'climate forcing' coincided with the most significant, and also the most controversial, historical phenomenon of the sixth
century: the advent of bubonic plague. Whether there is any direct or
causal link between the global climate event and the arrival of plague in
the Mediterranean basin has not yet been determined. Plague was first
reported in Pelusium, at the eastern edge of the Nile delta, from which
it spread rapidly to Alexandria and from there reached Constantinople
in the year 542. Procopius, who provides an account based on Thucydides's description of the plague in Athens a millennium earlier, states
that the disease originated in Egypt. Evagrius, who lived through the first

epidemic as a child, had contracted and survived the disease, but had lost many family members; he believed that it originated in Ethiopia. He, like Procopius, was well aware of Thucydides's ascription of the Athenian plague to that region, but also that the diseases were not identical. Modern scholars have long suggested that the disease originated in Asia, perhaps India, but the fact that it reached both China and Persia later than the Mediterranean, although both were more active trading partners with India than were the Romans and were closer to it, supports the views of contemporaries that the plague emerged from Africa, possibly the great lakes of Central Africa, where the disease remains endemic. Very recently, DNA analysis of plague victims from sixth-century cemeteries in Bavaria has suggested that the plague originated in Central Asia or China, despite an absence of historical evidence for a contemporary Chinese pandemic. Equally, there is no written evidence for the bubonic plague in southern Germany at this time, although the bodies of plague victims from which the aforementioned DNA was extracted demonstrate that it was there.[8]

The pathogen responsible for the epidemic disease known as bubonic plague is the bacterium *Yersinia pestis*. Bubonic plague is not spread by people, but requires a vector, the tropical rat flea (*Xenopsylla cheopis*). Fleas were borne on a host, believed to be the black or ship rat (*Rattus rattus*), since this was the case in the second plague pandemic of the fourteenth century AD. The bacterium *Y. pestis* is transmitted to humans when an infected flea leaves its rat host, generally when the rat dies, and, seeking another source of blood, bites a person. For this reason, bubonic plague generally spreads quite slowly and persists within a locality for some time. For bubonic plague to spread over great distances relatively quickly requires a functioning and efficient system for transporting rats and their fleas, and an environment in which fleas and their hosts can flourish. Rats do not otherwise move very far or very fast. The grain ships that sailed between Alexandria and Constantinople provided an ideal environment, as did the merchant ships that plied the coasts of North Africa, Syria, Lebanon and Asia Minor. However, the speed with which the pandemic travelled is still remarkable, far swifter than the third plague pandemic that began in China in 1855. Reports suggest that plague had reached Britain in 544, and DNA analysis has now identified plague victims in a sixth-century cemetery at Edix Hill near Cambridge in eastern England.[9]

The first symptom following infection by bubonic plague is a fever, followed a few days later by the most obvious diagnostic manifestation of the disease, that is buboes, inflammations of the lymphatic nodes in the armpits and groin, as well as on the thighs, neck and near the ears. These are reported by numerous sixth-century sources, including Procopius, who writes of doctors scrupulously performing autopsies and examining swollen glands. The same sources suggest that this outbreak of epidemic disease was immediately devastating. Procopius writes of the death rate rising to 10,000 deaths each day as the epidemic raged in Constantinople through three months in the spring of 542. John of Ephesus, who was travelling from Constantinople to Alexandria as the disease arrived in Syria and Palestine, offered a portrait of urban and rural devastation, with people devoting themselves entirely to transporting corpses and digging graves while fields lay abandoned and crops unharvested. 'Crops of wheat ... were white and standing, but there was no one to reap them and store the grain. Vineyards, whose picking season came and went, shed their leaves, since winter was severe.'[10]

In the longer term, archaeological evidence suggests that plague was not equally devastating across the empire, and that it affected rural areas rather less than the cities. Furthermore, there is an absence of documentary and material evidence to support the devastation of the initial outbreak reported by our writers. For example, since we know that plague entered the eastern empire through Egypt, the absence of direct references to bubonic plague in the Egyptian papyri is quite remarkable. Papyri do however reveal a rapid increase in taxation rates during the middle years of the sixth century, reaching three times the rate of the 520s under Justin II. This could be related to crop failures and loss of manpower, the results of both climate events and the onset of plague. More problematic is the absence of rat bones from excavations across the eastern Mediterranean. If hordes of rats were dying and passing their infected fleas to humans one would expect to find abundant archaeological evidence. Rat bones are fragile, certainly, but they do appear in archaeological contexts in some numbers, just not in those associated with the first plague pandemic. It may be, therefore, that rats were not the only or even the principal hosts for infected fleas, and that transmission during the first pandemic was quite different to that during the 'Black Death'. Even more problematic, pending further analysis of human remains, is the fact that there has not yet been a single body of a plague victim identified in the eastern Mediterranean, and only thirty in the western Mediterranean (southern France, northern Spain) and northern

Europe (Austria and France, as well as the aforementioned bodies in Britain and Germany).[11]

For two centuries after its advent, according to our written sources, waves of plague swept through the eastern empire and the broader Middle East, striking the same locations every six to twenty years. Evagrius noticed that outbreaks coincided with the first or second year of the fifteen-year indictional tax cycle. While not exact, this does agree with the frequency of outbreaks reported at Constantinople every eleven to seventeen years, or on average every fourteen years. Modern scientific analysis has demonstrated that the plague pathogen mutated frequently, retaining its potency. However, new spikes in mortality relied on the presence of large numbers of people without immunity. In truth the disease did not go away and return in waves, but lay dormant in the cities waiting for a population with immunity to die of other causes, and for a community without immunity to regenerate by birth and immigration, when it could explode again. Those who survived infection enjoyed robust immunity, but their children and new neighbours did not.[12]

After 542, Constantinople suffered contagion again in spring 558, in 573–4, in 599 and in 619. These are only the recorded episodes, which coincide with five of the first six known outbreaks. It is possible that Constantinople was spared during the fourth outbreak, which is reported in Rome and Ravenna in 591 and Antioch in 592, but is unmentioned in sources for New Rome. After 628, with the end of the *Easter Chronicle*, we lack an extant contemporary historical source for Constantinople, although records for lands to the east remain good. Further outbreaks of plague are reported in Syria or Palestine during 626–8, 639, 669–73, 683–7, 698–700, 704–6, 713–15, 718–19, 724–6, 732–5 and 745–7. The 698 outbreak is reported to have lasted for four months at Constantinople, and that of 747 to have endured for a year. This last recorded outbreak at Constantinople, which Theophanes attributed to the removal of icons by impious rulers, struck 'the imperial city in the fifteenth indiction'.

> All of a sudden, without visible cause, there appeared many oily crosslets on the men's garments, on the altar cloths of churches and on hangings. The mysteriousness of this presage inspired great sorrow and despondency among the people. Then God's wrath started destroying not only the inhabitants of the city, but also those of all its outskirts ... In the spring of the first indiction, the plague intensified and in the

summer it flared up all at once so that entire households were completely shut up and there was no one to bury the dead ... When all the urban and suburban cemeteries had been filled as well as empty cisterns and ditches, and many vineyards had been dug up and even the orchards within the old walls to make room for the burial of corpses, only then was the need satisfied.[13]

The use of mass graves and other extraordinary burial practices, including covering bodies with lime, may account in part for an oddity that had perplexed historians, namely the absence of plague-induced spikes in the record of funerary inscriptions. The epigraphic record may not reflect the scale of mortality because individual inscriptions were not made. In this scenario, if it is accurate, absence of evidence can be read as evidence of absence.[14]

Estimates of how many died of bubonic plague in the sixth to eighth centuries are all guesses and vary wildly, ranging from minimal disruption to the loss of 50 per cent or more of the empire's total population. A drop in total population of between 10 per cent and 25 per cent would have had an enormous impact on urban life and education, patterns of settlement, agricultural production and the means to raise taxes. A loss greater than 25 per cent would have been devastating. From then on, much of the eastern Roman world would have suffered from a chronic dearth of manpower, and its ability to sustain an effective army and to rebuild after natural disasters would have been greatly diminished. There is no evidence that this was the case. Indeed, the Roman ability to rebound and rebuild after invasions, fires and earthquakes remained remarkable for more than a century after the advent of bubonic plague. It is only for the later seventh century that there is solid evidence for population decline and the abandonment of cultivation.

If 'climate forcing' and plague were exceptional features of the sixth and seventh centuries, earthquakes were a more familiar torment. The fact that we may read a great deal about earthquakes in written sources of the period is not conclusive evidence that they were more frequent or of greater magnitude than at other times. Writers were inclined to report on disasters at home and abroad, both as sensational news and because, frequently, they wished to report the generosity of benefactors, notably emperors, in supporting rebuilding. However, the phrase 'Early Byzantine Tectonic Paroxysm' has been coined to describe an intense period of tectonic activity that commenced in the middle of the fourth century and lasted into the middle of the sixth. Through two hundred years, there appears to have been a narrowing of the Mediterranean basin, as the African and Eurasian tectonic plates collided, resulting in strong seismic activity and substantial coseismic uplift to coastlines in many places. Elevated coastlines are indicated by erosion notches and barnacles located significantly above sea level. As an extreme example, western Crete was raised by as much as eight metres above sea level, probably by a single massive earthquake in the fourth century that sent tsunami waves across the whole Mediterranean. At the northern end of the Adriatic, in contrast, there are signs of coseismic subsidence (land being forced down).[15]

Constantinople sits on the north Anatolian tectonic fault, one of the most active large faults, so there were (and are) frequent earthquakes, one at least with a concomitant tsunami. In our period, significant earthquakes are recorded at Constantinople in 400, 402, 417, 437, 447, 450, 477, 487, 525 and 526, 533, 542 and 543, possibly 546, 548, 554, 557, 580, 583 and 611, before our contemporary sources disappear. The earthquake of 557 brought down many buildings in Constantinople, including the dome of Hagia Sophia, and appears to have caused a tsunami that is identifiable in the sedimentary structures of the Theodosian harbour (Yenikapı). There were many equally devastating earthquakes elsewhere in the empire through these centuries. To cite just one example, an earthquake on 9 July 551 was felt strongly in Palestine, Mesopotamia, Arabia and Syria, and more weakly as far away as Alexandria. Destruction was reported in Tripolis, Byblos, Tyre, Beirut, Sidon and Petra, as were aftershocks and associated quakes and waves in Greece and Bithynia.[16]

Earthquake damage was frequently identified as evidence of God's wrath, on which matter Malalas, a native of Antioch, was insistent. He reported on a fire and devastating earthquake in that city in apocalyptic terms.

> Great was the fear of God that occurred then, in that those caught in the earth beneath the buildings were incinerated and sparks of fire appeared out of the air and burned anyone they struck like lightning. The surface of the earth boiled and foundations of buildings were struck by thunderbolts thrown up by the earthquakes and were burned to ashes by fire, so that even those who fled were met by flames. It was a tremendous and incredible marvel.[17]

Antioch was struck by devastating earthquakes in 526 and 528. In March 529 a band of Arabs – pre-Muslim Arabs, engaged by the Persians, then fighting Justinian's troops – attacked Antioch's suburbs, plundering and looting. It was only in 538, six years after the 'eternal peace' was signed between Justinian and Chosroes, that Antioch's cathedral rose again. An inscription from the same year attests that Flavius, the *comes orientis*, had restored the great public baths. Very soon afterwards Chosroes invaded Syria, sacking Aleppo and falling upon Antioch, which many of its wealthier inhabitants swiftly abandoned. As we have seen, Chosroes's army overwhelmed the local circus factions before many Antiochenes were transported to Persia and settled at 'Antioch of Chosroes', a living symbol of his great victory. Justinian's response was to send money and builders to the old Antioch, and from late in 540 a whole new city was constructed, above the ruins of the old, but rather smaller city. The later colonnaded central street was built atop the old, and although it was somewhat narrower, it was still 85 feet wide, lined with shops and residences. Evidently, Justinian agreed with Libanius on the significance of the *embolos*. However, the island where the imperial residences had stood was left abandoned, behind a new circuit wall, and excavations suggest that the theatres and baths were not fully restored.

Looking to a fresh start and a prosperous future, the city was devastated again when the plague arrived in 542. It continued to suffer through the remaining decades of the century, being struck by earthquakes in 553, 557 and 588, and was once again sacked and burned by the Persians in 573, still under an aged Chosroes. Ultimately, it fell to the Persians in 610, returning to Roman control briefly before being captured by the Arabs in 637. Still, the city was not always firmly held. As we have seen, in 684, Abd al-Malik once again 'subjugated Antioch'. The various destructions of Antioch have preserved a good deal of the city under the rubble for archaeologists to uncover. Most notable are the more than 300 mosaics discovered during excavations between 1932 and 1939 (see Plate 4). The mosaics of Antioch are well preserved largely because the site ceased to be a great city, unlike Constantinople, where constant redevelopment has destroyed most of them, and where what remains is largely beneath the modern city. Several panels from the 'House of the Bird Rinceau' date from *c.*525/6 and 540, between the devastating fire and earthquake and the sack of the city by the Persians (Plate 22). Fecund vines and diverse birds would have spoken to Hellenes and Christians alike of Dionysus the vintner and drinker and of Hera, whose peacock was a counterpart to Zeus's eagle, or of the wonders of resurrection and creation, the fruit of the vine becoming the blood of Christ and the feathers of the bird moulting each season only to grow back in equal beauty. For many, perhaps most Antiochenes, there was no contradiction.[18]

The Muslim capture of Antioch was, therefore, just the latest disaster that its citizens endured. The city centre remained in use and new structures were placed directly on to the pavement of Justinian's reconstructed *embolos*. While very few new Roman coins were put into circulation after the Muslim capture of the city, existing coins remained in circulation in Antioch alongside new 'Arab-Byzantine' coins. These continued to facilitate trade in the city, although over time the silting up of the river Orontes prevented easy access to the Mediterranean. The patriarch lived in exile in Constantinople until 742, but then was welcomed back. Later reports suggest that Abd al-Malik stationed Slavs (*saqaliba*) at Antioch, and that Marwan II preserved the city's walls, when he destroyed those of other cities, so that he might use the city as a refuge during the third *fitna*.[19]

Many of the administrative functions performed at Antioch were transferred to Emesa (modern Homs), including the minting of coins and the collection of taxes. Emesa was already a vibrant city. Life there at the end of the sixth century was vividly portrayed in the *Life of St Symeon the Fool*, which tells of the saint entering the walled city through its gate, a dead dog tied to his foot, to the jeers and blows of local children exiting school. In the *Life* we learn of the city's churches and baths, of the mimes still playing at the theatre and the activities of the circus factions, of the taverns and shops that lined the great colonnaded street. There is no evidence that this all ended when the city surrendered to the Arabs in 637. Indeed, a tradition emerged, often repeated but hard to prove, that the Church of St John the Baptist was divided, with one half of the building serving as the city's Friday mosque.[20]

From *polis* to *madina*

A great deal of recent work has elucidated the transition of urban life in Syria and Palestine, the region known to the Umayyads as Bilad al-Sham. Two mosaics found today in Jordan illustrate how this world of cities was transformed between the middle of the sixth century, when a remarkable map was created at Madaba, and the first decades of the eighth century, when a floor mosaic was installed at the Church of St Stephen at Mefaa, today Umm ar-Rasas. The Madaba mosaic map famously depicts Jerusalem as it appeared in the age of Justinian, a city among its neighbours, near and far, as if viewed from an airship approaching the coast of Palestine (Plate 23). Jerusalem is shown in remarkable detail and given the epithet 'Holy City'. To the left, the viewer sees the cities lining the River Jordan as it flows into the Dead Sea, which is placed centrally, above the dominant city of Jerusalem. To the right, the viewer sees the cities of the Nile delta, including Pelusium, where bubonic plague was identified a year before it swept all before it in Constantinople. Cities are identified by inscriptions and illustrated by walls and towers or a known building.[21]

In G. W. Bowersock's words, 'What survives for our contemplation at Madaba is the self-representation of the Christian near east in the sixth century ... in which cities saw themselves in relation to other cities'. This, he concludes, is a contemporary portrait of the landscape at the time of Justinian. The worshippers of Madaba who looked upon this map understood that Jerusalem was the centre of their world, but as citizens they also placed themselves and their city in a traditional relationship with the imperial capital. Elsewhere in the city, in the so-called Hippolytus Hall – named for the sixth-century mosaic depicting the ancient myth of Hippolytus – a personification of Madaba sits alongside two more *tychai*, the enigmatic 'Gregoria', and the more obvious 'Rome'. Gregoria may be Antioch, whose most famous sixth-century patriarch was named Gregory. Rome is certainly Constantinople (Plate 24).

It is hard to escape the conclusion that, however much had changed, cities continued to dominate the mental map of those who lived in greater Syria-Palestine two centuries later, when mosaicists were active at Umm ar-Rasas. In the Church of St Stephen two chains of cities line the borders of the central floor mosaic (Plate 25). On the left, when viewed facing the apse, are eight cities to the west of the Jordan; and to the right are seven cities to the east of the same river. Jerusalem has first place among the western cities, in the top left-hand corner. It is identified only as Hagia Polis. In the lower left corner are seen Ascalon and Gaza, coastal cities on the approach to Egypt. The upper right-hand corner is dominated by Mefaa, the city now identified as Umm ar-Rasas, followed by Philadelphia (modern Amman), the region's major city, and Madaba (Plate 26). Within these chains of cities is a border fully encompassing the central

Figure 24. Great oval piazza at Gerasa (Jerash), Jordan

panel of the mosaic, which represents the Nile delta, full of fish, fisher-men and their boats, sailing between important cities. The bottom right corner shows the greatest of these, the metropolis of Alexandria. Gerasa (Jerash, in Jordan), to the north of Philadelphia, was not considered part of this regional network and was not depicted at Umm ar-Rasas. It has its own collection of mosaics, however, for example at the Church of St John (dated to 531), where the Nile is shown teeming with life as it approaches the walled city identified as Alexandria.[22]

Gerasa, which retains the skeleton of its striking cityscape today, may serve as a case study in gradual urban transformation brought to a sudden stop. As antiquity ended, Gerasa's great public baths, on either side of the river, ceased to function and were replaced by a single, far smaller facility. The long, intersecting colonnaded streets were blocked off and smaller structures inserted, including in the great oval piazza (Figure 24). An industrial complex was inserted into the north theatre and its portico, and another in the abandoned Temple of Artemis. Part of the street at the entrance to the temple was given a roof and an apse was constructed, converting it into a church. This took place as early as 565, according to a mosaic inscription at its entrance. Gerasa's hippodrome, completed in the second century AD, was still operational at the end of the fourth century, when blocks were taken from it to reconstruct parts of the city walls. A mosaic inscription, installed in the city wall in AD 578, mentions the Blue

Figure 25. Hippodrome at Gerasa, Jordan

faction, which suggests that games and races persisted then, a decade after the Church of St Marianus, completed in AD 570, was built entirely from stones removed from the hippodrome. Today, the excavated hippodrome is used again to stages races for tourists (Figure 25).[23]

The great meat market (*macellum*), where shops were arranged around an octagonal, peristyled courtyard, had been expanded in the later fifth century, an indication of local prosperity. By the end of the sixth century, however, the shop fronts and market entrances were blocked up and the public space became an enclosed industrial area. Excavations have identified furnaces and kilns, associated with dye and pottery manufacturing, as well as storerooms and stables, perhaps for camels, whose urine was used as a fixative in the dying process. The entire building was destroyed in the early seventh century, perhaps during the Persian assault of 614 or an earthquake in 631/2. Afterwards, the octagonal courtyard was cleared of debris, which was reused as building materials. Two vaulted galleries and a staircase were built abutting the exterior of the courtyard in the second part of the seventh century, part of one of a large number of domestic buildings that were placed on the street to the south of the *macellum* that have been dated by early Umayyad pottery. A mint, part of a larger administrative complex, functioned at the site producing 'Arab-Byzantine' coins. A mosque was built immediately to the north of the former meat market, adjacent to the mint, although within a few decades of its construction the building was abandoned after a devastating earthquake in 749.[24]

That earthquake, which destroyed much of Gerasa, is attested by a small hoard of fifteen copper coins, ten of which were Roman ranging from the reign Anastasius to that of Heraclius. The same purse, found in a layer of ash and debris, also held five 'Arab-Byzantine' coins, including two struck in the reign of Abd al-Malik. The small hoard is indicative of the continued currency of very old, low-value Roman coins in cities and towns under Muslim rule. A large hoard found at Hama confirms evidence from across Syria that copper coins were newly struck by Constans II and imported into Syrian lands under Muslim rule, perhaps in part as payments to soldiers active in the region, but also to disseminate the emperor's image and message among Christian communities and to sustain local systems of exchange before 'Arab-Byzantine' coins were minted in great numbers. Large early Islamic copper mines, with associated smelting activities, have been identified in Nahal Amram, close to Ayla (today Aqaba), a port city at the northern end of the Red Sea to the east of Sinai.[25]

Alexandria also underwent the transition from *polis* to *madina*, and like Antioch it lost its position as regional administrative capital. Alexandria was supplanted by Fustat (now part of Old Cairo), an entirely new city that emerged on the site of a small camp established by the invading Arab army. Alexandria's cityscape developed to reflect the dominance of a new religion, but also to meet the needs of its majority Christian population. The Church of St Theonas was converted to the mosque 'of a thousand columns', but St Mark's church was restored. Egypt's largest Muslim garrisons were maintained at Fustat and Alexandria, with smaller detachments stationed elsewhere. Later reports suggest that some 40,000 troops were stationed at Fustat during the reign of Mu'awiya, and that he sent 15,000 troops from Syria and Medina to Alexandria, more than doubling the existing garrison. The great lighthouse of the Pharos stood and continued to function into the fourteenth century, earning the attention of numerous commentators in Arabic, and perhaps influencing minaret design. It remained important because Mediterranean trade, notably that with Cyprus, continued to be concentrated at Alexandria. The grain trade, however, was now diverted via Fustat to the Red Sea port of Clysma (al-Qulzum). Each year huge amounts of grain were despatched to support the faithful in Arabia. In 674, Alexandria's main shipyard was transferred to Rawda, a Nile island in the vicinity of Fustat, signalling a further decline in the city's fortunes. Limited excavations show that parts of Alexandria were abandoned, buried under layers of debris, and small shops began to crowd the ancient colonnaded streets.[26]

Documents travelled back and forth between Fustat, Alexandria and other cities and villages across Egypt, as well as to the Umayyad capital at Damascus. There are clues in several papyri that the communications system that linked cities across Egypt was sustained under Muslim rule, staffed by locals and Arabs. A document dated 22 October 669, issued by Flavius Petterius, 'pagarch' (a local official) of Arsinoe (Fayum), orders the villagers of Straton to provide salt and seasoning to the master of the local postal stables. The message was brought to them by the stable master of Psenyris, another village. We cannot establish, however, that a fully functioning postal system, such as would later exist as the *barid*, was operational.[27]

Situating major administrative centres further inland was carefully considered policy. The greatest challenge to Umayyad hegemony through-out the seventh century was posed by sea from the west. Control of the coastal cities of Syria-Palestine changed hands several times in the middle years of the seventh century. The largest of these, Caesarea Maritima, had been captured by both sides without obvious signs in the archaeological record of widespread destruction or abandonment. Sources report that the Arab capture required a seven-year siege, and in the later seventh century the south-western region of the city lay deserted, no longer sup-plied by the city's aqueducts and its harbour silted up, temporary signs of the city's diminished importance and prosperity.[28]

The status of regional capital was transferred from Caesarea to the far smaller inland city of Lod-Diospolis, which was in turn, from around 715, replaced by the new city of Ramla. Ramla was conceived by the new caliph, Sulayman, to rival Jersualem and Damascus, the two cities that had seen extensive construction by his father Abd al-Malik and brother al-Walid. Damascus was the least significant city of a group known as the Decapolis, the ten cities, the greatest of which were Gerasa and Scythopo-lis. Damascus was captured in summer 634, but its history thereafter is quite obscure, at least until the early years of the eighth century. We know that a large urban palace was constructed in the city, which also housed a mighty Umayyad garrison in its fortress. Between 706 and 715, the city's grandest structure, the Great Mosque (Figure 26), was constructed for

Figure 26. Great Mosque at Damascus, Syria

al-Walid, at around the time that the second Aqsa mosque was under construction at Jerusalem, and a mosque was constructed at Muhammad's house in Medina. Until this time, mosque construction was rare. The region around Damascus benefited economically from the city's growth.[29]

Carthage was among the last great cities to fall. It continued to strike imperial coins until *c.*692–5, including gold solidi of great fineness (more than 96 per cent gold) and of the same weight as those struck in Constantinople. Carthaginian solidi of the seventh century remained in circulation for about forty years, less than half the time of those minted in the sixth century, suggesting that taxation was efficient and that coins were regularly reminted. Because copper coins circulated principally within North Africa, a different weight standard could be maintained for the Carthage follis, which was worth a third more than one struck in Constantinople. However, there are many finds of Carthage coppers outside North Africa, including in Britain (Hamwic), which tend to follow established trade routes or troop movements. Although copper coins from Constantinople did circulate in North Africa during the sixth century, none dating later than the reign of Heraclius have been found. As the Muslim advance threatened Carthage, the minting of imperial coins appears to have been moved to Sardinia, where the fineness of solidi fell to below 95 per cent, and then in the reign of Justinian II to below 90 per cent. The Muslim capture of Carthage in 698 saw the introduction of Umayyad coins to North Africa, struck at the existing mint and emulating the nature of the existing currency. The first gold issues bore frontal portraits and inscriptions in Latin, not Arabic, although by then the reforms of Abd al-Malik had been in effect for several years. Purely epigraphic gold coins, without imperial busts, were first struck at Carthage in 703/4, but Latin remained the language of the inscriptions. The gold content of these coins was significantly lower than that of earlier Roman coins struck at Carthage, and even of those now struck in Sardinia, with a fineness of between 85 per cent and 89 per cent.[30]

The Umayyad mint was moved from Carthage to Cordoba, Spain, in 711 as the status of Carthage was supplanted by those of two new cities, Kairouan and Tunis, the only substantial new settlements in Muslim North Africa (Ifriqiya) to be established before AD 800. Tunis is a port city, situated immediately to the west of Carthage, which is today a suburb of the modern city of Tunis. It was founded as an arsenal and naval base atop the abandoned Roman town and bishopric of Thunes, and became the principal port of Ifriqiya. Kairouan was founded inland, 175 kilometres south of Carthage, offering security from attacks by sea. Like Fustat in Egypt, Kairouan was the principal garrison city (*misr*) and new provincial capital, and as such it was furnished with a fortress, a governor's palace and a large congregational mosque, as well as cisterns to guarantee its water supply. More strikingly, the new city was given a grand colonnaded street, the *simat*, but no circuit walls. These were added only when the Abbasids captured the region. Other cities retained their late Roman walls, many built in the 530s and 540s, and their ancient street plans.

Jerusalem, distinct from most Near Eastern *poleis*, had developed as a city of Christian pilgrimage. On the Madaba mosaic map, Jerusalem was the 'Holy City', the centre of the world, clearly circumscribed by its Herodian walls (Plate 23). These were punctuated by gates and linked by two prominent colonnaded streets lined with large and beautiful churches, including the complex sponsored by Constantine I at the site of Christ's tomb, the Holy Sepulchre. By the middle of the fourth century a rotunda, similar to Constantine's own mausoleum, enclosed the tomb. Originally, however, the tomb lay at the western end of an open colonnaded court-yard. Facing the tomb across the courtyard Constantine erected a grand basilica. Justinian restored the Holy Sepulchre complex, which now incorporated the site identified as Golgotha, the skull-shaped hill of the Crucifixion. Justinian added his own complex in 543, the New Church of the Mother of God, also visible on the Madaba map, which included both a hospital for the sick and a hostel for pilgrims. Excavations have revealed a remarkable number of structures offering accommodation to pilgrims.

In 427–8, as war escalated with the Persians, Theodosius II had sent to Jerusalem 'a golden cross studded with precious stones to be raised on the holy site of Golgotha', which symbolised Christ's victory over death at that place. It was to Golgotha that Heraclius returned the True Cross, which had been taken by the Persians during the storming of the city in 614. Written accounts emphasise destruction and slaughter at the city's capture, including the burning of the Lord's tomb. Archaeological evidence supports destruction at some sites, notably the Damascus gate, where a thick layer of ash and stones attests to the Persian assault that breached the wall and burned the gate. However, it also shows that new circuit walls were soon constructed, as were nearby palaces, and a large mosque, the original Aqsa mosque, was placed on the Temple Mount probably in the reign of Mu'awiya.[31]

In 691, the Dome of the Rock was placed on the Temple Mount (Figure 27). The date of construction is provided in the final words of an immensely long mosaic inscription around the octagonal arcade within the building. This is placed between the outer wall and a circular inner arcade that supports the high drum of the dome. The inscription, replete with the earliest recorded Quranic quotations, offers the first full profession of the Islamic faith (Plate 27). Its central idea is that the divine message present in both Judaism and Christianity has been fully revealed in Islam through Muhammad, the messenger of God. For this reason, it is as well to emphasise that the Dome of the Rock, whatever it was at its inception – today it is understood to mark Muhammad's 'night journey' and the site of his ascent into heaven – was not a mosque. The best clues we have to its original meaning are its form and decoration, not only the long inscription around the octagonal arcade, but also additional inscriptions on bronze plaques that were once fixed above the building's four entrances, and the whole complex of exquisite mosaics that adorn its interior. The form of the structure is that of a martyrium, a site of commemoration,

Figure 27. Dome of the Rock at Jerusalem

but also literally 'a place of witness'. The glittering mosaics depict paradise, an eternal verdant garden with trees and streams to shade and refresh. Together they declared the triumph of Islam, the final revelation that completed and supplanted those of Jews and Christians, and memorialised a moment soon to come, 'when God would return to earth to judge men and women. He would appear at the place from which He left the earth at the time of creation, the Rock over which the Dome was built'.[32]

Apocalypse

The imposition of the Dome of the Rock on one of the holiest sites in Christendom underlined the moment of its creation, but its eschatological message built on centuries-old Christian traditions. In Acts (1:3), Jesus is held to have appeared to his apostles prior to his Ascension, and 'over a period of forty days ... spoke about the kingdom of God'. Apocryphal writings emerged that set out what he had told them. Among these is the prologue to the so-called *Testament of Our Lord Jesus Christ*, a work of the fifth century.

After the famines and pestilences and tumults among the nations, then there shall rule and rise to power princes who love money and hate the truth, who kill their brethren, liars, haters of the faithful ... Then there shall be signs in heaven. A bow shall be seen, and a horn, and lights, and noises out of season, and sounds, and ragings of the sea and a roaring of the earth ... But they who in my name endure unto the end shall be saved ... Then shall come the Son of Perdition, who boasts and exalts himself, working many signs and miracles that he may deceive the whole earth and overcome the innocent. Blessed are they who endure in those days, but woe to those who are deceived. But Syria shall be plundered and shall weep for her sons ... Cappadocia, Lycia, Lycaonia shall bow down the back, for many multitudes shall be depraved by the corruption of their wickedness. And then shall be opened the camps of the barbarians, for many chariots shall go forth and cover the earth ... Judaea shall clothe herself with lamentation and shall be made ready for the day of destruction because of her uncleanness.[33]

The original Greek text has been lost, but because of this apocalyptic prologue, which appeared to capture the situation in which he lived, a monk named Jacob translated it into Syriac in 687, probably at Edessa in upper Mesopotamia.[34]

At the same time, also in northern Mesopotamia, John bar Penkaye concluded a fifteen-book summary of world history in Syriac with an apocalyptic chapter, convinced that he was living through the 'End Times'. He wrote of famine and earthquakes, solar eclipses and other terrible signs in the sky, of the coming of the Arabs, the Lord's punishment for the sin and heresy of his fellow Christians, and also of God's punishment of the Arabs, since he 'gave them two leaders from the beginning of their kingdom and divided them into two factions'. He wrote of the first victory of the Umayyads, in 657, and of the reign of Mu'awiya, a time of peace, when all religions were tolerated upon payment of tribute. He wrote of the sins and crimes of fellow Christians, who left the dead unburied, followed magicians and soothsayers, committed murder and adultery, and scorned the poor, pilgrims and even monks like John himself. Following an account of the second *fitna*, John dwells on the current situation of famine and pandemic, as the plague swept through the Near East once again between 683 and 687. 'Let us stop here', he concludes, 'because here begins the kingdom of the Lord'.[35]

Around the same time and in the same region a work known as the

Apocalypse of Pseudo-Methodius was written down. It is one of very few works written in Syriac during the Arab period to be translated into Greek, and this happened almost immediately, so urgently was the need to spread its revelation felt. Like John bar Penkaye, its author interprets recent history by reference to prophecy, relying on the schema of four kingdoms set out in the book of Daniel. The fourth kingdom was that of the Romans, now beset, but soon to rise again and clear the path for the Lord's return.

> After the kingdom of the Hebrews had been extirpated, in its place the children of Ishmael, son of Hagar, waged war on the Romans. These are the people whom Daniel [11:5] calls 'the arm of the south'. And he [Ishmael] will be waging war with it for seven weeks of years, for the end is known and there is no extension in between. In this last millennium, which is the seventh, during which the kingdom of the Persians will be extirpated, the children of Ishmael will come out from the desert of Yahrib ... and they too [the Romans] will be devastated by ... Ishmael, the 'wild ass of the desert', who will be sent in fury of wrath against mankind, against wild animals and domestic beasts, against trees and plants. It is a merciless punishment.

Pseudo-Methodius portrays the passage of the Arabs from desert to sea as a wave of devastation, such that 'deserted lands, which have been left like widows without cultivation, shall belong to them', until the 'chastisement of which the Apostle spoke [2 Thessalonians 2:3]' is complete. Then, 'there will be a great famine and many will die, their corpses thrown like mud into the streets ... plagues of wrath will be sent upon humanity'. When all hope is lost, the last emperor will 'go out against them in great wrath', the lands will be delivered to him, and the affliction of 'those who have denied Christ' will be seven times that which they have inflicted. 'The gates of the north shall be opened, and the armies of those people confined there shall come forth, the land quaking before them.' When the last emperor arrives in Jerusalem, the 'Son of Perdition shall appear', and the emperor will 'stand on Golgotha and the holy cross will be placed on that spot', so that he may place his crown upon it, bring an end to sovereignty, and await the return of the Lord in his glorious fury, when the Son of Perdition will be delivered into the 'Gehenna of fire'.[36]

The last emperor is called 'King of the Greeks' throughout the Syriac apocalypse, placing him in succession to Alexander the Great. In the *Alexander Romance*, it was Alexander who had built the 'gates of the north', often associated with the Caspian Gates, to hold back the barbarians beyond, the followers of Gog and Magog. In the Greek translation of Pseudo-Methodius, we meet instead the 'King of the Romans'. In the first iterations of the apocalypse, this last emperor is understood to be Justinian II, who will drive the Arabs back, a hope derived from the period of Arab civil wars. Justinian's association with Bulgars and Khazars, barbarians from beyond the gates of the north, seemed probative, until it was not. The Greek translation of Pseudo-Methodius was regularly updated to account for the enduring presence of Islam and the failure of each successive emperor to be the last. Updating was necessary also to fulfill a principal function of the text, which was to persuade Christians that the Arab hegemony was a temporary phenomenon, the eschatological trial promised in scripture that would separate the sheep from the goats, true believers from infidels.[37]

Within a decade another Syriac apocalypse was written, drawing heavily on Pseudo-Methodius and the *Testament of Our Lord*. The *Gospel of the Twelve Apostles* offers a prologue, where Jesus addresses his twelve apostles in life and again in death, instructing them on the signs of his 'Second Coming' through three revelations witnessed by different apostles on the mountain of the Transfiguration. The first two revelations, to Simeon Kepha and James, serve as preparation for the third, the vision offered to John 'the Little', who is identified with John the Evangelist, understood here to be identical with John the Divine. His revelation begins with a scene drawn from Revelation (4:4, 6:11, 7:13–14) and Daniel (7:9): 'And there was suddenly a great earthquake and John ... fell on his face on the earth and with great trembling he worshipped God the Lord of all. And our Lord sent to him a man in white raiment and mounted on a horse of fire, and his appearance was like the flashing of fire.' Again, drawing on Daniel – 'and then suddenly shall be fulfilled the prophecy of Daniel' – the Arabs are described as a mighty wind from the south, 'a people of deformed aspect ... and there shall rise up from among them a warrior, and one whom they call prophet'. The devastation they bring upon Christian communities echoes that in Pseudo-Methodius, but here it is implied that the fault lies with the Chalcedonians, or Duophysites, whose heretical division of the Lord is identified as the cause for God's wrath. The last emperor appears as an unnamed descendant of Constantine the Great, who will defeat the Muslims as Constantine defeated all enemies by the sign of the cross.[38]

This proliferation of apocalyptic texts in a very short period of time, all written in a region now firmly under Muslim rule, appears to correlate with a rapid increase in the number of Christian converts to Islam. While in the earliest Islamic conquests the Arabs enslaved enormous numbers of Christians, principally civilians whose lands had been overrun, whether they became Muslims was a matter of individual conscience or advantage, not of coercion. As a prestige religion, the victorious creed of the new rulers proved attractive to warriors and those whose attachment to settled Christian communities was tenuous. A great number of early Christian converts to Islam appear to have been from among the Bedouin.[39]

From the 680s, for the first time, the integrity of settled Christian communities was threatened by economic incentives to convert. According to the *Chronicle of Zuqnin*, in 691–2,

> Abd al-Malik carried out a census (*tadil*) on the Syrians. He issued a harsh order that everyone should go to his region, village, and father's house, so that everyone would register his name, his lineage, his crops and olive trees, his possessions, his children, and everything he owned. From this time, the poll tax (*jizya*) began to be levied per capita; from this time, all the evils were visited upon the Christians. Until this time, kings had taken tribute from land rather than from men. From this time, the sons of Hagar began to inflict on the sons of Aram servitude like the servitude of Egypt.[40]

Abd al-Malik's brother, Abd al-Aziz, was at this time governor of Egypt, where he undertook a census of Christian monks both to ensure that they paid the poll tax and also to fix the number of monks to the number assessed at that time. There is evidence for the widespread use of censuses and cadastral surveys aimed at calculating tax revenues due from Muslim and non-Muslim communities. Muslims paid a lower rate of land tax and no poll tax. A dossier of papyri from Nessana, Palestine, includes a register of households liable to pay poll tax, receipts for the payment of that tax and land taxes, and a letter to the governor in Gaza protesting the burden of taxation. Additionally, there are lists of Arab Muslims, soldiers, who are to be paid directly by the citizens of Nessana in money and grain.[41]

While conversion on an individual basis was now encouraged, Muslim authorities were anxious to maintain distinctions between faith communities, introducing sanctions against Christians who appeared too Muslim in their dress or behaviour – such as wearing the waistband, dyeing their hair and beards – and against Muslims who consorted too closely with Christians. Muslims were advised to avoid contact with Christian monks, and particular care was taken to avoid practices characteristic of monks, with an absolute rejection of celibacy. In the ninth century, when there is far more written evidence, Ibn Hanbal offered observations on how Muslims should interact with Christians on common public occasions such as during death rituals and funerals, and visits to monasteries were strongly discouraged, despite the example of the Prophet himself, whose encounter with the monk Bahira (Sergius) became a popular story at exactly this time. An admonitory tale emerged about al-Mutawwakil, a caliph who travelled to a Christian monastery, drank wine and listened to music, reducing his guard sufficiently that he seduced the daughter of the abbot, whom he then married.[42]

Muslims were not encouraged to take an interest in Christian writings, especially the apocalyptic works that animated the Christian communities, but it is clear that they did and that it informed their own developing eschatology. A story relates that Umar I beat an Arab scribe whom he caught copying the book of Daniel, with every blow repeating 'We have revealed to you an Arabic Quran'. The Quran is replete with apocalyptic imagery and the Last Judgement and Resurrection are among its most urgent themes. The second coming of Isa, Jesus, was identified as a major sign of the imminent coming of 'the hour' (Quran 43:61). That very imminence saw Islamic eschatology develop quickly in the *Hadith*, the sayings of the Prophet, where it is held that Isa would descend at Jerusalem, al-Quds, 'as a just ruler. He will break the cross, kill the swine and abolish the poll tax'. One encounters other familiar characters, including the Mahdi, or just ruler, who will reign at the time of Isa's coming, and the false messiah, the last deceiver.[43]

The theologians Maximus the Confessor (d. 662), whom we have met as a staunch opponent of Monotheletism, and John of Damascus (b. 675) were among the greatest minds of their age, and both engaged directly with the challenge posed by emergent Islam. Both men were for a time monks at Mar Saba, a monastery in the Judean Desert nine miles to the south-east of Jerusalem. Both men perceived the range of possible knowledge as lying within certain well-defined parameters, and in their writings they aimed at a complete synthesis of contemporary knowledge. Complete knowledge, however, was seen to be knowledge of God and his design for earth, and true insight could be granted only to those who devoted their lives to its pursuit. John's masterwork was *The Fountain of Knowledge*, a compendium that featured a refutation of heresies including Islam. This heresy he attributed to Muhammad's reading of the Old and New Testaments and a conversation with an Arian monk. John's close reading of the Quran and detailed knowledge of Muslim practices informs his critique, although his tone can be dismissive.[44]

> This Muhammad wrote many ridiculous books, to each one of which he set a title. For example, there is the book *On Woman*, in which he plainly makes legal provision for taking four wives and, if it be possible, a thousand ... in the book of *The Heifer*, he says some other stupid and ridiculous things, which, because of their great number, I think must be passed over. He made it a law that they be circumcised and the women, too, and he ordered them not to keep the Sabbath and not to be baptised. And, while he ordered them to eat some of the things forbidden by the Law, he ordered them to abstain from others. He furthermore absolutely forbade the drinking of wine.[45]

Maximus the Confessor articulated life's purpose as purification (*catharsis*) by purging a hierarchy of vices, and through illumination (*theoria*), to achieve union with God (*theosis*). Maximus's thought was profoundly influenced by the celestial and ecclesiastical hierarchies of Pseudo-Dionysius the Areopagite, who promised a path, a ladder of images, by which monastic initiates might achieve mystical union with God. The ultimate attainment of life, therefore, was achievable only by those who, like John and Maximus, were monks. The *Ladder of Divine Ascent*, written by John Climacus ('of the ladder'), a contemporary of Maximus and abbot of St Catherine's Monastery at Mt Sinai, presents thirty chapters on virtues and vices as thirty rungs on a ladder, an image that proved extremely popular. The top rung of the ladder, the last chapter, concerns 'Faith, Hope and Love', the ultimate virtues.[46]

Anastasius of Sinai, an abbot at St Catherine's, whose last works appeared shortly after the year 700, wrote for the learned and the humble, monks and lay people. His works included one of the most important medieval collections of 'Questions and Answers', as well as 'a two-volume treatise against the Jews ... a short tract on heresies and synods, a confession of faith comprising also an anti-heretical defence of Neo-Chalcedonian orthodoxy, a series of sermons on diverse themes, a series of homilies, and a dogmatic *tomos* intended as an intellectual tool in his life-long struggle against Monophysitism'.[47]

In his *Questions*, Anastasius offered observations on a range of arcane or magical practices concerning healing, astrology and prophecy, but he also supplied far more mundane reassurances, for example to questions about Christian life in general: can one go straight to church after having sex, or from dreaming in bed? Can one take communion having swallowed bathwater? In particular, he was interested in Christian life under Muslim rule. Answers are offered to those wondering whether their sins can be redeemed if they live in servitude or captivity and cannot attend church services; or if, as a female in captivity, they are subjected to sexual assaults and other sinful acts; or if they should pray for masters who are pagans, Jews or heretics; or whether God has willed the suffering of Christians at the hands of the Arabs. Several questions and answers related to oppression, persecution and even death at the hands of the Arabs. These were all very new concerns, indicative of the status of Christian communities that were no longer ruled by the emperor of New Rome.[48]

For Christians in the last decades of the seventh century, the world was not simply transformed, it seemed to be ending. For several centuries, earthquakes had been abnormally frequent and a large number were cataclysmic, destroying whole cities and causing tsunami waves. Massive volcanic eruptions threw up dust veils that blocked out the sun, causing cold summers and freezing winters after 536, 540 and 626. Crop failures and famines ensued. Waves of bubonic plague had struck every fifteen or so years after 541, returning to major cities where, through deaths, births and migration, herd immunity had weakened. Written sources suggest that the plague was devastating, but archaeological evidence proves that the disease was not equally destructive across the late Roman world. Most compelling is the absence of infected human remains. If tens of millions died of plague over two centuries, surely we would by now have found some mass graves and evidence of other extraordinary burial practices. Where are all the bodies? The deaths of millions would have left Rome without the manpower to rebuild, restore and resist; patterns and intensity of settlement and agriculture would have changed suddenly, and this would show up in the environmental record. It does not in the sixth century, but by the end of the seventh century it is clear that vast areas were no longer cultivated and were rapidly rewilded. Natural disasters alone did not bring this about, but even the most resilient communities could not endure this in combination with decades of existential warfare.

In the wake of the last brutal and protracted confrontation between the Romans and the Persians, which diminished both superpowers, the Arabs had launched a remarkable series of expeditions. They had overrun Persia, ending the Sasanian empire forever, and very nearly destroyed Rome after that, reducing it to a rump Christian state perched on the edge of a vast Eurasian-North African Islamic caliphate. It is hardly surprising, given the tone and nature of Roman–Persian warfare, that as this new force swept out of the Arabian peninsula, it was propelled by a developing notion of 'holy war'. It is clear that the two religious cultures, established Christianity and emergent Islam, conducted a debate on this matter even as their practitioners shed each other's blood. For Christians of all stripes, the remarkable eruption of Muslims could only be understood as God's punishment for sin, especially heresy, which had fractured the church for centuries.

The rise of apocalypticism, therefore, was a reasoned response to the calamities of the period and to the pressing need to maintain the integrity of Christian communities in the face of an increasingly rapid pace of conversions to Islam. It was also integral to a broader recalibration of what constituted knowledge and what was worthy to be written down. Under Heraclius, Theophylact Simocatta wrote secular military history and prefaced his work with a dialogue between history and philosophy, while George of Pisidia wrote florid secular poetry in praise of the emperor and his exploits. Other than the lost *History to 720*, secular historical writing in Greek then ends for nearly two centuries, during which interlude writers in Arabic offer no historical writing from the perspective of 'the victors'. At this time, Constantine the Great was removed from history to legend, becoming a saint, and the history of Constantinople became 'Brief Historical Notes', *Parastaseis Syntomoi Chronikai*, a semi-mythical account of the city and its landmarks. Collections of parables or 'beneficial tales', such as those collected in John Moschus's *Spiritual Meadow*, flourished, aimed at the spiritual education of less learned Christians, including monks, who were no longer rhetorically trained.[49]

12

EMPERORS OF
NEW ROME

Emperors equated their own interests with those of the Roman people. Frequently, those interests did coincide, and emperors who protected and enriched the greatest number of their subjects, and who upheld the rule of law, were the most successful in life and revered in death. Sustaining a robust economy, strong defences and an effective legal system required good government. Emperors who governed badly, who flouted the rule of law or interfered in matters better left to bureaucrats or clerics, or who otherwise lost the support of the people, knew they could be deposed and defamed. Just as they were acclaimed by the people, so they might be cast aside with the public cry of 'Unworthy!' or 'Dig up his bones!' There are key questions that we have not touched upon in this book, which go to the heart of the imperial system. How was the role of the emperor conceived constitutionally? What was the Christian emperor's role at New Rome? What was the relationship between the emperor and the law? What role was assigned to the people in choosing, creating and sustaining an emperor? Imperial power may appear to have been unfettered, given the actions of many emperors, but it was in theory delimited by a political system that sustained the Roman *res publica*, in Greek the *politeia*, which may be translated as 'republic'.

An argument can be made that Rome remained a republic because sovereignty resided ultimately with the people, who entrusted it to each absolute monarch until they wished to take it back and reassign it to another. The wealthiest and most powerful people were found in the senate, which was always present but is frequently invisible in our sources. Senators did not often become emperors, although emperors stacked the senate with their supporters, diluting old money and influence with new men. The senate represented the interests of the empire's elite families, protecting their own interests and checking those of rivals. This is clear on the occasions when the senate emerges from the darkness, for example in rejecting Heraclius's last will and testament, effecting a new constitutional settlement, and removing his second 'illegitimate' family from power. Rising stars rose more quickly in military circles and generals from humble backgrounds frequently became senators. The Roman people at war, the army (and later also the navy), could make or break emperors. Military rebellions began frequently in distant provinces and occasionally had an impact in Constantinople, if an army marched (or later sailed) against the city. Those who enjoyed military power within the city were effective at installing emperors, as is demonstrated by the number of emperors who had commanded the imperial guards. Finally, since they were bound to the city, emperors were sustained in power or were removed by the citizens of Constantinople and their most violent representatives, the circus factions. Popular uprisings, military rebellions and urban riots were not aberrations at New Rome, they were the means by which the people reclaimed its sovereignty, before it was reassigned to another.[1]

Imperial government and ceremony

A clear and concise articulation of the Christian emperor's role at New Rome is set out in Agapetus's *Ekthesis*, an 'Exposition' of advice to the emperor, which if followed might help him to remain in power. Its seventy-two short chapters are arranged so that their initial letters form an acrostic in Greek spelling out 'To our most sacred and most pious emperor Justinian, from Agapetus, the humble deacon'. It begins:

> Since you have a dignity above all other honour, Emperor, you render honour above all to God, who gave you that dignity; in so far as He gave you the sceptre of earthly power after the likeness of the heavenly

empire, to that end you should instruct men how to guard the cause of justice, and should punish the howling of those who rage against that cause; being yourself under the rulership of the laws and lawfully ruler of those who are subject to you.[2]

The emperor might be the 'living law', but he should also be constrained by the law and subject himself willingly to it. A law of 429 states that 'It is a pronouncement worthy of the majesty of the ruler that the emperor should declare himself bound by the laws, so much does our authority depend upon the authority of law. To submit our imperial office to the laws is in truth a greater thing than our imperial sovereignty.' Justinian, as we have seen, did not consider himself 'under the rulership of the laws', and would have preferred Agapetus's second chapter of advice, 'to keep a firm hold of the rudder of enforcement of the law'. It was widely understood that an emperor who placed himself above the law to promote his own interests at the expense of the common good was a tyrant and would be deposed. As such, the Nika riot in 532 was remarkable not in its magnitude, although it was impressively broad-based and destructive, but for the fact that it failed to overthrow a tyrant. Men recognised as emperors, such as Maurice and Phocas one after the other, and as usurpers, such as Basiliscus in the fifth century and Valentinus in the seventh, failed where Justinian had succeeded. Justinian killed an unprecedented number of his own citizens and survived, in effect ruling as a tyrant. Procopius recognised this and named him as such.[3]

Having commenced with justice, we may identify in the admonitions of Agapetus the remaining cardinal virtues and their Christian emanations with which emperors should be abundantly endowed. There is wisdom, temperance and courage; and also faith, hope and charity, the theological virtues (1 Corinthians 13:13). God is the source of all grace, Agapetus reminds the emperor, and 'he gives grace in return for grace ... as if it were a debt', repaying not good words but good works. Agapetus offers advice on keeping the soul wiped clean, 'like a mirror for divine illumination'. Faith is paramount, since 'above all the glories of rulership, the crown of piety adorns the emperor'. Agapetus addresses Justinian as a Platonic philosopher-king, 'Pursuing the study of philosophy, you were counted worthy of rulership, and now ruling you do not abandon philosophy', which is 'the love of wisdom'. And as Proverbs 1:7 has now revealed to Christians, 'the fear of God is the beginning of wisdom'.[4]

Truly you define yourself as emperor, because you dominate and rule your passions, and have been wreathed with the crown of temperance (*sophrosyne*) and bedecked with the purple robe of justice (*dikaiosyne*) ... Your rulership is rightly revered because it exhibits power to enemies and humanity (*philanthropia*) to your allies. When it has subdued the one group by force of arms, it yields to unarmed love (*agape*) of subjects.[5]

Agapetus expands upon these themes, often repeating himself. His work was not emulated until late in the ninth century, when the so-called *Hortatory Chapters* appeared, addressed to Leo VI, 'the Wise'. At that time, however, Agapetus's *Ekthesis* proved so popular that it was the first secular text to be translated using the newly created Slavonic alphabet, in *c.*900, for the Bulgarian khan Symeon. It proved to be the mostly widely translated, read and published later Roman work in both eastern and western Europe. It survives in more than eighty late manuscripts, and was printed in twenty editions in the sixteenth century, reflecting a passion for 'Mirrors for Princes', a corpus that now included the most famous of all such tracts, Machiavelli's *The Prince*.

For a readership as exclusive as Agapetus's was wide, the tenth-century *Book of Ceremonies* preserved records of ceremonies that had taken place centuries earlier. A ceremony that marked the elevation of two sons of Heraclius preserves the order of precedence at court as it was in July 638, where after the imperial family the patricians ranked first, followed by senators from those of consular rank down to those who were merely illustrious (*illoustrioi*). Those of consular rank included consuls (*hypatoi*), proconsuls (*anthypatoi*) and ex-consuls (*apo hypaton*). The praetorian prefectures had lost power long before that date. The last known holder of that office was Alexander in 626.[6]

The senate of Constantinople was also the city council, so the urban prefect (*eparchos*) served as its president. The imperial treasurer (*sakellarios*) also held senatorial rank, equivalent to that of the urban prefect, but the authority that both held derived more from their access to the emperor himself, as members of his consistory or, when that official body ceased to exist, its informal replacement, a group of advisers with 'unfettered access' (*parresia*) to the emperor, including those whose formal role was to manage the imperial residence within the palace. The power of office holders of all ranks was reduced as the authority of the emperor and his immediate advisers increased.[7]

In the long reign of Justinian, an insomniac emperor who rarely left Constantinople, government became increasingly concentrated in the Great Palace. Government bureaux were located there, giving the emperor and his coterie of advisers oversight, and the barriers between civilian and palatine administrative departments were elided. Close control was also exercised, so far as it could be, over the military administration and its most senior staff. Justinian was always wary of his generals, and his rivalry with Belisarius ensured that his successors frequently circulated their best men between commands and theatres of war, so that they were not able to build regional power bases or to threaten the sitting emperor. Justinian II and Tiberius managed this quite successfully, as did Maurice before he fell victim to a mere centurion, Phocas. Exarchs, who held both civilian and military authority in their regions, posed a major threat to a sedentary emperor's control. Although they were also usually sent out from the centre, several managed to establish themselves as more than local plenipotentiaries. Heraclius, the son of an exarch, employed North Africans in his successful bid for power, but his ability to command loyalty beyond Carthage evidently relied on his relationship with the family of Nicetas, his cousin from Cyrenaica. No exarch of Ravenna became emperor, but nor did every exarch recognise the authority of the man who ruled from Constantinople.

Just as government was increasingly focused on the palace and the person of the emperor, so the volume and importance of court ceremonial increased substantially over time. The first records of what was done and when were written down under Justinian by Peter the Patrician, some of whose stage directions are preserved in the *Book of Ceremonies*. Corippus's panegyric for Justin II elucidates aspects of various ceremonies that had developed, including an imperial funeral and a coronation, the reception of foreign dignitaries (an Avar embassy), and consular inaugurations. The Great Palace at Constantinople was transformed and expanded at Justin's initiative. According to John of Ephesus, space constraints obliged Justin to 'remodel the whole of the northern side', or upper palace, and 'build a palace for himself', a lower palace, 'for which purpose he demolished many spacious buildings and destroyed a beautiful garden that had existed in the middle of the palace and had been a great pleasure to former emperors'. In their place he 'erected larger and more magnificent buildings, including a splendid bath ... and spacious stabling for his horses'. The new lower palace was henceforth used for the daily imperial business, much of which was conducted in an octagonal Golden Throne Room (*Chrysotriklinos*). Its ceiling was decorated with scenes from the life of Christ and, above the throne itself, the figure of Christ. The old palace, or upper palace, gradually fell out of daily use.[8]

Evidently, Justin and his empress Sophia were extremely wealthy patrons and great builders in the manner of Justinian and Theodora. Some other projects had commenced prior to their accession, while others date between their accession and the death of Tiberius II Constantine. Among the largest projects was the dredging and refurbishment of the Julian harbour, which was renamed the 'Harbour of Sophia', second in size and importance only to the Theodosian harbour. The Sophiae Palace was built there. As emperor, Justin built the Deuteron Palace at the location where he had lived while serving as major-domo for Justinian. There he erected bronze statues of himself and his wife. Statues of Sophia and her relatives were erected at the Milion, Constantinople's central milestone, and at the new Harbour of Sophia. Justin also erected a colossal consular statue of himself. For Sophia a summer palace was built across the Bosphorus, the Sophianae, mentioned by Corippus as already under construction when the pair came to power.[9]

Justin is also credited by Theophanes with restoring the Aqueduct of Valens, supplying 'the city with abundant water'. Additionally, he built a public bath, named like so much else after his wife, at the Forum Tauri. An inscription attests to a partial restoration of the Theodosian land walls by Justin. Great attention was paid to the Church of the Holy Apostles, where the remains of emperors rested. A new patriarchal palace was erected for John Scholasticus adjoining Hagia Sophia, part of which is still preserved at the south-west ramp of the building. Elsewhere, shrines were built or restored to house relics brought to the city, which included the Camuliana – an *acheiropoietos* image of Christ's face 'not made by human hand' – the Virgin's girdle, and part of the True Cross. A precious fragment of the cross was sent west, to Poitiers, at the request of Abbess Radegund, and this act of pious generosity inspired a poem in praise of the imperial couple by Venantius Fortunatus.[10]

Like his predecessors, Maurice built in and around Constantinople, including the Church of the Forty Martyrs, placed on the Mese at the site of the old praetorium, the seat of the city prefect that had been burned more than once. This was perhaps part of a series of reconstructions near the Great Palace that followed a fire that started at the forum in the first spring of Maurice's reign. At the Magnaura, where the emperor met foreign dignitaries, a circular terrace was built around a courtyard at the centre of which was placed a statue of the emperor. The arsenal was also located there, adjacent to the upper palace. At the suburban Palace of Blachernae, next to the land walls at the Golden Horn, Maurice erected the so-called Carian portico, which was painted to depict 'all his deeds from his childhood until his reign'. This new facility included a public bath. The emperor demonstrated his piety by instituting a feast of the Assumption, celebrated with the *panegyris*, a new litany performed at Blachernae in honour of the Mother of God. The Blachernae would be incorporated by an extension of the land walls of Constantinople after the great siege of 626 (see Figure 9, p. 106). At the end of the seventh century, attention was paid to the city's sea defences. Leontius dredged the Neorion Harbour for the fleet, and Apsimar restored stretches of the sea walls.[11]

Justinian II was busier than most 'with constructions in the palace. He built Justinian's Hall, as it is called, and the circuit wall of the palace.' Additionally, he demolished a church near the palace dedicated to the Mother of God and rebuilt it near the Golden Horn. In its place, now within his palace walls, Justinian built a fountain and benches for the Blues, where they could acclaim and petition the emperor. The Greens also had a fountain courtyard where they could participate in regular ceremonies. The *Book of Ceremonies* preserves descriptions of ceremonies performed from the time of Justinian II until, two centuries later, the fountain courtyards were demolished to make way for a new church. The Hall of Justinian II, also called the Justinianus, and fountain courtyards were adjacent to each other and connected to the Lausiacus, a second grand reception hall, which led on to the Golden Throne Room. They were the locations for a complex set of court ceremonies to glorify the emperor on even quite mundane occasions, as is clear from a brief excerpt from the 'Reception of the Gold Hippodrome Festival on the Monday after Antipascha', which is to say the Monday after the Sunday after Easter Sunday.[12]

> After a faction cries out 'Holy!' then the organ is played in the fountain courtyard, and the officiating eunuch (*praipositos*) receives a sign from the emperor and he signals with his hand three times and the organ stops. Then the factions begin to do everything that is customary ... and they give acclamations and recite imperial eulogies, and

when these have been completed they begin the chant. When they begin the chant, the master of ceremonies receives a sign from the *praipositos* and signals to those of consular rank [the most senior senators], who process away to the Hall of Justinian to stand in ordered ranks to either side. When the chant has been completed the emperor stands up and ... goes through the Hall of Justinian while the senators ... stand to either side.

The Blues and Greens had increasingly formalised roles, representing the people in acclaiming emperors on this and all public occasions, including at imperial inaugurations and coronations. Beneath the formality, however, the factions could wield real power on behalf of the people and did so periodically.[13]

The emperor and the people

The age of Justinian was an extremely volatile period in the history of the empire's larger cities, where there were frequent riots and fires. Much of this focused on the theatres, notably the hippodromes, where large crowds gathered regularly.[14] In 493, there was 'Civil War in Constantinople against the rule of Anastasius: statues of the emperor and empress were dragged down and through the city on ropes'. Marcellinus, author of this terse notice, fails to mention a further hippodrome riot in 498, which the *Easter Chronicle* reports at length:

In the time of these consuls [John Scythopolites and Paulinus], while chariot races were being held, those of the faction of the Greens appealed to the emperor Anastasius that certain men be released who had been detained by the city prefect for stone-throwing. Anastasius refused the appeal and in anger he ordered an armed force to go out against them and there ensued a great riot ... [during which] a great Moor hurled a stone at the emperor, who avoided the stone, since he would have been killed by it.

The people then set fire to the hippodrome, which was burned from the gates as far as the imperial box, and the porticoes outside the stadium, which were burned as far as the Forum of Constantine.[15]

Civil unrest was associated with many public spectacles and entertainments, and Anastasius tried to ban events that facilitated or precipitated riots. In the Cynegium, Constantinople's amphitheatre, wild beast fights (*venationes*) were occasionally staged, although only as part of consular games and only by the richest and best-connected consuls. In 498, Anastasius recapitulated a ban on wild beast fights, a law that had been in place since 414, which also permitted the hunting and killing of lions in the wild. If wild beast fights were rare, pantomimes were ubiquitous and immensely popular across the eastern Roman provinces. The Greek term *pantomimos* captures well the fact that a single artist portrayed all (*panta*) the characters in a story, generally mythological, hence considered by some to be 'pagan'. The solo performer was a dancer (*orchestes*), accompanied by a chorus and musicians (hence 'orchestra'), showing 'in movement what the singer sends echoing through the theatre: he fights, he plays, he is in love, he is frenzied, he spins, he stops'. Each pantomime, male, with long flowing hair and a silk robe, was supported by a claque, who would lead an audience in their applause, and apparently could whip them into a frenzy. Riots were frequently attributed to pantomime performances and dancers were frequently banished from cities, for example from Rome by Tiberius, Nero, Domitian, Trajan and Commodus. In 502, Anastasius banned pantomime dancing from all the empire's cities, and pantomimes were banished from Constantinople. Apparently this was a consequence of spectacular water dances performed at an aquatic festival known as the Brytae, which ended in 499 in a riot, and again in 501 with the deaths of 3,000 participants, among them Anastasius's own illegitimate son.[16]

His bans earned Anastasius the approval of Christian writers who feared the challenge of the hippodrome. John Chrysostom had raged 'Against the games and the theatre', evidently to limited effect, some years after the Cappadocian father Amphilochius of Iconium had condemned those who enjoyed his demonic local hippodrome. Provincial circuses could not compete with the great cities for the number and qualities of its street artists, acrobats and conjurers, dancers and prostitutes. There was also gambling, including betting on the outcome of races, and even simulated races. Zealous Christians reviled charioteers, who were known to be in league with sorcerers who might send demons to trip rivals' horses or otherwise cause their vehicles to crash. An odd law confirmed by Justinian in 534 implies that charioteers who employed sorcerers were not permitted to murder them if they proved ineffectual. Preserved curse tablets attest to efforts by fans to counter magic employed by rivals.

It is unclear whether puritanical laws against the games were ever enforced. A *venatio* is shown taking place along with chariot racing, mimes and dramatic performance on the consular diptych of Flavius Anastasius, carved in ivory for the games of 517. This suggests that all of the familiar spectacles remained vital to consular entertainments, and certainly to how they were imagined. The emperor himself is shown presiding over all of it, his bust centrally placed above the enthroned consul. For his consular games in 521, Justinian is reported to have 'exhibited together in the amphitheatre twenty lions and thirty panthers, not counting other wild beasts', an attempt to surpass the spectacular games staged in 520 by Vitalian. Justinian was reluctant to allow anyone to compete with him and regularly failed to appoint consuls for the first ten years of his reign. On 28 December 537 he issued a law that restricted the number of games a consul could stage and limited the generosity each might show, imposing a fine of 100 pounds of gold for violations. Still, this reduced programme included seven days of events spaced through the year, including two sets of chariot races, a wild beast hunt in the Cynegium, and 'the so-called *pancarpon*, men fighting with beasts ... and the show called theatrical entertainments (*pornai*)'.[17]

The role of the circus factions in the transfer of imperial power is never clearer than at the end of the sixth century and the beginning of the seventh. The factions supported Maurice's efforts to defend the city, first against the Avars by manning the Long Wall, and then against Phocas on the city walls. They were instrumental in the removal of Maurice, when they abandoned the walls and rioted, allowing Phocas's entry. As Alan Cameron long ago noticed, 'It is notorious that the factions caused a lot of trouble during the reign of Phocas, but more of it than is usually realized stemmed directly from his crass and cavalier attitude to ceremonial'. On the day after his coronation, as he prepared to crown his empress, Phocas displayed this ignorance and was chided by the Blues with the cry 'Go away and learn the ceremonial'. On another occasion when he appeared drunkenly in the imperial box, Phocas was greeted with a variant on the regular greeting, 'At the bottle again, it's befuddled your brain'. In revenge, Phocas had many mutilated and their limbs hung at the end of the hippodrome. When the Greens rioted, he banned them from holding public office, denying a large and powerful segment of the citizenry access to ranks and titles with their associated stipends. The Greens exacted their revenge: although they had twice scorned the patrician Germanus and rejected the general Priscus in their bids to become emperor, they threw their support behind Heraclius. Knowing that the fate of an unpopular emperor was determined, more often that not, in the city of Constantinople, Heraclius avoided the city after his incestuous second marriage caused public outrage. He preferred villas in distant cities or suburban palaces, and perhaps he dreamt of ruling from Carthage. However, when in 623 he went out to Heraclea Lyncestis to meet the Avar chagan, he brought the city with him, seeking to impress with horse races and ceremonial performed as if in the capital, and for that reason took along contingents of the circus factions.[18]

Although Heraclius spent more time outside Constantinople than any emperor since Theodosius I, he was acutely aware that the city and its suburbs, their palaces and churches, were elements of a grand imperial stage. There is confusion about where and when Heraclius became emperor, with reports that he was acclaimed long before he reached the city, then again at the Hebdomon Palace's parade grounds, before he was crowned by Patriarch Sergius in a church, which may have been St Stephen's chapel in the Great Palace, the Church of St Thomas, or Hagia Sophia. It is possible that rituals took place at all those locations. Immediately after his coronation Heraclius was married to Fabia, who took the name Eudocia. Nine months later, Eudocia gave birth to a daughter, Epiphania, at the imperial summer villa at Hieria. Epiphania was baptised at another suburban palace, the Blachernae, a month later, in August 611. Nine months after that, in May 612, Eudocia gave birth to a son, Heraclius Constantine, at yet another seasonal palace, the Sophianae. A short while later, at Blachernae, Fabia Eudocia died. Her body was taken by royal barge along the Golden Horn to the dock at the Great Palace, then in procession through the city to be interred at the Holy Apostles. In October, two months after her mother's death, the young Epiphania was invested as Augusta. The child, 'who was also called Eudocia, was crowned in St Stephen's chapel in the Great Palace. Seated in a chariot ... she departed as is customary for the Great Church'. The following January, Heraclius Constantine was crowned emperor in the Great Palace, then 'ascended into the hippodrome and there, wearing the crown, received the obeisance of the senators as emperor, and was acclaimed by the factions'. It was presumably around this time that Heraclius erected a gilded equestrian statue of his cousin Nicetas in the Forum of Constantine, since the honorand had arrived in Constantinople in 612 and was made a patrician. All of these exceptional events took place within two years within a packed calendar of regular ceremonial activity.[19]

When Heraclius married his niece Martina, her coronation as Augusta took place in the Augustaeum, the plaza outside the Great Palace dedicated to empresses. If this took place as late as 622, it marks the end of the longest continuous period that Heraclius spent in Constantinople. On the next occasion when he spent an extended time in the city his tenuous hold on power led him to humiliate and imprison his brother, and then mutilate and exile both his bastard son and his nephew. Shortly after his death, his wife and sons would all be removed from office, mutilated, exiled and executed. Mutilating living emperors who were found wanting, as well as their relatives or others who aspired to the office, became commonplace in the seventh century. Exile, when granted, became a prelude to murder. The burning of imperial bodies denied them a Christian burial. At the same time, for those considered to have reigned justly, their bodies and those of their relatives were interred with ceremony and with their insignia of office in the imperial mausolea at the Church of the Holy Apostles. Reputations could be destroyed as effectively as bodies. Some emperors still sport the epithets their enemies fashioned for them, such as Leo 'the Butcher'. No emperor was considered 'Great' during his own lifetime, but several gained that epithet posthumously, to contrast their achievements, or more accurately the record of those achievements, with those of emperors who were truly atrocious or whom the winners traduced.

The end of the 'statue habit'

The image of the Roman emperor was assiduously cultivated throughout antiquity and each emperor played an active role in the articulation of his image. At the start of our period, harmony between east and west was projected by textual, visual and material media assiduously produced by Theodosius's sons: Arcadius, who ruled from Constantinople, and Honorius, whose court was in Milan then Ravenna. At the end of our period, emperors cast themselves as solitary Davidic priest-kings, fighting in single combat to defend the Christian world order. However, between Theodosius I and Heraclius, no eastern Roman emperor led his own army and most left Constantinople rarely, restricting travel to suburban palaces and hunting grounds or visiting shrines and holy persons and sites. Emperors and their families furnished Constantinople and other major cities with new churches and chapels, and packed them with relics and treasures. All emperors regarded themselves as divinely favoured; as instruments of God and mediators of grace. However, they shared those qualities with others – patriarchs and priests, holy men and monks, empresses and nuns – and competed to demonstrate the right to rule through ostentatious piety and humility, or through brutal acts that history would judge as atrocities. This is the stuff of politics and power, the building blocks for narrative history, which we have set out at length. It is also the subject matter of art history, for which the period under scrutiny was transformative.

Our period witnessed the decline (if not the complete disappearance) of luxury art forms such as metalworking and mosaic art, and the emergence of new forms, such as the portrait icon. The desire for realism employed in icon portraiture, and the quality of representation achieved, for example in the magnificent sixth-century icon of Christ preserved at Mt Sinai, stood in stark contrast to the impressionistic manner in which emperors were portrayed on contemporary coins (see Figure 23, p. 269). The rise of the icon also mirrored the fall of the statue, second only to numismatic art in impressing the imperial image across the empire, and throughout our period there was a broad and precipitous decline in the erection of new statues. This was the case even in cities where the 'statue habit' was strong and very well established. It is only at Constantinople that the habit persisted for a while longer, but it had ended there too before 700.

As we have seen, sources record a number of dedications of imperial statues in Constantinople in various materials, including silver, gilded bronze, bronze, marble and even iron. These were frequently raised on columns, high above the people, dominating a public plaza or square. The columns were fashioned from porphyry, or constructed from hollow white marble drums carved with a spiralling narrative frieze depicting the exploits of the imperial honorand. Perhaps most fantastically, Theodosius I mounted himself on a chariot pulled by elephants atop the Golden Gate. Constantinople was the site of the last known statue dedicated in the east to a western emperor, Valentinian III, who was raised to the throne by his uncle Theodosius II while resident at the eastern court. The last known statue dedicated to a Roman emperor in the west was of Phocas, raised in

Figure 28. Portrait bust of a young man, possibly Arcadius or Theodosius II

608 atop a column in the forum at Rome by the exarch Smaragdus. Two like it were raised in Constantinople.[20]

Imperial dynastic groups were placed at various locations across the city, including at the Forum Tauri, where in June 1949 the portrait head of a young man, perhaps twenty years old, was discovered. He is carved from Pentelic marble, his eyes turned towards the heavens, his hair neatly combed and full (Figure 28). He wears a diadem formed from two rows of pearls lining a broad band, a large jewel centred above the narrow face, which has a flat forehead, fine nose, well-defined lips and small round chin. The diadem has been identified as the imperial stemma as it appeared at the end of the fourth century, and the sovereign as Arcadius, although it could just as easily be his son Theodosius II. There were also statues of important dignitaries and commanders, aristocrats, intellectuals and, uniquely, charioteers, raised throughout both the fifth and sixth centuries. Women were not so well represented as men, but a marble bust of a woman, not an empress, but of sufficient wealth and distinction that she might once have attracted the interest of John Chrysostom, is today to be seen at the Metropolitan Museum of Art. She holds in her right hand a scroll, an indication that she had donated something of value to earn the portrait. A bonnet covers her elaborate hairstyle, a sign of her modesty as well as her sense of fashion (Plate 28).[21]

After the death of Arcadius, new statues of emperors were rarely erected in provincial cities, as once they had been in great numbers, dedicated by the wealthy to demonstrate their loyalty and to affirm the empire's unity. Some regions had never developed a 'statue habit', notably the north-western provinces where cities were sparse, but also Egypt and the Levant, where cities were many and rich. In contrast, in Italy, North Africa, the Aegean and Asia Minor, thousands of inscribed bases from before our period attest to the regular erection of honorific statues to local worthies and to emperors. For the period from the accession of Theodosius I to the death of Arcadius, only around 130 known statues were erected across the whole empire. Of these, thirty-one were statues of emperors raised in eastern cities other than Constantinople. For the two centuries after 450, the total number of new statues is only forty, and those of emperors fewer than a dozen. The very few new statues of dignitaries and emperors erected in the fifth century stood among hundreds of their ancestors in cityscapes that were visually and graphically rich. The disappearance of new inscribed bases and plaques, identifying those honoured by statues, busts and portraits, was part of a broader waning across the empire of epigraphic culture, the practice of writing on and inscribing in stone or bronze. The vast majority of Roman inscriptions are funerary, directed at posterity in the same manner as those far fewer inscriptions, but still many thousands, that commemorate the generosity of individuals, families, societies and emperors.[22]

Between earth and heaven

After Theodosius I it was not until the seventh century that eastern emperors again led their own armies into battle. Generals like Marcian, who led armies before ascending the throne, did not do so as emperors. Without the opportunity to command the personal loyalty of their troops, emperors were vulnerable to strongmen and warlords like Stilicho and Aspar, generals who ruled from behind the throne, and Odoacer who dispensed with an emperor altogether. Some strongmen stepped forward to seize the throne by intrigue and negotiation, including three emperors in succession, Marcian, Leo and Zeno, and later Justin. Some generals marched on Constantinople and seized the throne by threat of violence, including Phocas, who murdered Maurice, and Heraclius, who murdered Phocas. There are more examples of generals marching on Constantinople but failing to oust the sitting emperor, although some gained concessions, like Gainas of Arcadius and Vitalian of Anastasius.

Lacking military standing, fifth-century emperors were portrayed by their supporters as divinely rewarded for their outstanding piety and humility. The most successful emperors were able to project their power over the army without taking personal risks, working through loyal commanders whom they rewarded generously and replaced regularly, not allowing any one to become a threat. Theodosius II managed to avoid military coups through a fifty-year reign despite never leading his army, although the absence of great military victories never saw a true rival emerge. Socrates Scolasticus (7.22) offers a chapter on the virtues of Theodosius II, who is said to have turned the palace into a monastery. The emperor fasted very frequently, and always on Wednesdays and Fridays; rose early each day to pray with his virginal sisters; collected holy books and learned scripture by heart, citing it accurately and often in conversation with bishops; and wore as a cloak the filthy, hairy cassock of a holy man who had died at Constantinople, hoping it would act as a contact relic and imbue him with sanctity. More substantively, Theodosius was said to have mastered 'anger, grief and pleasure', so that 'he never took revenge on anyone by whom he was injured, nor has anyone ever seen him annoyed'. This was written shortly before Theodosius had his lifelong friend Paulinus executed, but that act appears to have been exceptional, and it is further noted that he often pardoned criminals sentenced to death as they were led to the gallows. According to Socrates, Theodosius II's power was that he was "'Meek above all the men which are upon the face of the earth" (Numbers 12:3). It is because of this meekness that God subdued his enemies without martial conflicts.' Rather than lead his army in war, Theodosius prayed, since 'like David he had recourse to God, knowing he is the arbiter of battles'. His father was more Davidic, of course, in having prayed most potently on the battlefield.[23]

The Theodosian court was a hotbed of ostentatious piety and the imperial family engaged in competitive relic collecting. The empress Eudocia, emulating Constantine's mother Helena, travelled to the Holy Land and returned with a range of trophies including the dirty cloak of the terrifying holy man Barsauma, who had earned an impressive reputation for violence against Jews, pagans and bishops he considered heretics. Pulcheria, not to be outdone, engineered the return of the remains of John Chrysostom, and towards the end of her life located the bones of the Forty Martyrs of Sebaste, their resting place in Constantinople revealed to her in a vision. This set a standard that future emperors and empresses, their families and courtiers, sought to meet, gathering relics from all corners of the empire and beyond and depositing them in established or newly dedicated shrines. Relics did not always remain in Constantinople. Many, as we have seen, were sent forth from the city to the battlefield, to be paraded before the troops and to sanctify them.[24]

The emperor moved between secular and sacred hierarchies in a manner denied to all others. This was demonstrated on each occasion that he came before his people in his cathedral church, where graduated architecture created degrees of sacred space that mirrored the progression of the Christian revelation. The altar was fenced off within the sanctuary, the holy of holies, where only those ordained might enter. Alone of lay people, the emperor was permitted inside. When he entered Hagia Sophia on the major feasts of the Christian calendar, he removed his crown, signifying that for this moment he had set aside his earthly dominion. Once and briefly during the liturgical entrance, the emperor was permitted to enter the sanctuary, led by the patriarch, to kiss the altar cloth. Afterwards, this area was off limits to him, and he conducted his role in proceedings from behind the chancel barrier. The initial entry into the holy of holies was understood to signify the emperor's quasi-priestly status, and this is reflected in a famous mosaic portrait of Justinian at the Church of San Vitale in Ravenna. On the north apse wall, Justinian is depicted as the central figure in a processional entry holding a large paten. To his left, before him in the procession, are priests, most conspicuously Bishop Maximian, who is labelled and holds a large cross, and two deacons holding a bejewelled Gospel book and a censer. Behind him, to his right, are his guards, who hold shields marked with the chi-rho and spears (Plate 29).

Justinian rarely left Constantinople and he entrusted his many foreign wars to others, notably the charismatic Belisarius. He still wished, however, to be considered a master of victory. To mark his first success over the Persians a great column topped with an equestrian statue of him was raised in Constantinople. For almost a thousand years this stood both as a dedication to Justinian and a general image of the victorious emperor mounted on his horse wearing a peacock-feathered tiara, the *toupha*, which was adopted from Persian triumphal regalia. According to a contemporary inscription, the statue was cast from captured Persian arms and armour: 'The bronze from the Assyrian spoils moulded the horse and the monarch and Babylon perishing. This is Justinian, whom Julian, holding the balance of the east, erected, his own witness to the slaying of the Persians.'[25]

The image of the Roman emperor between earth and heaven, protected and rewarded by Christ, is nowhere better captured than on the Barberini ivory, a combination of five ivory plaques carved in high relief (Plate 14). Once identified as Constantine, the emperor is now imagined to be either Anastasius or Justinian. Neither man led an army into battle. The emperor is shown as the master of victory, the main figure on the central panel, mounted on a rearing horse that he is coaxing capably to his right with a left hand lightly holding reins. The emperor wears no helmet and carries no weapons, but is distinguished by his diadem, studded with pearls. His horse's bridle is also jewelled. Both emperor and horse face forwards, the mount glancing upwards, directing a docile and loving gaze at the rider. The emperor holds a sceptre in his right hand, which is supported by the right hand of an obscured, bearded man in a distinctive cap, whose large left hand is raised above the horse's rump. The emperor's riding cloak is fixed at his shoulder with a brooch. A personification of the earth, holding a basket of its fruits, supports the emperor's foot as she reclines beneath the horse's raised front hooves. The emperor's toes project from a greave, the top of which is decorated with the head of a lion, beneath a muscular calf and bare knee. Above the emperor's left shoulder, standing on a *globus cruciger*, is a winged victory, holding a palm frond and offering the emperor a crown or garland, now lost. The panel to the left of the emperor has also been lost.

To the emperor's right, the viewer's left, a carved ivory panel depicts a soldier, surely a victorious general, lightly bearded and unhelmeted. His scabbard projects behind his legs as he stands offering the emperor a statuette of a victory holding a garland. Two column capitals are visible above his shoulders and a bag of money sits by his advancing feet. He wears the same cavalry uniform as the emperor, but his greave lacks the lion motif. Naturally, he has no diadem. In the register above the emperor two winged figures, both victories and angels, support the firmament, within which is the bust of Christ raising his right hand in benediction. Christ holds a cruciform sceptre in his left hand. Representations of the sun, moon and a star are scratched above his shoulders. The lowest register depicts embassies arriving to honour the emperor with gifts and tribute, demonstrating his worldly dominion. They both approach a central figure, yet another winged victory, who looks and gestures up towards the emperor. To his right is a group wearing short tunics and Phrygian caps, bringing the emperor a lion, to which the emperor's lion-greaved leg happens to point. Their capes billowing out behind them, the ambassadors also carry gifts. An Indian embassy approaches from the emperor's left. Two men, nude from the waist up and wearing turbans, bow low to the emperor and offer him a tiger and a rather small elephant. The tiger symbolises strength and ferocity, and the Indian ambassadors knew that Roman emperors conceived of wild beasts as akin to the barbarians they defeated in war. One man is weighed down by an enormous tusk, reminding us that the Romans relied on such imports and gifts for the production of the very item we behold. In fact, the Barberini ivory is too large to have come from an Indian elephant, unless it was among the largest Indian tusks ever recorded. At 20.1 cm long, 13.4 cm wide and 3.6 cm thick, with a maximum depth of relief of 2.8 cm, with no evident curvature, the central plaque would probably have been sliced from a larger African forest elephant tusk. The mighty African bush elephant (*Loxodonta africana*) was unknown in antiquity.[26]

Conquering cross, new David

After Justinian, and despite the lengthy campaigns in the west that marked his reign, there was a parting of the ways between eastern and western Christians. The militant Christ wielding his cross as portrayed in the sixth century at Ravenna (see Figure 20, p. 124) continued to appeal to the Franks, as is clear from Venantius Fortunatus's sixth-century hymns *Vexilla regis* and *Pange lingua*. At Charlemagne's court in Aachen, from 800, Christ trampling the lion and serpent was painted into several books, including the Stuttgart and Utrecht Psalters, and carved into at least three ivory book covers. The Douce ivory, produced perhaps for Charles himself, places the Old Testament motif centrally, surrounded by New Testament scenes of Christ's life. Between Venantius and Charlemagne, the Franks had established an empire and had stopped the advance of Islam in the west. In contrast, the eastern Romans had suffered the shock of defeat, but as a recovery began the cross returned to the fore.[27]

The centrality of the victory-bringing cross, *stauros nikopoios*, to military thought becomes clear in the reigns of Maurice and Heraclius. In 591, the emperor had ridden out behind a fragment of the True Cross raised

Figure 29. Gold coin (solidus) of Heraclius and two sons

on a golden spear. The recovery of the wood upon which Christ had been crucified, and which had been captured in the Persian sack of Jerusalem in 614, was not a motivation for Heraclius's wars, and does not feature in literature composed between 622 and 628 (notably the *Persian Expedition* and *Heraclias* of George of Pisidia). However, after the negotiated recovery of the cross, its return to Jerusalem and its transfer to Constantinople, 'New Jerusalem', it became the central motif of the Heraclius story. The cross was presented as the symbol of Christian victory in various historical works, in the later poems of George of Pisidia, and on Heraclius's coins and those of Constans II. Indeed, Heraclius's removal of the cross-on-steps from the reverse of his coins, and its replacement with 'the figures of the three emperors, that is himself and his two sons' was believed by some, according to John of Nikiu, to have led to the emperor's demise (Figure 29). 'After the death of Heraclius they obliterated these figures.'[28]

Andrew of Crete (*c.*660–740) wrote at least three homilies on the exaltation of the cross, in which its military function played a minor but significant role, for example: 'Besides the cross is the victory of emperors and pious generals and armies, the defeat of opponents and the weapon of truth shielding the faithful according to the great David – for Christ is the truth and the cross is the weapon of Christ.' The homily employs the new imperial style, *basileus*, introduced by Heraclius to emphasise the sacred, Davidic nature of the office, and it alludes to Psalm 90 (91).[29]

The Davidic nature of the imperial office was implicit in Constantine's retention of the title *pontifex maximus*, chief priest of all the empire's cults, including Christianity. It had been made explicit since the sixth century, for example in the Sinope Gospels, an imperial commission written in gold letters on purple-dyed parchment. David is depicted on four of five illustrated leaves in imperial costume, remarkably similar to that in a contemporary apse mosaic at Sinai. At the base of an apse mosaic of the Transfiguration at St Catherine's Monastery at Mt Sinai, built on Justinian's orders as a divine fortress and a powerhouse of prayer deep in the desert, one finds a rare surviving sixth-century representation of David (Plate 30). He is shown in the purple robe (*chlamys*) and diadem of the emperor and appears to share Justinian's features. David's bust is immediately below the depiction of Christ, alluding to the genealogy outlined in Romans 1:3: 'Concerning his Son Jesus Christ our Lord, which was made of the seed of David according to the flesh.' Above the standing Christ is another roundel within which one sees the cross. The emperor is David, but so is Christ, for that is his human form; the cross reveals Christ's divine nature, which distinguishes him from the emperor, and is also the symbol of Christ's victory over death. The visual rhetoric of the scene is as compelling as anything one finds in contemporary writings, including the complex imagery conjured by Paul the Silentiary in his description of Hagia Sophia, where David is the only Old Testament figure to be named. It is David's psalms that rise up in the church's narthex, 'a melodious sound pleasing to the ears of Christ'. 'David the meek, whom the divine voice praised, a glorious light', is identified as the progenitor of Christ, David's 'much-sung scion' born of Mary.[30]

Nine silver plates with scenes from the life of David were discovered near the northern coast of Cyprus in 1902. They bear imperial silver stamps, fixing their date of manufacture to between 613 and 630, during the reign of Heraclius. It is likely that the plates were hidden within a few decades, along with other treasures, when the Arabs attacked Cyprus between 649 and 654. The plates are of three sizes: four smaller, four larger, and one very large. Three of the smaller plates show David fighting a lion, fighting a bear, and playing a harp while seated as a messenger approaches him. A fourth small plate (Figure 30) shows David speaking to a soldier, possibly his confrontation with his brother Eliab (1 Samuel 17:28–30), or his encounter with an Egyptian (1 Samuel 30:11–15). The latter two plates show the sun and moon in the heavens above the humans meeting. The medium plates strongly suggest an imperial connection, placing a scene from David's life before an arcaded lintel, an architectural setting used frequently to depict the imperial court. In two scenes, David

Figure 30. David Plate, showing David speaking to a soldier; his confrontation with Eliab

is placed centrally, receiving anointment from Samuel and his arms and armour from Saul. In two other 'court scenes' the place of honour goes to Saul, who is seated on a throne receiving David in one of them, and presiding at the wedding of his daughter to David in the other. The largest plate offers a scene in three registers, the story at 1 Samuel 17:41–51 of David fighting Goliath (Plate 18). In the uppermost register, David and Goliath approach a seated figure, the river from which David collected the stones for his sling. All are beneath the sun and moon in the heavens, shown as on two of the small plates. Here, however, the hand of God projects beyond the firmament, indicating that David will be victorious. This is made manifest in the lower register, the denouement, where David is standing behind the fallen Goliath, cutting his head from his shoulders.

The middle register of the largest silver plate is a scene of single combat, David to the viewer's left confronting Goliath to the right. David stands alone, surrounded by helmeted Philistines, holding only his slingshot and his cloak in place of a shield. In fact, he needs no arms and armour, because he is protected by God. David is shown here, as on all the plates, nimbate, a halo of divine light around his head. There are compelling reasons to associate David with the emperor Heraclius, who at the battle of Nineveh in December 627 had engaged a Persian general, Razates, in single combat. Stories were regularly circulated of Heraclius's remarkable courage and how he was divinely protected. A year before he fought Razates, 'a giant of a man confronted the emperor in the middle of a bridge and attacked him, but the emperor struck him and threw him into the river'. The Persians looked on in wonder and remarked at 'how boldly the Caesar stands in battle, how he fights alone against such a multitude and wards off blows like an anvil'. It was at exactly this time, when he had achieved his greatest victories, that Heraclius began consistently to use the new imperial style, *basileus*, the Greek word used for Old Testament kings. In November 630, Heraclius named a newborn son David.[31]

The image of Heraclius, sacred king and cosmic conqueror, is captured in George of Pisidia's *Heraclias*, where he asks, 'How then did you pass through the deserts like cities, unless it was by passing through the spiritual gate itself?' Describing the emperor's boot, George observes that it is not stained with Persian blood, but with the emperor's own blood, sweat and tears.

> O now you show the true purple
> For it is reddened with immortal dye
> Piously drenched with your sweat
> It remains white although it is purple
> And gleaming with your new and great achievements,
> The more it is worn the more radiant it becomes
> Hail, general of the cosmic rebirth![32]

After Heraclius, as New Rome suffered the shock of defeat, emphasis began to shift from the emperor as divinely inspired to individual soldiers, whose spiritual purity became essential to the Christian empire's survival. Ideas developed for, and discipline demanded of, spiritual warriors – monks and holy men – were transferred to soldiers. The pure, penitent and chaste were prepared to receive spiritual rewards for their sacrifice, and ultimately would wear the martyr's crown. This idea was adopted into Islam, forming the kernel of the earliest formulation of jihad, even as it was debated, and ultimately rejected, in Constantinople.

Christians and Muslims in the seventh century basically agreed on the sanctity of warfare, even as they disagreed on the power of the cross and related imagery. A key issue of disagreement that emerged as the seventh century ended involved the depiction of living things, and a debate raged within eastern Christianity about holy images or icons (*eikones*). The debate over icons, generally known as 'Iconoclasm' forced New Rome to confront once again Christological issues many thought resolved. The veneration of icons, or their destruction, were alternative responses to the issue of authority and knowledge, which had undergone a radical intellectual realignment, in tandem with a total economic and political transformation. A new conception of what constituted knowledge emerged, and within that system the issue of authority, orthodoxy, became paramount.

This fight over icons took place not in a vacuum, but in a political space between emperors, clerics and monks. It set patriarchs against abbots, with monasteries serving as centres for resistance. Christians understood that images were connected to their prototypes, and therefore that ancient idols served as repositories for the supernatural, demonic forces that might be animated by triggers, known and unknown. Surviving statues, since there were no new ones, were branded with the sign of the cross, not as an attempt to Christianise an idol, but rather as an *apotropaion*, a mark to prevent demonic forces from entering or exiting. The *Parastaseis Syntomoi Chronikai*, a semi-mythical account of Constantinople's landmarks, interpreted statues as tellers of fortunes, good and bad, fulfilled and unfulfilled. It evinced no interest in history. The people of Constantinople, fewer than at any time since the city's foundation, lost touch with their past, creating fantastic fictions to explain the cityscape. The Theodosian Obelisk and the Column of Arcadius were imagined to foretell the end of the world and the Last Judgement. This was the world we now call Byzantium.[33]

BIBLIOGRAPHY

All works are cited in full on the first occasion they appear in the notes for each chapter and are abbreviated thereafter. Commonly used abbreviations are listed below, as are works that have been cited more than once in a widely available English translation.

Abbreviations

CAH	*Cambridge Ancient History*
CJ	*Codex Justinianus*
CTh.	*Codex Theodosianus*
DOP	*Dumbarton Oaks Papers*
EHB	*Economic History of Byzantium, from the Seventh through the Fifteenth Century*, ed. A. Laiou (Washington, DC, 2002), 3 vols
OCD	*Oxford Classical Dictionary*, third edn, eds S. Hornblower and A. Spawforth (Oxford, 2005)
ODB	*Oxford Dictionary of Byzantium*, eds A. Kazhdan and A.-M. Talbot (Oxford, 1991), 3 vols
PLRE	*The Prosopography of the Later Roman Empire*, eds A. H. M. Jones, J. Morris and J. R. Martindale (Cambridge, 1971–92), 3 vols
PNAS	*Proceedings of the National Academy of Sciences of the USA*
JRS	*Journal of Roman Studies*

Sources in English translation

Cassiodorus, *Variae The Letters of Cassiodorus*, trans. T. Hodgkin (London, 1886)

Daniel the Stylite 'Life of St. Daniel the Stylite', in *Three Byzantine Saints*, trans. E. Dawes and N. Baynes (London, 1948), pp. 1–84

De Cerimoniis Constantine Porphyrogennetos, The Book of Ceremonies, trans. A. Moffatt and M. Tall (Canberra, 2012), 2 vols

Easter Chronicle Chronicon Paschale 284–628 AD, trans. M. Whitby and Mary Whitby (Liverpool, 1989)

Evagrius *The Ecclesiastical History of Evagrius Scholasicus*, trans. M. Whitby (Liverpool, 2000)

John of Ephesus *The Third Part of the Ecclesiastical History of John of Ephesus*, trans. R. Payne Smith (Oxford, 1860)

John of Nikiu *The Chronicle of John of Nikiu*, trans. R. H. Charles (London, 1916)

Libanius, *Oration* G. Downey, 'Libanius' Oration in Praise of Antioch (Oration XI)', *Proceedings of the American Philosophical Society* 103 (1959), 652–86

Malalas *The Chronicle of John Malalas*, trans. E. Jeffreys, M. Jeffreys and R. Scott (Melbourne, 1986)

Mango, *Art* C. Mango, *The Art of the Byzantine Empire, 312–1453: Sources and Documents* (Englewood Cliffs, NJ, 1972)

Marcellinus Comes *The Chronicle of Marcellinus Comes*, ed. and trans. B. Croke (Sydney, 1995)

Menander Protector *The History of Menander the Guardsman*, trans. R. C. Blockley (Cambridge, 1985)

Nicephorus *Nikephoros, Patriach of Constantinople: Short History*, trans. C. Mango (Washington, DC, 1990)

Philostorgius *Philostorgius, Church History*, trans. P. R. Amidon (Atlanta, GA, 2007)

Procopius, *Wars Prokopios: The Wars of Justinian*, trans. H. B. Dewing, revised A. Kaldellis (Indianapolis, 2014)

Procopius, *Secret History Prokopios: The Secret History with Related Texts*, trans. A. Kaldellis (Indianapolis, 2010)

Sebeos *The Armenian History attributed to Sebeos*, trans. R. Thomson and J. Howard-Johnston (Liverpool, 1999), 2 vols

Simocatta *The History of Theophylact Simocatta*, trans. M. Whitby and Mary Whitby (Oxford, 1986)

Socrates *The Ecclesiastical History of Socrates Scholasticus*, trans. A. C. Zenos, in P. Schaff and B. Wace, eds, *Nicene and Post-Nicene Fathers* (New York, 1890), vol. 2

Sozomen *The Ecclesiastical History of Sozomen*, trans. C. D. Hartranft, in P. Schaff and B. Wace, eds, *Nicene and Post-Nicene Fathers* (New York, 1890), vol. 2

Theodoret, *Letters The Ecclesiastical History, Dialogue, and Letters of Theodoret*, trans. B. Jackson, in P. Schaff and H. Wace, eds, *Nicene and Post- Nicene Fathers* (New York, 1890), vol. 3

Theophanes *The Chronicle of Theophanes Confessor: Byzantine and Near Eastern History, AD 284–813*, trans. C. Mango and R. Scott (Oxford, 1997)

Theophilus of Edessa *Theophilus of Edessa's Chronicle*, trans. R. Hoyland (Liverpool, 2011)

Zosimus *Zosimus, New History*, trans. R. T. Ridley (Canberra, 1982)

NOTES

Introduction

1. The quotations are from J. H. W. G. Liebeschuetz, *The Decline and Fall of the Roman City* (Oxford, 2001), pp. 248–9.

1. Life at the End of the 'Lead Age'

1. M. Johnson, 'On the burial places of the Valentinian dynasty', *Historia* 40 (1991), 501–6; B. Brenk, *The Apse, the Image and the Icon* (Wiesbaden, 2010), pp. 20–22, 52–3.
2. S. Manning, 'The Roman world and climate: context, relevance of climate change, and some issues', in W. V. Harris, ed., *The Ancient Mediterranean Environment between Science and History* (Leiden, 2013), pp. 103–70. A powerful synthetic argument by K. Harper, *The Fate of Rome: Climate, Disease, and the End of an Empire* (Princeton, 2017), has met with some criticism, since it offers certainty that troubles some scholars, but it is replete with fascinating data presented lucidly. An excellent collection of essays established the current state of the art (the science) and highlights further opportunities for historians: W. Scheidel, ed., *The Science of Roman History: Biology, Climate, and the Future of the Past* (Princeton, 2018). Initiatives at both Harvard and Princeton universities, led by Michael McCormick and John Haldon, continue to expand our field of vision.
3. U. Büntgen et al., 'Cooling and societal change during the Late Antique Little Ice Age from 536 to around 660 AD', *Nature Geoscience* 9 (2016), 231–6, with ongoing discussion; A. Izdebski et al., 'The environmental, archaeological and historical evidence for regional climatic changes and their societal impacts in the Eastern Mediterranean in Late Antiquity', *Quaternary Science Reviews* 136 (2016), 189–208; K. Harper and M. McCormick, 'Reconstructing the Roman climate', in Scheidel, ed., *Science of Roman History*, pp. 11–52; J. Haldon et al., 'Plagues, climate change,

and the end of an empire: a response to Kyle Harper's *The Fate of Rome*', *History Compass* 16:12 (2018), in three parts: e12506, e12507, e12508.

4. A. Izdebski, 'Why did agriculture flourish in the late antique East? The role of climate fluctuations in the development and contraction of agriculture in Asia Minor and the Middle East from the 4th till the 7th c. AD', *Jahrbuch zu Kultur und Geschichte des ersten Jahrtausends n. Chr.* 8 (2011), 291–312; N. Roberts, 'Revisiting the Beyşehir Occupation Phase: land-cover change and the rural economy in the eastern Mediterranean during the first millennium AD', *Late Antique Archaeology* 11 (2018), 53–68.

5. Local variations in the lead content of ores produce an isotopic signature, which allows us to distinguish between some sources of lead (although not always precisely, and there are many overlaps). See J. R. McConnell et al., 'Lead pollution recorded in Greenland ice indicates European emissions tracked plagues, wars and imperial expansion during antiquity', *PNAS* 115 (2018), 5726–31, offers a worryingly precise correlation between specific historical events and the environmental record. K. Rosman et al., 'Lead from Carthaginian and Roman Spanish mines isotopically identified in Greenland ice dated from 600 BC to 300 AD', *Environmental Science and Technology* 31 (1997), 3413–16, identifies Spanish lead in the Greenland cores. Generally, on the state of the art and limitations of isotope analysis, see F. Cattin et al., 'Lead isotopes and archaeometallurgy', *Archaeological and Anthropological Sciences* 1 (2009), 137–48.

6. M. Wagreich and E. Draganits, 'Early mining and smelting lead anomalies in geological archives as potential stratigraphic markers for the base of an early Anthropocene', *The Anthropocene Review* 5 (2018), 177–201; G. Schettler and R. L. Romer, 'Atmospheric Pb-pollution by pre-medieval mining detected in the sediments of the brackish karst lake An Loch Mór, western Ireland', *Applied Geochemistry* 21 (2006), 58–82; I. Renberg et al., 'Preindustrial atmospheric lead contamination detected in Swedish lake-sediments', *Nature* 368/6469 (1994) 323–6; W. A. Marshall et al., 'The isotopic record of atmospheric lead fall-out on an Icelandic salt marsh since AD 50', *The Science of the Total Environment* 407 (2009), 2734–48; W. Shotyk et al., 'Accumulation rates and predominant atmospheric sources of natural and anthropogenic Hg and Pb on the Faroe Islands', *Geochimica et Cosmochimica Acta* 69 (2005), 1–17; S. Hong et al., 'Greenland ice evidence of hemispheric lead pollution two millennia ago by Greek and Roman civilizations', *Science* 265 (1994), 1841–3; S. Hong et al., 'History of ancient copper smelting pollution during Roman and Medieval times recorded in Greenland ice', *Science* 272 (1996), 246–9; S. Preunkert et al., 'Lead and antimony in basal ice from Col du Dome (French Alps) dated with radiocarbon: a record of pollution during antiquity', *Geophysical Research Letters* 46 (2019), 4953–61.

7. K. Butcher and M. Ponting, *The Metallurgy of Roman Silver Coinage: From the*

Reform of Nero to the Reform of Trajan (Cambridge, 2015), which is the first of three projected volumes.

8. A. Hillman et al., 'Lead pollution resulting from gold extraction in northwestern Spain', *The Holocene* 27 (2017), 1465–74. The Roman peak of lead pollution was twice that of the modern peak, related to the use of leaded fuels, which is evident in the same lake sediment core. Since the modern peak globally in lead pollution in around 1990 is significantly higher than that associated with Roman metallurgy, it is clear that the contamination of the Spanish lake bed reflects local smelting activity. A. Martínez Cortizas et al., 'Atmospheric Pb deposition in Spain during the last 4600 years recorded by two ombrotrophic peat bogs and implications for the use of peat as archive', *Science of the Total Environment* 292 (2002), 33–44. An ombrotrophic peat bog receives all its moisture from precipitation and not from streams or springs, so offers an extremely reliable record of environmental pollution.

9. H. Delile et al., 'Lead in ancient Rome's city waters', *PNAS* 111 (2014), 6594–9, by examining sediment cores from the Trajanic harbour basin, observed that piped water in the city of Rome contained one hundred times the amount of lead as there was in local spring water. See now also H. Delile et al., 'A lead isotope perspective on urban development in ancient Naples', *PNAS* 113 (2016), 6148–53; H. Delile et al., 'Rome's urban history inferred from Pb-contaminated waters trapped in its ancient harbor basins', *PNAS* 114 (2017), 10059–64.

10. J. Boulakia, 'Lead in the Roman World', *American Journal of Archaeology* 76 (1972), 139–44; M. Still, *Roman Lead Sealings*, 2 vols, unpublished PhD dissertation, University College London, 1995; E. Welcomme et al., 'Investigation of white pigments used as make-up during the Greco-Roman period', *Applied Physics A* 83 (2006), 551–6; S. Stewart, '"Gleaming and deadly white": toxic cosmetics in the Roman world', in P. Wexler, ed., *Toxicology in Antiquity* (London, 2018), pp. 301–11. More than two hundred lead coffins have been discovered in Britain, and even more in the Levant, but very few in comparison in Spain.

11. R. Homsher et al., 'From the Bronze Age to the "Lead Age": observations on sediment analyses at two archaeological sites in the Jezreel Valley, Israel', *Mediterranean Archaeology and Archaeometry* 16 (2016), 203–20; A. Véron et al., 'Pollutant lead reveals the pre-Hellenistic occupation and ancient growth of Alexandria', *Geophysical Research Letters* 33 (2006), L06409 (four pages); Y. Kahanov and D. Ashkenazi, 'Lead sheathing of ship hulls in the Roman period: archaeometallurgical characterisation', *Materials Characterization* 62 (2011), 768–74; B. Rosen and E. Galili, 'Lead use on Roman ships and its environmental effects', *International Journal of Nautical Archaeology* 36 (2007), 300–307; A. Mosyak et al., 'Thermodynamics of a brazier cooking system modeled to mimic the lead brazier of a Roman ship', *Journal of Archaeological Science: Reports* 16 (2017), 19–26.

12. J. Montgomery et al., 'Gleaming, white, and deadly: using lead to track human exposure and geographic origins in the Roman period in Britain', *Journal of Roman*

Archaeology, supplementary series, 78 (2010), 199–226, 210. Unlike lead itself, galena and mine waste are not soluble in water and therefore cannot be absorbed into the body through contamination of food or water supplies. A. T. Hodge, 'Vitruvius, lead pipes and lead poisoning', *American Journal of Archaeology* 85 (1981), 486–91. There is an established literature on lead poisoning in antiquity and popular theories have been advanced by medics and environmental scientists, but largely rejected by historians, classicists and archaeologists. See S. C. Gilfillan, 'Lead poisoning and the fall of Rome', *Journal of Occupational Medicine* 7 (1965), 53–60; T. Waldron, 'Lead poisoning in the ancient world', *Medical History* 17 (1973), 391–8; J. Nriagu, *Lead and Lead Poisoning in Antiquity* (New York, 1983). More recent research posits the question: P. Charlier et al., 'Did the Romans die of antimony poisoning? The case of a Pompeii water pipe (79 CE)', *Toxicology Letters* 281 (2017), 184–6. Vitruvius, *De Architectura*, 8.6.10–11, 'Cerussite in particular is said to be injurious to the human system'. Strabo, *Geography*, 3.2.10, identified 40,000 slaves at a single Spanish mine, Cartagena. J. Eisinger, 'Lead and wine. Eberhard Gockel and the *colica Pictonum*', *Medical History* 26 (1982), 279–302.

13. G. Müldner, 'Stable isotopes and diet: their contribution to Romano-British research', *Antiquity* 87 (2013), 137–49; T. Waldron, 'The exposure of some Romano-British populations to lead', *Anthropologie* 26 (1988), 67–73; J. Rogers and T. Waldron, 'Lead concentrations in bones from a Neolithic long barrow', *Journal of Archaeological Science* 12 (1985), 93–6. S. R. Scott et al., 'Elevated lead exposure in Roman occupants of Londinium: new evidence from the archaeological record', *Archaeometry* 62 (2020), 109–29, which compares bones from Neolithic and Roman age London to determine that lead "concentrations of lead in the Roman/Londinium-era femora were more than 70-fold greater than those from pre-Roman populations". Inhumation replaced cremation as the principal method of burial in Britain only in the third century, after the peak of lead pollution related to smelting, so one might posit higher concentrations in the first and second centuries. However, the remains of cremation – typically crushed burnt bone in a pottery vessel – do not lend themselves to comparable analysis. A typical cremation of an adult woman might leave 1.5 kg of bone fragments and ashes, of a man somewhat more. Far less than this is generally found in well-sealed contexts.

14. J. Longman et al., 'Exceptionally high levels of lead pollution in the Balkans from the early Bronze Age to the Industrial Revolution', *PNAS* 115 (2018) E5661–8; K. Aslihan Yener and A. Toydemir, 'Byzantine silver mines: an archaeometallurgy project in Turkey', in S. Boyd and M. Mundell Mango, eds, *Ecclesiastical Silver Plate in Sixth-Century Byzantium* (Washington, DC, 1993), pp. 155–67.

15. The laws are *CTh.* 10.19.15 and *CTh.* 10.9.5. A related law suggests that miners had absconded to Sardinia by ship, and that ships' captains were liable for a fine of five solidi for each miner they helped to escape. Evidently, the practice continued, and a further law of 378 returned to the matter of miners absconding to Sardinia, notably gold miners.

16. Malalas 18.54, trans. Jeffreys et al., p. 267. C. Meyer, 'Bi'r Umm Fawâkhir: gold mining in Byzantine times in the Eastern Desert', in J.-P. Brun et al., eds, *The Eastern Desert of Egypt during the Greco-Roman Period: Archaeological Reports* (Paris, 2018), online (no pagination). See also K. P. Matschke, 'Mining', in *EHB*, pp. 115–20; C. Entwhistle and A. Meek, 'Early Byzantine glass weights: aspects of function, typology and composition', *British Museum Technical Research Bulletin* 9 (2015), 611–14.

17. W. V. Harris, 'Defining and detecting Mediterranean deforestation, 800 BCE to 700 CE', in W. V. Harris, ed., *The Ancient Mediterranean Environment between Science and History* (Leiden, 2013), pp. 173–94, and other papers from this fine collection, notably the critical conclusion by A. Wilson, and the paper on charcoal by R. Veal.

18. J. Grattan et al., 'Analysis of copper and lead mineralizations in human skeletons excavated from an ancient mining and smelting centre in the Jordanian desert: a reconnaissance study', *Mineralogical Magazine* 69 (2005), 653–66; J. Grattan et al., 'The local and global dimensions of metalliferous pollution derived from a reconstruction of an eight thousand year record of copper smelting and mining at a desert-mountain frontier in southern Jordan', *Journal of Archaeological Science* 34 (2007), 83–110; M. Perry et al., 'Condemned to *metallum*? The origin and role of 4th–6th century A.D. *Phaeno* mining camp residents using multiple chemical techniques', *Journal of Archaeological Science* 38 (2011), 558–69.

19. However, there is some evidence that lead poisoning remained a risk for those who inherited Roman infrastructure and habits or where mining continued: R. Paine et al., 'A health assessment of high status Christian burials recovered from the Roman-Byzantine archaeological site of Elaiussa Sebaste, Turkey', *Journal of Comparative Human Biology* 58 (2007), 173–90. D. Brothwell, 'The human bones', in R. M. Harrison, ed., *Excavations at Saraçhane in Istanbul* (Princeton and Washington, DC, 1986), I, pp. 374–98, offered an average life expectancy at birth of twenty-eight in sixth-century Constantinople, which is relatively high compared with other Mediterranean contexts (and perhaps due to the fragmentary nature of much of the evidence, which was hard to interpret). More comprehensive are material and analysis from Byzantine Crete, presented by Chryssi Bourbou in several articles and her excellent monograph *Health and Disease in Byzantine Crete (7th–12th centuries AD)* (Farnham, 2010), which suggests life expectancy at birth on Crete ranged from twenty-three at Stylos to twenty-nine at Gortyn.

20. K. Harper, *The Fate of Rome: Climate, Disease, and the End of an Empire* (Princeton, 2017), pp. 65–118, is compelling and comprehensive. On malaria, see R. Sallares, *Malaria and Rome: A History of Malaria in Ancient Italy* (Oxford, 2002) and his subsequent studies.

21. See W. Jongman et al., 'Health and wealth in the Roman empire', *Economics and Human Biology* 34 (2019), 138–50, currently the largest analysis of Roman skeletal data.

22. C. Radu et al., 'Bioanthropological data for a skeletal sample retrieved from the extra-mural necropolises at Histria', in *Histria: histoire et archéologie en Mer Noire, Pontica* 47, supplementum III (Constanţa, 2014), pp. 283–97.

23. J. Koder, 'Fresh vegetables for the Capital', in C. Mango and G. Dagron, *Constantinople and its Hinterland* (Aldershot, 1995), pp. 49–56; J. Koder, 'Stew and salted meat – opulent normality in the diet of every day?', in L. Brubaker and K. Linardou, eds, *Eat, Drink and Be Merry (Luke 12:19): Food and Wine in Byzantium* (Aldershot, 2007), pp. 59–72.

24. G. Jansen, 'The toilets of Ephesus: a preliminary report', in G. Wiplinger, ed., *Cura Aquarum in Ephesus* (Leuven, 2006), pp. 95–8; M. L. Ledger et al., 'Intestinal parasites from public and private latrines and the harbour canal in Roman period Ephesus, Turkey (1st *c.* BCE to 6th *c.* CE)', *Journal of Archaeological Science Reports* 21 (2018), 289–97; J. Baeten et al., 'Faecal biomarker and archaeobotanical analyses of sediments from a public latrine shed new light on ruralisation in Sagalassos, Turkey', *Journal of Archaeological Science* 39 (2012), 1143–59; M. Flohr and A. Wilson, 'The economy of ordure', in A. Kolowski-Ostrow et al., eds, *Roman Toilets: Their Archaeology and Cultural History* (Leuven, 2011), pp. 147–56.

25. R. J. A. Wilson et al., 'Funerary feasting in Early Byzantine Sicily: new evidence from Kaukana', *American Journal of Archaeology* 115 (2011), 263–302; J. Ramsay and R. J. A. Wilson, 'Funerary dining in Early Byzantine Sicily: archaeobotanical evidence from Kaukana', *Mediterranean Archaeology* 25 (2012), 81–93; R. J. A. Wilson, 'An unusual early Byzantine "thrice-holy" inscription and accompanying design from Punta Secca, Sicily', *Phoenix* 67 (2013), 163–81.

26. E. Galili et al., 'Anchors bearing apotropaic inscriptions and symbols from Israel', in A. Zemer, ed., *Seafarers' Rituals in Ancient Times* (Haifa, 2016), pp. 122–9.

27. *The Letters of Synesius of Cyrene*, trans. A. Fitzgerald (London, 1926), p. 157. See A. Cameron and J. Long, *Barbarians and Politics at the Court of Arcadius* (Berkeley, 1993), pp. 17–19.

28. See generally E. Dauterman Maguire, H. Maguire and M. Duncan-Flowers, *Art and Holy Powers in the Early Christian House* (Urbana and Chicago, 1989). On this image, and on Sisinnius more generally, see M. Immerzeel, 'Holy horsemen and crusader banners: equestrian saints in wall paintings in Lebanon and Syria', *Eastern Christian Art* 1 (2004), 29–60.

29. P. Hatlie, *The Monks and Monasteries of Constantinople, ca. 350–850* (Cambridge, 2007), pp. 102–10.

30. 'Life of St. Daniel the Stylite', in *Three Byzantine Saints*, trans. E. Dawes and N. Baynes (Oxford, 1948), pp. 1–84.

31. *The Miracles of St. Artemios: A Collection of Miracle Stories by an Anonymous Author of Seventh-Century Byzantium*, trans. V. Crisafulli and J. Nesbitt (Leiden, 1997).

32. *ODB*, pp. 58 (Alexander of Tralles), pp. 1327–8 (Medicine), p. 1473 (Nicander), pp. 1607–8 (Paul of Aegina).

33. Now that each map has been drawn to the reader's attention, I shall refer to a map only when it offers insight into the narrative.

2. Family and Faith

1. *P. Oxy.* I. 129, published and translated at *The Oxyrhynchus Papyri*, I, ed. and trans. B. Grenfell and A. Hunt (London, 1898), pp. 200–201. The Greek text of *P. Oxy.* I. 129 can also be seen at papyri.info/ddbdp/p.oxy;1;130

2. Cameron and Long, *Barbarians and Politics at the Court of Arcadius* (Berkeley, 1993), pp. 18–19; A. Laiou, 'Family structure and transmission of property', in J. Haldon, ed., *The Social History of Byzantium* (Oxford, 2009), pp. 51–75. For the division of an estate valued at 360 solidi between the son and two daughters of a certain Paulus, see *Oxyrhynchus Papyri*, I, pp. 205–6. *P. Oxy.* XVI. 1895. Greek text at: papyri.info/ddbdp/p.oxy;16;1895

3. See B. Porten, *The Elephantine Papyri in English: Three Millennia of Cross-Cultural Continuity and Change* (Leiden, 1996), pp. 396, 503–5, for the fuller context of the will, including family disputes and property transactions, and an English translation. The Greek text of *P. Lond.* 1727 can be seen at papyri.info/ddbdp/p.lond;5;1727

4. The papyrus is *P. Münch.*14. See Porten, *Elephantine Papyri*, p. 509. Also, G. Husson, 'Houses in Syene in the Patermouthis archive', *Bulletin of the American Society of Papyrologists* 27 (1990), 123–37; J. Keenan, 'Evidence for the Byzantine army in the Syene papyri', *Bulletin of the American Society of Papyrologists* 27 (1990), 139–50, at 143–4.

5. See J. Evans, *The Age of Justinian: The Circumstances of Imperial Power* (Abingdon, 1996), pp. 207–8.

6. G. Prinzing, 'On slaves and slavery', in P. Stephenson, ed., *The Byzantine World* (London), pp. 92–102, discusses the terms *therapon* and *therapaina*, *oiketes, oikogenes, pais, paidiske* and *threptos, paidarion, psycharion, soma* and *threptarion*.

7. *P. Oxy.* I. 130. The translation is developed from that at *Oxyrhynchus Papyri*, I, pp. 201–3. See also P. Sarris, *Economy and Society in the Age of Justinian* (Cambridge, 2006); P. Sarris, 'Social relations and the land: the early period', in Haldon, ed., *Social History*, pp. 92–111, at p. 104. The Greek text of *P. Oxy.* I. 130 can be seen at papyri.info/ddbdp/p.oxy;1;130

8. The Greek text of *P. Oxy.* XVI. 1918 can be seen at papyri.info/ddbdp/p.oxy;16;1918 G. Ruffini, *Social Networks in Byzantine Egypt* (Cambridge, 2008), p. 105, cites a second document of AD 580 that reveals remarkably similar figures for revenue and tax payments. However, it is suggested elsewhere that a portion of the revenue here was simply processed by the Apions as tax farmers, so the rate of taxation on their own lands was closer to 48 per cent. See G. Bransbourg, 'Capital in the sixth

century: the dynamics of tax and estate in Roman Egypt', *Journal of Late Antiquity* 9 (2016), 305–414, at 314–15.

9. Bransbourg, 'Capital in the sixth century', 361, estimates that the estates owned or managed by the Apions grew by *c.*30 per cent between 540 and 587. However, he later (395–400) posits bursts of rapid growth before 540 and after 580 and is unable to identify any substantial boost to the estates from land abandoned due to population decline.

10. Bransbourg, 'Capital in the sixth century', 319, summarises the system well: 'What sixth-century Egyptian papyri often display on the ground are two distinct channels of tax collection, the grain of the *embole* on one hand, and everything else, combining gold *kanonika* [the general tax levies, assessed in gold] and local *annona* in kind [i.e. grain and produce] or gold, on the other ... The *embole* grain ended up with local boatmen responsible for its transportation to Alexandria, while the gold and local payments in kind were handled by specific officials at the village, city, or large estate level, irrespective of these payments' final destination.'

11. Bransbourg, 'Capital in the sixth century', 314: 'The Apions contribute 25% of the total or 30% once [an additional levy for the baths at the village of] Takona is added ... the Apions seem invariably to be assessed at a rate about 1/3 of the overall figure.' See also Ruffini, *Social Networks*, pp. 102–4, 117; P. Van Minnen, 'The other cities in later Roman Egypt', in R. Bagnall, ed., *Egypt in the Byzantine World 300–700* (Cambridge, 2007), pp. 207–25, at p. 221. The Greek text of *P. Oxy.* XVI. 2040 can be seen at papyri.info/ddbdp/p.oxy;16;2040

The Greek text of *P. Oxy.* I. 135 can be seen at papyri.info/ddbdp/p.oxy;1;135

12. Aphrodite was a village that had won and jealously guarded its right to pay its own taxes (*autopragia*), which were collected by a council of leading villagers (*syntelestai*). When this was threatened the villagers sought the patronage of the empress Theodora, and it was only after her death that their 'autopract' status was lost.

13. J. Keenan, 'Aurelius Phoibammon, son of Triadelphus: a Byzantine Egyptian land entrepreneur', *Bulletin of the American Society of Papyrologists* 17 (1980), 145–54; J. Keenan, 'Byzantine Egyptian villages', in Bagnall, ed., *Egypt in the Byzantine World*, pp. 226–43; Ruffini, *Social Networks*, pp. 147–97.

14. Libanius, *Oration* 11.19–21; trans. Downey, p. 677. For commentary and archaeological evidence in support of this passage, see C. Morrisson and J.-P. Sodini, 'The sixth century economy', in *EHB*, pp. 171–220, at pp. 179–81.

15. Libanius, *Oration* 11.230; trans. Downey, p. 657.

16. Libanius, *Oration* 47.4, interpreted in context by J. Banaji, *Agrarian Change in Late Antiquity: Gold, Labour and Aristocratic Dominance* (Oxford, 2007), pp. 10–12.

17. Sarris, *Economy and Society*, p. 119: 'Throughout the work, Theodore can be seen maintaining and restoring peace and harmony to agrarian social relations through

his interventions, exorcisms and cures. This picture, should not, however, be taken at face value.'

18. 'The Life of St. Theodore of Sykeon', trans. E. Dawes and N. Baynes, in *Three Byzantine Saints* (Oxford, 1948), pp. 88–192. One is reminded of Peter Brown's observation on the character of Monica in Augustine's *Confessions* that few mothers can survive being presented to us exclusively in terms of what they have come to mean to their sons.

19. 'The Life of St. Theodore of Sykeon', pp. 117–18. See also R. Leader-Newby, *Silver and Society in Late Antiquity: Functions and Meanings of Silver Plate in the Fourth to Seventh Centuries* (Aldershot, 2004), pp. 66–72.

20. Chrysostom, *Homily 7 on Colossians*, perhaps alluding to the episode in Petronius's *Satyricon* where Trimalchio relieves himself into such a vessel while playing a ball game at the public baths. Leader-Newby, *Silver*, p. 112.

21. S. Boyd, 'A metropolitan treasure from a church in the provinces: an introduction to the study of the Sion Treasure', in S. Boyd and M. Mundell Mango, eds, *Ecclesiastical Silver Plate in Sixth-Century Byzantium* (Washington, DC, 1992), pp. 5–38; Leader-Newby, *Silver*, pp. 82–97. Repoussé is the technique of tamping sheet metal from the back to produce a decorative raised surface.

22. M. Mundell Mango, *Silver from Early Byzantium: The Kaper Koraon and Related Treasures* (Washington, DC, 1986).

23. N. Lenski, 'Valens and the monks: cudgeling and conscription as a means of social control', *DOP* 58 (2004), 93–117; P. Hatlie, *The Monks and Monasteries of Constantinople, ca. 350–850* (Cambridge, 2007), pp. 62–132. On the life of Hypatius, see now R. Kosinski, *Holiness and Power: Constantinopolitan Holy Men and Authority in the Fifth Century* (Berlin, 2016). We shall return to Chrysostom and Olympias in a later chapter.

24. Malalas, 18.18, trans. Jeffreys et al., p. 253.

25. British Library Papyrus 2019, also SB VI 8986, TM 17839. See papyri.info/ddbdp/sb;6;8986

As this book went to press, the author became aware of a wonderful collection that captures many stories told in the papyrus archive. See S. Huebner et al., eds, *Living the End of Antiquity* (Berlin, 2020).

3. An Empire of Cities

1. On Pausanias, see F. Millar, *The Emperor in the Roman World* (Bristol, 1992), pp. 394–5; C. Ando, 'The Roman city in the Roman period', in S. Benoist, ed., *Rome, a City and Its Empire in Perspective: The Impact of the Roman World through Fergus Millar's Research* (Brill, 2012), pp. 113–14.

2. Libanius, *Oration* 11.214–15; trans. Downey, pp. 675–6.

3. J. C. Russell, 'Late ancient and medieval population', in *Transactions of the American Philosophical Society* 48 (1958), 1–152; A. Laiou, 'Human resources', in *EHB*, pp. 47–55.

4. K. Hopwood, 'The links between the coastal cities of western Rough Cilicia and the interior during the Roman period', *Anatolia Antiqua* 1 (1991), 305–10.

5. G. de Kleijn, 'Fistulae and water fraud in late antique Constantinople', in B. Shilling and P. Stephenson, eds, *Fountains and Water Culture in Byzantium* (Cambridge, 2016), pp. 55–67.

6. A. Wilson et al., 'Urination and defecation Roman-style', in A. O. Koloski-Ostrow et al., eds, *Roman Toilets: Their Archaeology and Cultural History* (Leiden, 2011), pp. 95–111. Separate latrines for women have been identified, for example at the Asclepius sanctuary at Pergamum. However, women would mostly be expected to go at home.

7. G. Downey, *A History of Antioch in Syria from Seleucus to the Arab Conquest* (Princeton, 1961); S. Ashbrook Harvey, 'Antioch and Christianity', in C. Kondoleon, ed., *Antioch: The Lost Ancient City* (Princeton, NJ, 2000), pp. 38–49.

8. Libanius, *Oration* 11.244–5; trans. Downey, pp. 678–9. See also Kondoleon, *Antioch*, pp. 144–5.

9. Nonnus, *Dionysiaca*, 9.25–6, trans. W. H. D. Rouse (Cambridge, MA, 1940), I, pp. 306–7. On the mosaic and this quotation see R. Molholt, 'Mosaic of Hermes carrying the infant Dionysos', in L. Becker and C. Kondoleon, eds, *The Arts of Antioch* (Worcester, MA, 2005), pp. 190–95. See also Alan Cameron, 'Poets and pagans in Byzantine Egypt', in his *Wandering Poets and other Essays on Late Greek Literature and Philosophy* (Oxford, 2016), pp. 147–62, at p. 157.

10. Malalas, 12.38; trans. Jeffreys, Jeffreys and Scott, p. 167. See F. Yegül, 'Baths and bathing in Roman Antioch', in Kondoleon, ed., *Antioch*, pp. 146–51.

11. Libanius, *Oration* 11.207; trans. Downey, p. 675. Shortly after his death, the older Symeon was emulated by a namesake, a native of Antioch who set up his pillar even closer to the city.

12. K. Weitzmann, *Age of Spirituality: Late Antique and Early Christian Art, Third to Seventh Century* (New York, 1979), pp. 176–7, no. 155; Kondoleon, *Antioch*, pp. 116–17.

13. R. Bagnall, *Early Christian Books in Egypt* (Princeton, 2009), pp. 6–7. On population see P. Sarris, *Economy and Society in the Age of Justinian* (Cambridge, 2006), pp. 10–11. D. Delia, 'The population of Roman Alexandria', *Transactions of the American Philological Association* 118 (1988), 275–92, suggested a population of 500,000; more than the 300,000 estimated by Diodorus Siculus in his day, *c.* 60 BC.

14. *Eusebius, Life of Constantine*, II. 64–72, trans. Averil Cameron and S. G. Hall (Oxford, 1999), pp. 116–19, 250–53. See P. Stephenson, *Constantine: Unconquered Emperor, Christian Victor* (London, 2009), pp. 264–9.

15. *Theodosian Code*, 16.10.10–12; Sozomen 7.12 and 7.20, quoted by Alan Cameron,

The Last Pagans of Rome (Oxford and New York, 2011), pp. 59–74, which is in fact a nuanced critique of scholarship on Theodosius's anti-pagan legislation. See also E. Watts, *Hypatia: The Life and Legend of an Ancient Philosopher* (New York and Oxford, 2017), pp. 45–6, 56–61.

16. Alan Cameron, 'Hypatia: life, death, and works', in his *Wandering Poets*, pp. 185–203, at p. 199.

17. *Letters of Synesius*, trans. Fitzgerald, p. 174.

18. John of Nikiu, 84.88, cited in Watts, *Hypatia*, p. 157, n. 4; Socrates Scholasticus, 7.15.

19. A. H. M. Jones, *The Greek City from Alexander to Justinian* (Oxford, 1940); A. H. M. Jones, *The Later Roman Empire, 284–602*, 3 vols (Oxford, 1964), I, pp. 712–66.

20. T. D. Barnes, 'Synesius in Constantinople', *Greek, Roman, and Byzantine Studies* 27 (1986), 93–112; A. Cameron and J. Long, *Barbarians and Politics at the Court of Arcadius* (Berkeley, 1993), pp. 71–102; N. Lenski, *Constantine and the Cities: Imperial Authority and Civic Politics* (Philadelphia, 2016), pp. 158–9.

21. Lenski, *Constantine and the Cities*, pp. 96–113, 150–64. See also Millar, *Emperor*, p. 410.

22. Jones, *Later Roman Empire*, II, p. 748.

23. I have followed the concise account of J. G. Keenan, 'Egypt', in *CAH* 14, pp. 612–37, at pp. 625–8. T. M. Hickey, *Wine, Wealth, and the State in Late Antique Egypt: The House of Apion at Oxyrhynchus* (Ann Arbor, 2012), offers a fuller account, and at pp. 14–16 suggests that our Flavius Apion and the consul are different men, with the consul remaining in Constantinople. This would require that Anoup's petition was sent rather later, perhaps in the 590s to Apion III. For the Apion diptych, see Alan Cameron and D. Schauer, 'The last consul: Basilius and his diptych', *JRS* 72 (1982), 126–45, at p. 134 and plate VIII. On the riot, see Malalas, 18.135, trans. Jeffreys et al., p. 300.

24. K. Dunbabin, *Mosaics of the Greek and Roman World* (Cambridge, 1999), pp. 116–17; J. N. Adams, ed., *An Anthology of Informal Latin, AD 200–900* (Cambridge, 2017), pp. 354–66.

25. D. Parrish, 'Two mosaics from Roman Tunisia: an African variation of the season theme', *American Journal of Archaeology* 83 (1979), 279–85; Dunbabin, *Mosaics*, pp. 118–21. On the house of Hesychius, see Cameron and Long, *Barbarians and Politics*, pp. 15–16, for references.

26. Dunbabin, *Mosaics*, pp. 116, 130–43; A. Castrorao Barba, 'Sicily before the Muslims. The transformation of the Roman villas between late antiquity and the early middle ages, fourth to eighth centuries CE', *Journal of Transcultural Medieval Studies* 3 (2016), 145–89; P. Pensabene et al., 'A quantitative and qualitative study on marble revetments of service area in the Villa del Casale at Piazza Armerina', in P. Pensabene and E. Gasparini, eds, *Interdisciplinary Studies on Ancient Stone* (Rome, 2015), pp. 641–9; P. Pensabene, 'Il contributo degli scavi 2004–2014 alla

storia della Villa del Casale di Piazza Armerina tra IV e VIII secolo', in C. Giuffrida and M. Cassia, eds, *Silenziose rivoluzioni: la Sicilia dalla tarda antichità al primo medioevo* (Catania, 2016), pp. 223–71.

27. I. Baldini, 'Early Byzantine churches in Crete and Cyprus: between local identity and homologation', *Cahiers du Centre d'Études Chypriotes* 43 (2013), 31–49; R. Kolarik, 'Mosaics of the early church at Stobi', *DOP* 41 (1987), 295–306; B. Shilling, 'Fountains of paradise in early Byzantine art, homilies, and hymns', in Shilling and Stephenson, eds, *Fountains and Water Culture*, pp. 208–28.

28. R. Kolarik, 'Seasonal animals in the narthex mosaic of the large basilica, Heraclea Lyncestis', *Niš and Byzantium* 10 (2012), 105–18.

29. *CTh.* 15.9.1, 15.9.2; A. Cameron, 'The origin, context and function of consular diptychs', *JRS* 103 (2013), 174–207.

30. A. Cameron, *Porphyrius the Charioteer* (Oxford, 1973), pp. 180–87.

31. Cameron, *Porphyrius*, pp. 7–8, 12–58, 96–116, 121–2, 209–11; *Greek Anthology* XV. 41–50; XVI. 335–87. Twenty-seven of the epigrams are also recorded in the Laurentian version of the anthology, and ten are preserved in a later addendum to the Palatine version of the anthology, MS Parisinus gr. 23, 707–9.

32. This notion is set out by H. Inglebert et al., *Histoire de la civilisation romaine* (Paris, 2005), pp. 31–3, and developed critically by J. Arnason, 'The Roman phenomenon: state, empire, and civilization', in J. Arnason and K. Raaflaub, eds, *The Roman Empire in Context* (London, 2010), pp. 351–86.

4. Culture, Communications and Commerce

1. *Edict*, 7.66–71, for which see R. Rees, *Diocletian and the Tetrarchy* (Edinburgh, 2004), p. 146.

2. See most recently the papers in W. A. Johnson and H. N. Parker, eds, *Ancient Literacies: the Culture of Reading in Greece and Rome* (Oxford, 2009).

3. T. Habinek, *The World of Roman Song: From Ritualized Speech to Social Order* (Baltimore, MD, 2005), pp. 1, 61–2, explains that ritualised speech was 'speech made special through the use of specialized diction, regular meter, musical accompaniment, figures of sound, mythical or religious subject matter, and socially authoritative performance context'.

4. *Menander Rhetor*, eds D. Russell and N. Wilson (Oxford, 1981). Menander Rhetor's entry (M590) in the Byzantine encyclopaedia, *Suda*, reveals his place of birth and that he was a sophist, often a synonym for rhetor, who wrote commentaries on Hermogenes and Minucianus. See also G. Kustas, *Studies in Byzantine Rhetoric* (Thessalonike, 1973); H. Maguire, *Art and Eloquence in Byzantium* (Princeton, 1981); and T. Habinek, *Ancient Rhetoric and Oratory* (Oxford, 2005), p. 44. In terms of popularity and influence, Menander's treatise must be placed beside five

books by Hermogenes (fl. later second century AD), on which he offered commentary, and somewhere behind the works of Aphthonius (fl. later fourth century AD), whose rhetorical exercises were the best known and most popular throughout the following millennium. There was also a tradition of rhetorical instruction in Latin, most importantly Cicero's treatise in three books *On Oratory* and the twelve-book instructional handbook by Quintilian, *Institutes of Oratory*.

5. *CTh.* 14.9.3; Averil Cameron, 'Education and literary culture', *CAH* 13, pp. 665–707, at pp. 676, 682–4; A. Sheppard, 'Philosophy and philosophical schools', *CAH* 14, pp. 835–54; A. Cameron and J. Long, *Barbarians and Politics at the Court of Arcadius* (Berkeley, 1993), pp. 39–62.

6. R. Browning, 'Education in the Roman Empire', *CAH* 14, pp. 855–83.

7. P. Rousseau, *Basil of Caesarea* (Berkeley, 1994), pp. 41–2, quoting also in translation from Gregory of Nyssa, *Life of Macrina*, 6. See also Cameron, 'Education and literary culture', p. 668.

8. Synesius, Letter 11, translated in Cameron and Long, *Barbarians and Politics*, p. 22.

9. C. Rapp, *Holy Bishops in Late Antiquity: the Nature of Christian Leadership in an Age of Transition* (Berkeley, 2005).

10. F. Millar, *The Emperor in the Roman World* (Bristol, 1992), p. 364: 'Letters sent by provincial governors, other officials, or exceptionally privileged private persons, could be carried by messengers, but otherwise any form of communication to the emperor necessitated a journey by the interested parties to wherever he currently was. The process thus involved journeys of anything up to a couple of thousand miles, and delays in time which could amount to weeks or months. None the less ... such journeys to the emperor were frequently undertaken ... [and] they were, in a number of different contexts, a standard and accepted part of the way the empire worked.'

11. D. Drakoulis, 'The Synekdemos of Hierokles and his cartographic representation', in A. Tsorlini, ed., *Proceedings of the 10th National Cartographic Conference* [in Greek] (Thessalonike, 2010), pp. 199–224.

12. See now J. Beresford, *The Ancient Sailing Season* (Leiden, 2013).

13. V. von Falkenhausen, 'Reseaux routiers et ports dans l'Italie meridionale byzantine (VIe-Xe s.)', in *Everyday Life in Byzantium* (Athens, 1989), pp. 711–31.

14. A. Avramea, 'Land and sea communications, fourth–fifteenth centuries', in *EHB*, pp. 57–90, at pp. 68–72.

15. D. French, *Roman Roads and Milestones of Asia Minor, IV: the Roads* (Ankara, 2016), electronic monograph.

16. M. Hendy, *Studies in the Byzantine Monetary Economy, c. 300–1450* (Cambridge, 1985), pp. 602–13. The only comparable modern service, the short-lived Pony Express, had stations every ten miles and averaged ten miles per hour across the western states of the US. The riders, on horseback, not in carriages, and weighing

no more than 125 pounds; carried up to forty pounds of mail and provisions in a pannier.

17. Lucian, *The Ship, or the Wishers*, trans. Harmon, pp. 434–7.

18. *The Letters of Synesius of Cyrene*, trans. A. Fitzgerland (Oxford, 1926), pp. 81–2. See also H. Casson, *Travel in the Ancient World* (Baltimore, 1994), pp. 159–62, and passim.

19. S. Leach et al., 'A lady of York: migration, ethnicity and identity in Roman Britain', *Antiquity* 84 (2010), 131–45. There is now a good deal of scientific data identifying individuals who travelled, and migrants. See M. Perry et al., 'Strontium isotope evidence for long-distance immigration into the Byzantine port city of Aila, modern Aqaba, Jordan', *Archaeological and Anthropological Sciences* 9 (2017), 943–64; M. Wong et al., 'Pursuing pilgrims: isotopic investigations of Roman and Byzantine mobility at Hierapolis, Turkey', *Journal of Archaeological Science: Reports* 17 (2018), 520–28. See also H. Shaw et al., 'Identifying migrants in Roman London using lead and strontium stable isotopes', *Journal of Archaeological Science* 66 (2016), 57–68, which identifies Spitalfields woman, a fourth-century resident of Roman London, who is very likely to have grown up in the city of Rome. Strontium in her tooth enamel reveals that she travelled to London after her adult teeth had erupted. A black gingival line around her teeth suggests that this woman suffered from lead poisoning as a young adult. The Spitalfields woman was buried in a coffin of British lead that was quite different from that in her tooth enamel, and which stained her bones brown.

20. A. Wilson, 'Saharan trade in the Roman period: short-, medium- and long-distance trade networks', *Azania: Archaeological Research in Africa* 47 (2012), 409–49; M. Bonifay, 'Africa: patterns of consumption in coastal regions versus inland regions: the ceramic evidence (300–700 AD)', in L. Lavan, ed., *Local Economies? Production and Exchange of Inland Regions in Late Antiquity* (Leiden, 2013), pp. 529–66; W. van Zeist and S. Bottema, 'Palaeobotanical studies at Carthage', *CEDAC Carthage Bulletin* 5 (1983), 18–22.

21. E. Fentress, 'An island in transition: Jerba between the fifth and ninth centuries', in S. Panzram and L. Callegarin, eds, *Entre Civitas y Madīna: El Mundo de las Ciudades en la Península Ibérica y en el Norte de África (Siglos IV–IX)* (Madrid, 2018) pp. 241–52.

22. A. Wilson, 'Cyrenaica and the late antique economy', *Ancient West and East* 3 (2004), 143–54.

23. F. Chevrollier, 'From Cyrene to Gortyn: notes on the relationship between Crete and Cyrenaica under Roman domination (1st century BC – 4th century AD)', in J. Francis and A. Kouremenos, eds, *Roman Crete: New Perspectives* (Oxford, 2016), pp. 11–26.

24. E. Fentress et al., 'Accounting for ARS: fineware and sites in Sicily and Africa', in S. Fontana et al., eds, *Side-by-Side Survey: Comparative Regional Studies in the*

Mediterranean World (Oxford, 2004), pp. 147–62; P. Bes and J. Poblome, 'African red slip ware on the move: the effects of Bonifay's *Études* for the Roman east', in J. H. Humphrey, ed., *Studies on Roman Pottery of the Provinces of Africa Proconsularis and Byzacena (Tunisia)* (Portsmouth, RI, 2009), pp. 73–91.

25. The partial quotation is from the fourth-century *Expositio totius mundi et gentium*. See P. Sarris, *Economy and Society in the Age of Justinian* (Cambridge, 2006), pp. 11–12; M. McCormick, *Origins of the European Economy: Communications and Commerce, AD 300–900* (Cambridge, 2001), p. 97. See also McCormick, *Origins*, p. 87, 'a state rate of 7 percent of the value of the grain compared with a private rate of 12 percent looks about right for the early 4th c.' It has elsewhere been estimated that Rome, at its height, imported 200,000 tonnes of grain to support a population of one million. See generally J. Teall, 'The grain supply of the Byzantine Empire, 330–1025', *DOP* 13 (1959), 87–139; C. Haas, *Alexandria in Late Antiquity: Topography and Social Conflict* (Baltimore, 1987).

26. D. Stone, 'Africa in the Roman empire: connectivity, the economy, and artificial port structures', *American Journal of Archaeology* 118 (2014), 565–600.

27. The Yenikapı ships are still being studied, but important preliminary reports include U. Kocabaş, 'The Yenikapı Byzantine-era shipwrecks, Istanbul, Turkey: a preliminary report and inventory of the 27 wrecks studied by Istanbul University', *International Journal of Nautical Archaeology* 44 (2015), 5–38; Z. Kızıltan, ed., *Stories from the Hidden Harbour: Shipwrecks of Yenikapı* (Istanbul, 2013), and on YK22 and its gold coin, pp. 65, 137; Ü. Akkemik, *Woods of Yenikapı Shipwrecks* (Istanbul, 2015), pp. 85–91, 124–35; R. Ingram, 'The hull of Yenikapı shipwreck YK11: a 7th-century merchant vessel from Constantinople's Theodosian Harbour', *International Journal of Nautical Archaeology* 47 (2018), 103–39.

28. N. Poulou, 'Transport amphorae and trade in the Aegean from the 7th to the 9th century AD: containers for wine or olive oil?', *Byzantina* 35 (2017), 195–216; F. H. van Doorninck Jr, 'The seventh-century Byzantine ship at Yassıadda and her final voyage: present thoughts', in D. Carlson et al., eds, *Maritime Studies in the Wake of the Byzantine Shipwreck at Yassıadda, Turkey* (College Station, TX, 2015), pp. 205–16; A. Sazanov, 'Cretan amphorae from the northern Black Sea region: contexts, chronology, typology', in N. Poulou-Papadimitriou et al., eds, *Late Roman Coarse Wares, Cooking Wares and Amphorae in the Mediterranean* (Oxford, 2014), pp. 399–409; E. C. Portale, 'The sunset of Gortyn: amphorae in 7th–8th centuries AD', in *Late Roman Coarse Wares*, pp. 477–89.

29. G. Hoffmann Barford et al., 'Geochemistry of Byzantine and early Islamic glass from Jerash, Jordan: typology, recycling and provenance', *Geoarchaeology* 33 (2018), 1–18; I. Freestone et al., 'Raw glass and production of glass vessels at late Byzantine Apollonia-Arsuf, Israel', *Journal of Glass Studies* 50 (2008), 67–80; Th. Rehren et al., 'Changes in glass consumption in Pergamon (Turkey) from Hellenistic to late Byzantine and Islamic times', *Journal of Archaeological Science* 55 (2015),

266–79; M. Phelps et al., 'Natron glass production and supply in the late antique and medieval Near East: the effect of the Byzantine-Islamic transition', *Journal of Archaeological Science* 56 (2016), 57–71.

5. Constantinople, the New Rome

1. E. Albu, *The Medieval Peutinger Map: Imperial Roman Renewal in a German Empire* (New York and Cambridge, 2014), pp. 95–103; R. Talbert, *Rome's World: The Peutinger Map Reconsidered* (Cambridge, 2010), pp. 10–11, 83–4 (palaeographical analysis by M. Steinmann), 163–4, problematises but does not reject this suggestion.

2. R. Taft, *The Byzantine Rite: A Short History* (Collegeville, Minnesota, 1992), p. 28.

3. Zosimus, 2.38, trans. Ridley, pp. 40–41; *Eusebius, Life of Constantine*, IV. 1, trans. Averil Cameron and S. G. Hall (Oxford, 1999), pp. 153–4, 310–11.

4. 'The Origin of Constantine', trans. J. Stevenson, in S. Lieu and D. Montserrat, eds, *From Constantine to Julian: Pagan and Byzantine Views: A Source History* (London and New York, 1996), pp. 39–62. See also P. Stephenson, *Constantine: Unconquered Emperor, Christian Victor* (London, 2009), pp. 190–211.

5. Symeonius was a native of Kaper Koraon and a donor of church silver, as we saw in an earlier chapter.

6. A. H. M. Jones, *The Later Roman Empire*, 3 vols (Oxford, 1964), I, pp. 448–9.

7. M. Hendy, *Studies in the Byzantine Monetary Economy, c. 300–1450* (Cambridge, 1985), pp. 602–13.

8. F. Millar, *The Emperor in the Roman World* (Bristol, 1992), p. 190; Stephenson, *Constantine*, pp. 256–78.

9. G. Fowden, 'Constantine's porphyry column: the earliest literary allusion', *JRS* 81 (1991), 119–31 at 124.

10. J. Bardill, 'The Golden Gate in Constantinople: a triumphal arch of Theodosius I', *American Journal of Archaeology* 103 (1999), 671–96; C. Mango, 'The triumphal way of Constantinople and the Golden Gate', *DOP* 54 (2000), 173–88.

11. S. Bassett, *The Urban Image of Late Antique Constantinople* (Cambridge, 2004), pp. 79–97; Z. Kızıltan, ed., *Stories from the Hidden Harbour: Shipwrecks of Yenikapı* (Istanbul, 2013).

12. C. Pearson et al., 'Dendroarchaeology of the mid-first millennium AD in Constantinople', *Journal of Archaeological Science* 39 (2012), 3402–14; G. Bony et al., 'A high-energy deposit in the Byzantine harbour of Yenikapı, Istanbul (Turkey)', *Quaternary International* 266 (2012), 117–30.

13. V. Onar, 'Animal skeletal remains discovered at the Metro and Marmaray excavation', in Kızıltan, ed., *Stories from the Hidden Harbour*, pp. 139–43, lists species, mostly domestic. See also V. Onar et al., 'Skull typology of Byzantine dogs from

the Theodosius Harbour at Yenikapı, Istanbul', *Anatomia Histologia Embryologia* 41 (2012), 341–52. The quotation is from Malalas, 18.51, trans. Jeffreys et al., p. 266.

14. The association between the two obelisks from Karnak is sustained in their inscriptions, for the Latin inscription in Constantinople echoes a longer dedication on the base of the Lateran Obelisk, alluding to the defeat by Constantius II of 'the vilest of tyrants', meaning the pretender Magnentius.

15. On Theodosius's claim to descend from Trajan, see now A. Kaldellis, 'The politics of classical genealogies in the Late Antique Roman East', in I. Tanaseanu-Döbler and L. von Alvensleben, eds, *Athens II: Athens in Late Antiquity* (Tübingen, 2020), pp. 259–77, at pp. 269–72.

16. I owe to Anthony Kaldellis the observation that the knotted columns of Theodosius represent the club of Hercules.

17. K. Dark and A. Harris, 'The last Roman forum: the Forum of Leo in fifth-century Constantinople', *Greek, Roman, and Byzantine Studies* 48 (2008), pp. 57–69.

18. A recent report has highlighted the lead pollution associated with the devastating fire at Notre Dame, Paris, in April 2019. See G. Le Roux et al., 'Learning from the past: fires, architecture and environmental lead emissions', *Environmental Science and Technology* 53 (2019), 8482–4.

19. P. Stephenson and R. Hedlund, 'Monumental waterworks in late antique Constantinople', in B. Shilling and P. Stephenson, eds, *Fountains and Water Culture in Byzantium* (Cambridge, 2016), pp. 36–54, at pp. 48–51.

20. A. Kaldellis, 'Christodoros on the statues of the Zeuxippos Baths: a new reading of the *Ekphrasis*', *GRBS* 47 (2007), 361–83; and Bassett, *Urban Image*, p. 162, for a photo (plate 8).

21. J. Crow, J. Bardill and R. Bayliss, *The Water Supply of Byzantine Constantinople* (London, 2008), pp. 127, 224 (Themistius, *Oration* 11.151c), 225 (Jerome, *Chronicle*, s.a., 373).

22. C. Mango, 'The water supply of Constantinople', in C. Mango and G. Dagron, eds, *Constantinople and its Hinterland* (Aldershot, 1995), pp. 9–18; *CJ* 11.42.9, for the edict of Zeno. It is possible that region XIV was the town of Regium, which although at some distance from the walls of Constantinople was regarded administratively as part of the city, in a manner similar to Sycae (Pera).

23. *CJ* 11.43.6; Crow et al., *Water Supply*, pp. 15–17, 125, 128–32, 138, 148. The Mocius reservoir is located in such a place that it cannot have been supplied directly either by the Hadrianic or Valens lines, as established by Crow and his team, and as such it is posited that it was supplied by a branch from the Valens line.

24. *CJ* 11.42.10; Crow et al., *Water Supply*, pp. 142–3, 231; Mango, 'The water supply', p. 14.

25. S. Runciman, 'The country and suburban palaces of the emperors', in A. Laiou-Thomadakis, ed., *Charanis Studies* (New Brunswick, NJ, 1980), pp. 219–28; A. Littlewood, 'Gardens of the Byzantine world', in H. Bodin and R. Hedlund, eds,

Byzantine Gardens and Beyond (Uppsala, 2013), pp. 30–113. John of Ephesus, 47, trans. E. W. Brookes, *Patrologia Orientalis* 17 (1923), 1–307; P. Hatlie, *The Monks and Monasteries of Constantinople*, ca. *350–850* (Cambridge, 2007), pp. 47, 143–4.

26. *De Cerimoniis* I, 18, trans. Moffatt, pp. 108–12.

27. R. Janin, 'Les processions religieuses à Byzance', *Revue des Études Byzantines* 24 (1966), 69–88; J. Baldovin, *The Urban Character of Christian Worship: The Origins, Development, and Meaning of Stational Liturgy* (Rome, 1987), pp. 205–26; R. Taft, *The Byzantine Rite: A Short History* (Collegeville, MN, 1992), pp. 30–34.

28. W. Mayer, 'The sea made holy: the liturgical function of the waters surrounding Constantinople', *Ephemerides liturgicae* 112 (1998), 459–68; C. Mango, 'Constantinople's Mount of Olives and Pseudo-Dorotheus of Tyre', *Nea Rhome* 6 (2009), 157–70; I. Kimmelfield, 'Defining Constantinople's suburbs through travel and geography', *Scandinavian Journal of Byzantine and Modern Greek Studies* 2 (2016), 97–114.

29. The *Easter Chronicle* provides details of these relic transfers, including that of Samuel, s.a., 406 and 411, trans. Whitby, pp. 60–62. See also B. Croke, 'Reinventing Constantinople: Theodosius I's imprint on the imperial city', in S. McGill et al., eds, *From the Tetrarchs to the Theodosians: Later Roman History and Culture, 284–450 CE* (Cambridge, 2010), pp. 241–64, at pp. 255–6; H. Klein, 'Sacred and imperial ceremonies at the great palace of Constantinople', in F. Bauer, ed., *Visualisierungen von Herrschaft Byzas* 5 (Istanbul, 2006), pp. 79–99; W. Mayer, 'The sea made holy: the liturgical function of the waters surrounding Constantinople', *Ephemerides Liturgicae* 112 (1998), 459–68.

30. K. Holum and G. Vikan, 'The Trier Ivory, *adventus* ceremonial, and the relics of St. Stephen', *DOP* 33 (1979), 113–33; J. Wortley, 'The Trier ivory reconsidered', *Greek, Roman, and Byzantine Studies* 21 (1980), 381–94, accepts the 439 translation, but questions the authenticity of that in 421, which is reported only far later by Theophanes, rather than in earlier sources that record around fifteen relic transfers between 350 and 450, including that of 439. The ivory is now held to represent the restoration of the relics of Euphemia by the empress Irene, in around 800. See A. Cutler and P. Niewöhner, 'Towards a history of Byzantine ivory carving', *Travaux et mémoires* 20 (2016), 89–107, at 93–8.

31. G. Downey, 'The tombs of the Byzantine emperors at the Church of the Holy Apostles in Constantinople', *Journal of Hellenic Studies* 79 (1959), 27–51; P. Grierson, 'The tombs and obits of the Byzantine emperors (337–1042)', *DOP* 16 (1962), 1–63; M. Johnson, 'On the burial places of the Theodosian dynasty', *Byzantion* 61 (1992), 330–39; Croke, 'Reinventing Constantinople', pp. 252–4. The origin of the Church of the Holy Apostles in Constantinople is discussed most convincingly by C. Mango, 'Constantine's mausoleum and the translation of relics', *Byzantinische Zeitschrift* 83 (1990), 51–61, who argues for the attribution of the mausoleum to Constantine, and of the adjacent basilica to Constantius. This develops the

detailed study by G. Downey, 'The builder of the original Church of the Apostles at Constantinople: a contribution to the criticism of the *Vita Constantini* attributed to Eusebius', *DOP* 6 (1951), 51–80, esp. 69–71.

32. A. Cameron, *The Last Pagans of Rome*, pp. 93–131, with translated quotations from John Chrysostom and Rufinus.

33. Marcellinus Comes, trans. Croke, p. 12; Philostorgius, 11.7–8, trans. Amidon, pp. 148–9. Philostorgius considered earthquakes to be God's wrath for sin, going so far as to refute natural explanations advanced by 'pagans'.

6. The Theodosian Age, AD 395–451

1. P. Stephenson, *Constantine: Unconquered Emperor, Christian Victor* (London, 2009), pp. 279–302.

2. D. Potter, 'The unity of the Roman empire', in S. McGill et al., eds, *From the Tetrarchs to the Theodosians: Later Roman History and Culture, 284–450 CE* (Cambridge, 2010), pp. 13–32, at p. 13; J. W. Drijvers, 'The *divisio regni* of 364: an end to unity?' in R. Dijkstra et al., eds, *East and West in the Roman Empire of the Fourth Century* (Leiden, 2015), pp. 82–96.

3. A. Cameron, *The Last Pagans of Rome*, pp. 93–131.

4. Gregory of Nyssa, *On the Deity of the Son*, PG 96, 557b, here in the translation of Kallistos Ware.

5. A. Cameron, *Claudian: Poetry and Propaganda at the Court of Honorius* (Oxford, 1970), pp. 63–92, 156–88.

6. Cameron, *Claudian*, pp. 37–45; A. Cameron and J. Long, *Barbarians and Politics at the Court of Arcadius* (Berkeley, 1993), pp. 1–11; E. Burrell, 'Claudian's *in Eutropium liber alter*: fiction and history', *Latomus* 62 (2003), 110–38. The law against Eutropius is *CTh.* 9.40.17.

7. Cameron and Long, *Barbarians and Politics*, pp. 337–98, offers a full English translation of *On providence*. Saturninus, a powerful general and opponent of the Goths, consul in 383, was also handed over to Gainas.

8. *Chronicle of Marcellinus*, s.a., 399–401, trans. Croke, pp. 7–8; *Easter Chronicle*, s.a., 401–2, trans. Whitby and Whitby, pp. 58–9; Zosimus, 4.51, 5.18, trans. Ridley, pp. 95, 108, 194 (n. 100 on Promotus). On the comet, which is also recorded in Chinese sources, see Cameron and Long, *Barbarians and Politics*, pp. 168–9. Arcadius's daughters with Aelia Eudoxia were named Flacilla (17 June 397), Pulcheria (19 January 399), Arcadia (3 April 400), and Marina (10 February 403).

9. W. Mayer, 'John Chrysostom as crisis manager: the years at Constantinople', in D. Sim and P. Allen, eds, *Ancient Jewish and Christian Texts as Crisis Management Literature* (London, 2012), pp. 129–43, at p. 140, translation modified. See also W. Mayer, 'Doing violence to the image of an empress: the destruction

of Eudoxia's reputation', in H. Drake, ed., *Violence in Late Antiquity* (Aldershot, 2006), pp. 205–14. Sozomen, 8.8, on processions.

10. Socrates, 8.20. The inscribed base of the column that supported the statue survives, currently in the Hagia Sophia museum collection. See *Last Statues of Antiquity (LSA) database*: laststatues.classics.ox.ac.uk/: LSA-27.

11. J. H. W. G. Liebeschuetz, 'Friends and enemies of John Chrysostom', in A. Moffatt, ed., *Maistor* (Canberra, 1984), pp. 85–111.

12. *Funerary Speech for John Chrysostom*, trans. T. Barnes and G. Bevan (Liverpool 2013), p. 77.

13. W. Mayer, 'Constantinopolitan women in Chrysostom's circle', *Vigiliae Christianae* 53 (1999), 265–88; W. Mayer, 'John Chrysostom and women revisited', in W. Mayer and I. J. Elmer, eds, *Men and Women in the Early Christian Centuries* (Strathfield, 2014), pp. 211–25.

14. Zosimus, 5.24–5; trans. Ridley, pp. 111–12.

15. Comparing Theodosius II poorly with his grandfather, A. H. M. Jones, *The Later Roman Empire, 284–602,* 3 vols (Oxford, 1964), I, p. 175: 'None of the male descendants of Theodosius the Great inherited his ability or force of character: they reigned rather than ruled the empire ... Theodosius II received high praises from contemporary Christian writers for his devout piety ... Such amiable qualities did not make a good emperor.' Two recent works have changed our view of Theodosius II: C. Kelly, 'Rethinking Theodosius', in C. Kelly, ed., *Theodosius II: Rethinking the Roman Empire in Late Antiquity* (Cambridge, 2013), pp. 3–64; F. Millar, *A Greek Roman Empire: Power and Belief under Theodosius II, 408–450* (Berkeley, 2006). Additionally, A. Cameron, 'The empress and the poet', in his *Wandering Poets and Other Essays on Late Greek Literature and Philosophy* (Oxford, 2016), pp. 37–80, offers a compelling, contrarian account of the reign through a careful examination of the career of Cyrus of Panopolis and, to a lesser extent, Eudocia.

16. Sozomen, 9.16; Socrates, 7.1, who adds that Anthemius sought good advice, notably from Troilus the sophist, who had probably taught Socrates. Anthemius and Troilus were correspondents of Synesius of Cyrene. Sozomen, who completed his work later than Socrates, failed to mention Anthemius.

17. Procopius, *Wars*, 1.2.1; trans. Dewing and Kaldellis, pp. 4–5. Theophanes Confessor in the ninth century and Nicholas Mysticus in the tenth both report the same story, adding new details. See G. Greatrex and J. Bardill, 'Antiochus the *Praepositus*: a Persian eunuch at the court of Theodosius II', *DOP* 50 (1996), 171–97, including a quotation from Synesius, letter 110.

18. Zosimus, 6.12, trans. Ridley, p. 131. On this period in the west, see now M. Kulikowski, *Imperial Tragedy: From Constantine's Empire to the Destruction of Roman Italy* (London, 2019), pp. 123–64.

19. J. B. Bury, *A History of the Later Roman Empire*, 2 vols (Cambridge, 1889), I, p. 122.

20. Sozomen, 9.11; Zosimus, 6.10, trans. Ridley, p. 130. *Orosius, Seven Books Against the*

Pagans, trans. A. T. Fear (Liverpool, 2010), pp. 409–11. Since Honorius's letter to the Britons is mentioned in the middle of an excursus on Italy, some have considered it to mean that a letter was sent to the cities of Bruttium, an Italian region, rather than to the cities of Britannia. However, this idea is now discredited. See Kulikowski, *Imperial Tragedy*, pp. 150–53, discussing Britain, Spain and Gaul.

21. Kulikowski, *Imperial Tragedy*, pp. 153–6.
22. Sozomen, 9.16; Philostorgius, 12.12, trans. Amidon, pp. 161–2. Theodosius I had ruled that Sundays were to be free of games, unless the emperor's birthday happened to fall on that day, which his own did in 393.
23. Quodvultdeus, *Liber promissionum et praedictorum Dei*, ed. R. Braun (Paris, 1964), II, pp. 558–60. See G. Greatrex and S. Lieu, *The Roman Eastern Frontier and the Persian Wars Part II, AD 363–630* (Abingdon, 2002), pp. 36–43; K. Holum, 'Pulcheria's crusade, AD 421–22 and the ideology of imperial victory', *Greek, Roman, and Byzantine Studies* 18 (1977), 153–72.
24. Philostorgius, 12.13–14, trans. Amidon, pp. 162–3. See also Marcellinus Comes, trans. Croke, pp. 13–14; Socrates, 7.23; Theophanes, trans. Mango and Scott, pp. 132–3; Kulikowski, *Imperial Tragedy*, pp. 164, 175–6.
25. P. van Nuffelen, 'Olympiodorus of Thebes and eastern triumphalism', in Kelly, ed., *Theodosius II*, pp. 130–52; W. Treadgold, 'The diplomatic career and historical work of Olympiodorus of Thebes', *International History Review* 26 (2004), 709–33.
26. Cameron, 'The empress and the poet', pp. 62–3. Others are more sceptical about the birth of Arcadius.
27. Malalas, 14.3–8, trans. Jeffreys et al., pp. 191–5; *Easter Chronicle*, trans. Whitby, pp. 66–8, 73–5; Theophanes, trans. Mango and Scott, pp. 155–6; Marcellinus Comes, trans. Croke, p. 17.
28. John of Nikiu, trans. Charles, pp. 104–5. K. Chew, 'Virgins and eunuchs: Pulcheria, politics, and the death of Theodosius II', *Historia* 55 (2006), 207–27, explores this and more salacious stories, including the notion advanced by Nestorius that Paulinus was also Pulcheria's lover, one of seven the virgin empress is said to have taken. Chew also notes that in the Old Church Slavonic version of Malalas, generally less full than the extant Greek version, it is recorded that Theodosius loved Paulinus greatly as his friend, just as in John of Nikiu.
29. The exchange between Nestorius and Pulcheria is preserved in a later Syriac translation of a Greek text, the so-called *Letter to Cosmas*, published in *Patrologia Orientalis* 13 (Paris, 1907), pp. 275–86, at p. 279.
30. The fullest account is by Evagrius 1. 2–7, trans. Whitby, pp. 8–25. See also Socrates 7.34; T. Graumann, 'Theodosius II and the politics of the first council of Ephesus', in Kelly, ed., *Theodosius II*, pp. 109–29. The 'formula of reunion' between Cyril and John is preserved in letter 39, dated 433, in Cyril's corpus. Cyril's 'twelve anathemas' against Nestorius is contained in letter 3.
31. See now E. Fournier, 'The Vandal conquest of North Africa: the origins of a

historiographical persona', *Journal of Ecclesiastical History* 68 (2017), 687–718; Theodoret, *Letters,* 29–36.

32. A. Leone, *Changing Townscapes in North Africa from Late Antiquity to the Arab Conquest* (Bari, 2007), pp. 168–95; R. Miles, 'Byzantine Carthage and its Vandal legacy', in É. Wolff, ed., *Littérature, politique et religion en Afrique vandale* (Paris, 2015), pp. 73–92.

33. Kulikowski, *Imperial Tragedy*, pp. 184–6.

34. Kelly, 'Rethinking Theodosius', pp. 9–10; *CTh.,* 16.8.22, 16.5.57–8, 16.9.3, 16.10.21. The *Theodosian Code* (*CTh.*16.5.56) also records a law of Honorius, directed in August 415 to Heraclian, count of Africa, which commanded the confiscation of the property of Donatists, North African heretics, should they seek to assemble or practise their rites.

35. *CTh.* 6.27.18, 6.27.23, 8.4.29.

36. Malalas, 14.16, trans. Jeffreys et al., pp. 197–8; Cameron, 'The empress and the poet', pp. 38, 56–64.

37. See J. Harries, 'Men without women: Theodosius' consistory and the business of government', in Kelly, ed., *Theodosius II*, pp. 67–89, at pp. 78–9.

38. D. Lee, 'Theodosius and his generals', in Kelly, ed., *Theodosius II*, pp. 90–108.

39. Jones, *Later Roman Empire*, III, pp. 347–80 (appendix II); M. Kulikowski, 'The *Notitia Dignitatum* as a historical source', *Historia* 49 (2000), 358–77; A. M. Kaiser, 'Egyptian units and the reliability of the *Notitia Dignitatum, pars Oriens*', *Historia* 64 (2015), 243–61.

40. The practice of indicating the number of Augusti with the number of Gs had commenced under Theodosius I and continued for around forty years, well into the reign of Theodosius II. See J. Kent, '"Concordia" *solidi* of Theodosius I: a reappraisal', *Numismatic Chronicle* 153 (1993), 77–90.

41. J. Kent, 'The coinage of Arcadius (395–408)', *Numismatic Chronicle* 151 (1991), 35–57; P. Guest, 'The production, supply and use of late Roman and early Byzantine copper coinage in the eastern empire', *Numismatic Chronicle* 172 (2012), 105–31, collates finds from numerous excavation sites showing that bronze coins dated to 379–408 appear in far greater numbers than for the rest of the fifth century, regardless of total numbers found, including at Athens (AD 379–408 = 2,929 coins, versus AD 408–491 = 570); Sardis (1816 v. 710); Caesarea Maritima (383 v. 171); Nicopolis ad Istrum (92 v. 29).

42. See the *Last Statues of Antiquity (LSA) database*: laststatues.classics.ox.ac.uk/: Corinth, LSA-52; Aphrodisias, LSA-165, 167, 168; Side, LSA-267; Erythrae, LSA-285; Heraclea Pontica, LSA-528, 529; Ephesus, LSA-711; Roma, LSA-1306. On the equestrian statues at the Forum of Theodosius, see B. Croke, 'Reinventing Constantinople: Theodosius I's imprint on the imperial city', in McGill et al., eds, *From the Tetrarchs to the Theodosians*, pp. 241–64, at p. 259.

43. R. Grigg, '"Symphōnian Aeidō tēs Basileias": an image of imperial harmony on the

base of the Column of Arcadius', *Art Bulletin* 59 (1977), 469–82; A. Taddei, 'La Colonna di Arcadio a Costantinopoli: profilo storico di un monumento attraverso le fonti documentarie dalle origini all'età moderna', *Nea Rhome* 6 (2009), 37–103.

44. Malalas, 14.13, trans. Jeffreys et al., pp. 196–7; G. Traina, 'Mapping the world under Theodosius II', in Kelly, ed., *Theodosius II*, pp. 155–71, at p. 165.

45. E. A. Thompson, 'The foreign policies of Theodosius II and Marcian', *Hermathena* 76 (1950), 58–75, at 59; Bury, *Later Roman Empire*, I, pp. 216–17, drew attention to Priscus's terminology, but remarked that 'it does not imply that there was any idea afloat at the time of a western Roman empire'. He offers a full translation of the relevant passages from Priscus.

46. Theodoret, *Letters*, 42, 43, 79, 92, 111, 119, 121, 139; Harries, 'Men without women', in Kelly, ed., *Theodosius II*, pp. 67–89; D. Lee, 'Theodosius and his generals' in Kelly, ed., *Theodosius II*, pp. 90–108, at pp. 93–4, 97–8.

7. Soldiers and Civilians, AD 451–527

1. C. Zaccagnino et al., 'The *missorium* of Ardabur Aspar: new considerations of its archaeological and historical contexts', *Archaeologia Classica* 63 (2012), 419–54. The dish is 95 per cent silver, 4 per cent copper, with traces of lead. See also R. Leader-Newby, *Silver and Society in Late Antiquity* (Aldershot, 2004), pp. 11–59.

2. R. W. Burgess, 'The accession of Marcian in the light of Chalcedonian apologetic and Monophysite polemic', *Byzantinische Zeitschrift* 86–7 (1993–4), 47–68. Evagrius contradicts this, stating that Pulcheria remained chaste and that Marcian's sovereignty was the prize of his virtue.

3. Malalas 14.28–34, trans. Jeffreys et al., pp. 201–2; Marcellinus Comes, trans. Croke, pp. 20–22; *Daniel the Stylite*, 35, trans. Dawes and Baynes, pp. 27–8. See also Procopius, *Wars*, 3.6.3, trans. Dewing and Kaldellis, p. 157, where it is reported that Aspar believed that an Arian would not be readily accepted as emperor, which was certainly the case in the sixth century. However, this was not perhaps so clear in the middle years of the fifth century. For this period in the west, see now Kulikowski, *Imperial Tragedy*, pp. 192–213.

4. Evagrius, 2.1, trans. Whitby, pp. 56–60. See also S. Barnish, 'A note on the *collatio glebalis*', *Historia* 38 (1989), 254–6; C. Humfress, 'Poverty and Roman law', in M. Atkins and R. Osborne, eds, *Poverty in the Roman World* (Cambridge, 2006), pp. 183–203. The translation, from Priscus fr. 19, is from G. Greatrex and S. Lieu, eds, *The Roman Eastern Frontier and the Persian Wars Part II, AD 363–630* (Abingdon, 2002), p. 46.

5. G. Bevan and P. Gray, 'The trial of Eutyches: a new interpretation', *Byzantinische Zeitschrift* 101 (2008), 617–57.

6. Kulikowski, *Imperial Tragedy*, pp. 214–26.

7. *Easter Chronicle*, trans. Whitby, pp. 85–6; Malalas 14.37, trans. Jeffreys et al., p. 203.

8. Malalas 14.39, trans. Jeffreys et al., p. 204. See P. Wood, 'Multiple voices in chronicle sources: the reign of Leo I (457–74) in book fourteen of Malalas', *Journal of Late Antiquity* 4 (2011), 298–314, on Leo's creating an anti-Arian platform that allows him to create distance from Aspar. See also M. MacEvoy, 'Becoming Roman: the not-so-curious case of Aspar and the Ardaburii', *Journal of Late Antiquity* 9 (2016), 483–511, on roles the Ardaburs continued to play after Aspar's fall, notably Areobindus, who married Julia Anicia, the last Theodosian princess.

9. *The Chronicle of Pseudo-Joshua the Stylite*, trans. F. Trombley and J. Watt (Liverpool, 2000), p. 12.

10. Malalas 14.44, trans. Jeffreys et al., p. 206.

11. Marcellinus Comes, s.a., 470, trans. Croke, p. 25; Malalas 14.40, trans. Jeffreys et al., pp. 204–5; *Easter Chronicle*, trans. Whitby, p. 89, misdates the episode to 467.

12. *De Cerimoniis* I, 94, trans. Moffatt, pp. 431–2; A. Boak, 'Imperial coronations ceremonies of the fifth and sixth centuries', *Harvard Studies in Classical Philology* 30 (1919), 37–47, offers a concise summary of the various ceremonies.

13. Mango, *Art*, pp. 34–5. The manuscript is Cod. Paris gr. 1447, fols 257–8.

14. *Pseudo-Joshua the Stylite*, trans. Trombley and Watt, p. 12; N. Lenski, 'Assimilation and revolt in the territory of Isauria, from the 1st century BC to the 6th century AD', *Journal of the Economic and Social History of the Orient* 42 (1999), 413–65, is a rich and insightful study complementing B. Shaw, 'Bandit highlands and lowland peace: the mountains of Isauria-Cilicia', *Journal of the Economic and Social History of the Orient* 33 (1990), 199–233. See also W. D. Burgess, 'Isaurian factions in the reign of Zeno the Isaurian', *Latomus* 51 (1992), 874–80.

15. Candidus, fr. 1.56–65. See *Fragmenta Historicorum Graecorum*, ed. K. Müller, (Paris, 1851), IV, p. 136. Translations of these fragmentary sources are provided in C. D. Gordon, *The Age of Attila* (Ann Arbor, MI, 1960). An introduction to, and commentary on, the work of Malchus, and therefore also Candidus, is presented in R. Blockley, *The Fragmentary Classicising Historians of the Later Roman Empire*, 2 vols (Liverpool, 1981).

16. *Daniel the Stylite*, 70, trans. Dawes and Baynes, pp. 134–5; Evagrius, 3.4, trans. Whitby, pp. 134–7.

17. D. Pingree, 'Political horoscopes from the reign of Zeno', *DOP* 30 (1976), 135–50, at 136–8. Pingree reminds us that in 486, according to Malalas (trans. Jeffrey et al., p. 219), Zeno consulted an astrologer named Maurianus. Another horoscope relating to the promotion of Theoderic by Zeno, and probably written in Greek by the same person, has been transmitted with the name Palchus. The most famous 'sorcerer' of the period, whose name is associated with horoscopes, was Pamprepius.

18. Malchus, fr. 9; Gordon, *Age of Attila*, pp. 146–7; Blockley, *Fragmentary Classicising Historians*, I, p. 74, assigns this passage to Candidus. For references and commentary, see B. Baldwin, 'Malchus of Philadelphia', *DOP* 31 (1977), 91–107, at 92–3.

19. Photius's summaries in Greek are supplied with Latin translations in *Fragmenta Historicorum Graecorum*, IV, pp. 111–12, 135–7, which also includes a fragment preserved in later sources, including the *Suda*, a Byzantine encyclopaedia, and Constantine Porphyrogenitus's compilation on embassies, the *Excerpta de Legationibus*.

20. John of Antioch, fr. 210. See Gordon, *Age of Attila*, pp. 144–5. An excellent potted biography of Verina can be read online: G. Greatrex, 'Aelia Verina (Wife of Leo I)', *De Imperatoribus Romanis: an online encyclopaedia of Roman emperors* (updated 11 August 2004).

21. R. McCail, 'P. Gr. Vindob. 29788C, hexameter encomium on an un-named emperor', *Journal of Hellenic Studies* 98 (1978), 38–63; *Pseudo-Joshua the Stylite*, trans. Trombley and Watt, p. 14.

22. R. Asmus, 'Pamprepios, ein byzantinischer Gelehrter und Staatsmann des 5. Jahrhunderts', *Byzantinische Zeitschrift* 22 (1913), 320–47; Gordon, *Age of Attila*, pp. 150–51, offers a translation of Malchus's fr. 20, concerning Pamprepius.

23. The translation from Zacharias's *Life of Severus* is quoted by A. Cameron, *Wandering Poets and Others Essays on Late Greek Literature and Philosophy*, 2nd edn (Oxford, 2015), pp. 156–7. See also S. Brock and B. Fitzgerald, ed. and trans., *Two Early Lives of Severos, Patriarch of Antioch* (Liverpool, 2013), pp. 37–56.

24. The translation is adapted from C. Mango, 'The conversion of the Parthenon into a church: the Tübingen Theosophy', *Deltion of the Christian Archaeological Society* 18 (1995), 201–3. See also Cameron, *Wandering Poets*, p. 248. On the work, which is a Byzantine epitome of the last four books of the treatise, the first seven on 'correct faith' having been lost, see P. F. Beatrice, 'Pagan wisdom and Christian theology according to the Tübingen Theosophy', *Journal of Early Christian Studies* 3 (1995), 403–18. For the Syriac tradition, see S. Brock, 'Some Syriac excerpts from Greek collections of pagan prophecies', *Vigiliae Christianae* 38 (1984), 77–90.

25. Mango, 'The conversion of the Parthenon', 202. See also A. Kaldellis, *The Christian Parthenon: Classicism and Pilgrimage in Byzantine Athens* (Cambridge, 2009), p. 48.

26. According to the *Life of Proclus*, written by his pupil Marinus immediately after his tutor's death in 485, the cult statue of Athena had recently been removed. See Mango, 'Conversion of the Parthenon', 202–3; Kaldellis, *Christian Parthenon*, pp. 33–4, 52, fig. 12 (the inscription from Icaria).

27. 'The Henotikon of Zeno, 482', trans. H. Bettenson and C. Maunder, *Documents of the Christian Church* (Oxford, 2011), pp. 93–5.

28. M. Redies, 'Die Usurpation des Basilikos im Kontext der aufsteigenden monophysitischen Kirche', *Antiquité tardive* 5 (1997), 211–21.

29. Malchus, fr. 18; trans. Blockley, *Fragmentary Classicising Historians*, II, pp. 426–35.

30. See Kulikowski, *Imperial Tragedy*, pp. 244–59, 260–76, for a less detailed but complementary treatment of Zeno's reign, and for a fuller analysis of western affairs than is offered here.

31. The Barberini ivory is considered in detail in our last chapter.

32. D. Westberg, *Celebrating with Words: Studies in the Rhetorical Works of the Gaza School*, Uppsala University PhD dissertation, 2010, pp. 52–62; P. Coyne, *Priscian of Caesarea's De Laude Anastasii Imperatoris* (Lewiston, NY, and Lampeter, 1991), a revision of a 1988 McMaster University PhD dissertation, offers a full English translation of Priscian's panegyric with commentary.

33. *De Cerimoniis* I, 92, trans. Moffatt, pp. 417–25, preserves the record of Anastasius's selection and inauguration ceremonies.

34. John of Antioch, fr. 239, translation modified from that provided at *Iohanni Antiocheni fragmenti quae supersunt Omnia*, ed. S. Mariev (Berlin, 2008).

35. A. Cameron, 'The House of Anastasius', *Greek, Roman, and Byzantine Studies* 19 (1978), 259–76; A. Kaldellis, 'Christodoros on the statues of the Zeuxippos Baths: a new reading of the ekphrasis', *Greek, Roman, and Byzantine Studies* 47 (2007), 361–83, at 377–82; A. Kaldellis, 'The politics of classical genealogies in the Late Antique Roman East', in I. Tanaseanu-Döbler and L. von Alvensleben, eds, *Athens II: Athens in Late Antiquity* (Tübingen, 2020), pp. 259–77, at pp. 272–4.

36. T. Birch et al., 'From *nummi minimi* to *fulūs* – small change and wider issues: characterising coinage from Gerasa/Jerash (Late Roman to Umayyad periods)', *Archaeological and Anthropological Sciences* 11 (2019), 5359–76. This is very recent research, with observations restricted to only a very few coins from one location, but the implications are fascinating.

37. The diverse sources for the Anastasian War and the construction of Dara are collected at Greatrex and Lieu, *The Roman Eastern Frontier*, pp. 62–81. See also B. Croke and J. Crow, 'Procopius and Dara', *JRS* 73 (1983), 143–59. See also E. Key Fowden, *The Barbarian Plain: Saint Sergius between Rome and Iran* (Berkeley, CA, 1999), pp. 60–67.

38. B. Croke, 'The date of the "Anastasian Long Wall" in Thrace', *Greek, Roman, and Byzantine Studies* 23 (1982), 59–78, has argued that references to a 'long wall' before Anastasius refer to a structure across the Gallipoli peninsula, the so-called Thracian Chersonese. See also J. Crow, 'The long walls of Thrace', in C. Mango and G. Dagron, eds, *Constantinople and its Hinterland* (Aldershot, 1995), pp. 109–24.

39. *Two Early Lives of Severos*, p. 2.

40. J. Dijkstra and G. Greatrex, 'Patriarchs and politics in Constantinople in the reign of Anastasius (with a reedition of *O.Mon.Epiph.* 59)', *Millennium* 6 (2009), 223–65 at 241–2.

41. Dijkstra and Greatrex, 'Patriarchs and politics', 230–39. On the ouster of Macedonius, see also G. Greatrex et al., ed., *The Chronicle of Pseudo-Zachariah Rhetor* (Liverpool, 2011), pp. 251–64. A sympathetic account of Macedonius's career, a partial counterweight to the writings of Zacharias and Severus, was written by Theodore, a lector at Hagia Sophia, who probably joined his patriarch in exile. Fragments of Theodore Lector's work are preserved by Theophanes Confessor.

42. Malalas, 16.19, trans. Jeffreys et al., p. 228.

43. M. Hendy, *Studies in the Byzantine Monetary Economy* (Cambridge, 1985), pp. 157–73, 475–9; *Pseudo-Joshua the Stylite*, trans. Trombley and Watt, p. 30.

44. Michael the Syrian, II, 170, noted at Theophanes, trans. Mango and Scott, p. 252; Malalas, 17.1, trans. Jeffreys et al., p. 230.

45. Marcellinus Comes, trans. Croke, p. 41. See also Cameron, *Porphyrius*, p. 229. Averil Cameron, 'Justin I and Justinian', in *CAH* 14, pp. 63–85, at pp. 63–5, offers an excellent summary of Justin's reign and its implications for Justinian's early years.

8. The Age of Justinian, AD 527–602

1. A. Kaldellis, *Procopius of Caesarea: Tyranny, History, and Philosophy at the End of Antiquity* (Philadelphia, PA, 2004).

2. Malalas, 18.1, trans. Jeffreys et al., p. 245.

3. Marcellinus Comes, trans. Croke, p. 42.

4. Procopius, *Wars*, 1.11; trans. Dewing and Kaldellis, pp. 24–6.

5. Hermogenes is 'Hermogenes 1' in *PLRE* 3, pp. 590–93; Mundo is 'Mundus 1', pp. 903–5; Rufinus is Rufinus 1, pp. 1097–8; Sittas is 'Sittas 1', pp. 1160–63.

6. G. Greatrex and S. Lieu, *The Roman Eastern Frontier and the Persian Wars Part II, AD 363–630* (London and New York, 2002), pp. 82–101, at p. 97.

7. Procopius, *Wars*, 1.24; trans. Dewing and Kaldellis, p. 61.

8. Malalas, 18.71, trans. Jeffreys et al., p. 278.

9. Procopius, *Wars*, 1.24; trans. Dewing and Kaldellis, p. 64. This is reported in a set-piece speech set out by Procopius, who slyly alludes to a more familiar interpretation, that tyranny is a good burial shroud. See also Malalas, 18.71, trans. Jeffreys et al., pp. 280–81. For a comprehensive reinterpretation of the riot see G. Greatrex, 'The Nika riot: a reappraisal', *Journal of Hellenic Studies* 117 (1997), 60–86.

10. B. Pentcheva, *Hagia Sophia: Sound, Space, and Spirit in Byzantium* (University Park, PA, 2017).

11. A. Kaldellis, 'The making of Hagia Sophia and the last pagans of New Rome', *Journal of Late Antiquity* 6 (2013), 347–66; Mary Whitby, 'The St Polyeuktos Epigram (*AP* 1.10): a literary perspective', in S. F. Johnson, ed., *Greek Literature in Late Antiquity: Dynamism, Didacticism, Classicism* (Aldershot, 2006), pp. 159–88.

12. The translation is by Cyril Mango, recorded with commentary at Croke, 'Church of Saints Sergius and Bacchus', 47–8. See also J. Bardill, 'The Church of Sts. Sergius and Bacchus in Constantinople and the Monophysite Refugees', *DOP* 54 (2000), 1–11; B. Croke, 'Justinian, Theodora, and the Church of Saints Sergius and Bacchus', *DOP* 60 (2006), 25–63.

13. Malalas, 18.24, trans. Jeffreys et al., p. 255. An edict on gambling issued at this time by Justinian to John the Cappadocian, as praetorian prefect, is preserved as *CJ* 3.43.

It prohibits gambling and named games of chance but prescribes no punishment, entrusting execution of the law to bishops.

14. Malalas, 18.18, trans. Jeffreys et al., p. 253. While Antioch was officially renamed 'City of God', several rebuilt garrison towns were called Justinianoupolis, and the fortress of Ansarthon was called Theodorias. See Malalas, 18.30, trans. Jeffreys et al., p. 259.

15. Quoted by C. Humfress, 'Law and legal practice in the age of Justinian', in M. Maas, ed., *The Cambridge Companion to the Age of Justinian* (Cambridge, 2005), pp. 161–84, at p. 168.

16. See *ODB*, s.v. Stobaios, p. 1958.

17. Humfress, 'Law and legal practice', p. 167; Procopius, *Wars*, 3.9; trans. Dewing and Kaldellis, p. 164; A. H. Merrills, 'The secret of my succession: dynasty and crisis in Vandal North Africa', *Early Medieval Europe* 18 (2010), 135–59.

18. Procopius, *Wars*, 3.15; trans. Dewing and Kaldellis, pp. 177–8.

19. Procopius, *Wars*, 4.5; trans. Dewing and Kaldellis, p. 201.

20. Procopius, *Wars*, 4.9; trans. Dewing and Kaldellis, pp. 208–10.

21. Procopius, *Wars*, 4.14–18; trans. Dewing and Kaldellis, pp. 220–30. Germanus is 'Germanus 4' in *PLRE* 2, pp. 505–7; Solomon is 'Solomon 1', *PLRE* 3, pp. 1167–7; Stotzas is at *PLRE* 3, pp. 1199–1200.

22. Procopius, *Wars*, 5.6; trans. Dewing and Kaldellis, p. 263.

23. Procopius, *Wars*, 5.18; trans. Dewing and Kaldellis, p. 294. See Kaldellis, *Procopius of Caesarea*, p. 192.

24. Procopius, *Wars*, 6.13–15, 23; trans. Dewing and Kaldellis, pp. 344–51, with a long digression on the Heruls, p. 366.

25. Procopius, *Wars*, 5.18; trans. Dewing and Kaldellis, p. 294, 'by some chance'. But see Kaldellis, *Procopius of Caesarea*, pp. 189–98.

26. Procopius, *Secret History*, 4.32; trans. Kaldellis, pp. 20–21.

27. Procopius, *Wars*, 7.1; trans. Dewing and Kaldellis, pp. 107.

28. Cassiodorus, *Variae*, 12.25. See *The Letter of Cassiodorus*, trans. T. Hodgkin (London, 1886), p. 519. We explore this matter fully in a later chapter.

29. Procopius, *Wars*, 2.22–3; trans. Dewing and Kaldellis, pp. 120–23.

30. Procopius, *Wars*, 2.14–21; trans. Dewing and Kaldellis, pp. 104–19. See also Marcellinus Comes, 544–5, trans. Croke, p. 50, where Belisarius falls into 'great danger' and is 'exposed to envy'.

31. John had a long and distinguished military career, details of which are recorded under 'Ioannes 46' at *PLRE* 3, pp. 652–61. See also 'Areobindus 2', at *PLRE* 3, pp. 107–9; 'Artabanes 2', pp. 125–30.

32. *The Iohannis or De Bellis Libycis of Flavius Cresconius Corippus*, trans. G. W. Shea (Lewiston, 1998), pp. 187–207. Also 'Ioannes qui est Troglita 36', at *PLRE* 3, pp. 644–9.

33. Procopius, *Secret History*, 18; trans. Kaldellis, pp. 81–2.

34. Procopius, *Wars*, 7.32; trans. Dewing and Kaldellis, pp. 440–44.
35. Germanus is 'Germanus 4' in *PLRE* 2, pp. 505–7, also *PLRE* 3, p. 527. His sons are 'Iustinianus 3', *PLRE* 3, pp. 744–7; 'Iustinus 4', *PLRE* 3, pp. 750–54; and 'Germanus 3', *PLRE* 3, p. 528.
36. Procopius, *Wars*, 8.21–35; trans. Dewing and Kaldellis, pp. 510–44; P. Rance, 'Narses and the Battle of Taginae (Busta Gallorum) 552: Procopius and sixth-century warfare', *Historia* 30 (2005), 424–72; J. Wood, 'Defending Byzantine Spain: frontiers and diplomacy', *Early Medieval Europe* 18 (2010), 292–319.
37. D. Neary, 'The image of Justinianic orthopraxy in monastic literature', *Journal of Early Christian Studies* 25 (2017), 119–47, with quotation from Novel 133 at 126.
38. See J. Meyendorff, 'Justinian, the empire and the Church', *DOP* 22 (1968), 43–60, at 48.
39. R. Price, 'The development of Chalcedonian identity in Byzantium (451–553)', *Church History and Religious Culture* 89 (2009), 307–25.
40. C. Trypanis, *Fourteen Early Byzantine Cantica* (Vienna, 1968), no. 12, pp. 139–47, at p. 143; translated by A. Palmer, 'The inauguration anthem of Hagia Sophia in Edessa: a new edition with historical and architectural notes and a comparison with a contemporary Constantinopolitan *kontakion*', *Byzantine and Modern Greek Studies* 12 (1988), 117–68, at 141.
41. Matthew 17:2. Cf. Mark 9:3; Luke 9:29.
42. Paul the Silentiary, lines 786–9, trans. Mango, *Art*, p. 89. The altar cloth of Justinian's Hagia Sophia has not survived, but similar scenes of churches and Christ's miracles are depicted on the hems of two textiles now preserved in Berlin, the Danielstoff and the Petrusstoff, neither of which are as splendid or made of silk. See M. Stein, 'Die Inschriften auf dem Daniel- und dem Petrus-Stoff in Berlin', *Jahrbuch für Antike und Christentum* 34 (2002), 84–98.
43. Trypanis, *Fourteen Early Byzantine Cantica*, p. 137; Palmer and Rodley, 'Inauguration anthem', 144.
44. Procopius, *Wars*, 7.33–4; trans. Dewing and Kaldellis, pp. 444–9; A. Sarantis, 'War and diplomacy in Pannonia and the northwest Balkans during the reign of Justinian: the Gepid threat and imperial responses', *DOP* 63 (2009), 15–40.
45. Menander Protector, fr. 6, trans. Blockley, pp. 55–91.
46. *Agathias Histoires: Guerres et malheurs du temps sous Justinien*, trans. P. Maraval (Paris, 2007). See also Malalas, 18.119–52, trans. Jeffreys et al., pp. 294–307.
47. *Flavius Cresconius Corippus: In Laudem Iustini Augusti Minoris*, ed. and trans. Averil Cameron (London, 1976). A law issued in *c.*566 (Novel 148) sets out the fiscal policy and its relationship to military affairs.
48. Averil Cameron and A. Cameron, 'Anth. Plan. 72: a propaganda poem for the reign of Justin II', *Bulletin of the Institute of Classical Studies* 13 (1966), 101–4. I have slightly modified their translation.
49. Menander Protector, fr. 13, trans. Blockley, pp. 144–7. See M. Whitby, *The Emperor*

Maurice and his Historian: Theophylact Simocatta on Persian and Balkan Warfare (Oxford, 1988), pp. 250–75. On Roman relations with the Turks see D. Sinor, 'The establishment and dissolution of the Türk empire', in D. Sinor, ed., *The Cambridge History of Inner Asia* (Cambridge, 1990), pp. 285–316, at pp. 301–5.

50. J. Kroll and B. Bachrach, 'Justin's madness: weak-mindedness or organic psychosis?', *Journal of the History of Medicine and Allied Sciences* 48 (1993), 40–67.

51. The relationships between characters with similar names can be teased out by careful consultation of the *PLRE*. The general Justinian is 'Iustinianus 3' in *PLRE* 3, pp. 744–7. His late brother Justin is Iustinus 4, pp. 750–54. Germanus, who will become Caesar in 582, does not appear to be their half-brother Germanus Postumus, but is possibly the son of Justinian ('Iustinianus 3'), recorded as 'Germanus 5', *PLRE* 3, p. 529.

52. Simocatta, 3.13, trans. Whitby, pp. 92–4. See also D. Frendo, 'The Armenian and Byzantine foundations of the concept of jihad', *Byzantine Studies* 13 (1986), 241–50.

53. Averil Cameron, 'The empress Sophia', *Byzantion* 45 (1975), 5–21; Averil Cameron, 'The artistic patronage of Justin II', *Byzantium* 50 (1980), 62–84.

54. John of Ephesus, trans. Payne Smith, p. 352.

55. Simocatta, 2.3, trans. Whitby, pp. 46–7. Also, Theophanes, AM 6078, trans. Mango and Scott, pp. 377–9. Excellent coverage and commentary can be found at Greatrex and Lieu, *Roman Eastern Frontier*, pp. 167–81.

56. Evagrius, 1.13, trans. Whitby, p. 38; *Maurice's Strategikon: Handbook of Byzantine Military Strategy*, trans. G. Dennis (Philadelphia, 1984), pp. 33–4. G. Dennis, 'Religious services in the Byzantine army', in E. Carr et al., eds, *Eulogema: Studies in Honor of Robert Taft, S.J.* (Rome, 1993), pp. 107–17; R. Taft, 'War and peace in the Byzantine divine liturgy', in T. S. Miller and J. Nesbitt, eds, *Peace and War in Byzantium: Essays in Honor of George T. Dennis, S.J.* (Washington, DC, 1995), pp. 17–32.

57. Evagrius 6.4–6, trans. Whitby, pp. 294–6; Simocatta, 3.5–6, trans. Whitby, pp. 76–80; Theophanes, AM 6080, trans. Mango and Scott, pp. 384–5.

58. Evagrius 6.7–15, trans. Whitby, pp. 296–308; Simocatta, 3.1–3, trans. Whitby, pp. 72–6; Theophanes, AM 6079, trans. Mango and Scott, pp. 381–4.

59. Simocatta, 3.1–3, trans. Whitby, pp. 72–6; Sebeos, trans. R. W. Thomson and J. Howard-Johnston (Liverpool, 1999), pp. 14–18.

60. Sebeos, trans. Thomson and Howard-Johnston, pp. 18–23; Simocatta, 4.12–13, trans. Whitby, pp. 118–24.

61. See Greatrex and Lieu, *Roman Eastern Frontier*, II, pp. 172–5, for references and a clear summary.

62. W. Pohl, *The Avars: A Steppe Empire in Central Europe, 567–822* (Ithaca, 2018), pp. 89–94; John of Ephesus, trans. Payne Smith, pp. 449–50.

63. F. Curta, *The Making of the Slavs: History and Archaeology of the Lower Danube Region* (Cambridge, 2001), pp. 90–119; B. Croke, 'Hormisdas and the late Roman

walls of Thessalonika', *Greek, Roman, and Byzantine Studies* 19 (1978), 251–8. A recent account of the siege of Thessalonica, dated to 586, is offered by Pohl, *The Avars*, pp. 127–30.

64. Maurice, *Strategikon*, trans. Dennis, pp. 120–26.
65. Theophanes, AM 6087 and AM 6092, trans. Mango and Scott, pp. 396–8, 403–6. For the whole period, see Whitby, *Maurice*, pp. 138–83.
66. Theophanes, AM 6093, trans. Mango and Scott, p. 408.
67. D. M. Metcalf, 'The Slavonic threat to Greece ca. 580: some evidence from Athens', *Hesperia* 31 (1962), 134–57; F. Curta, *The Edinburgh History of the Greeks, c. 500–1050: the Early Middle Ages* (Edinburgh, 2011), pp. 53–6, 84–5.
68. I. Popović et al., eds, *Sirmium à l'époque des grandes migrations* (Leuven, 2017); I. Popović, 'Marble sculptures from the imperial palace in Sirmium', *Starinar* 56 (2006), 153–66. For a general account of the siege, see Pohl, *The Avars*, pp. 83–9.
69. L. Mordechai, 'Antioch in the sixth century: resilience or vulnerability?', in A. Izdebski and M. Mulryan, eds, *Environment and Society in the Long Later Antiquity* (Leiden, 2018), pp. 25–41.
70. *The Third Part of the Ecclesiastical History of John of Ephesus*, trans. R. Payne Smith (Oxford, 1860), p. 141.

9. The Heraclians, AD 602–*c*.700

1. D. Sinor, 'The establishment and dissolution of the Türk empire', in D. Sinor, ed., *The Cambridge History of Inner Asia* (Cambridge, 1990), pp. 285–316, at pp. 305–8.
2. Theophanes, AM 6095–9, trans. Mango and Scott, pp. 418–24; Sebeos, trans. Thomson and Howard-Johnston, pp. 57–8, and xxii-xxv for a concise account of the 'Last Great War of Antiquity'. See Greatrex and Lieu, *Roman Eastern Frontier*, II, pp. 182–228, for key sources in translation.
3. *Easter Chronicle*, trans. Whitby, pp. 145–9; Theophanes, AM 6099–6101, trans. Mango and Scott, pp. 421–5. The chronology is here confused, and Theophanes is wrong in stating that the Persians advanced 'as far as Chalcedon' at this time. On the African revolt, see Kaegi, *Heraclius*, pp. 37–57. On the Persian invasions of Anatolia, the seminal works are C. Foss, 'The Persians in Asia Minor and the end of antiquity', *English Historical Review* 90 (1975), 721–47; C. Foss, 'Archaeology and the "twenty cities" of Byzantine Asia', *American Journal of Archaeology* 81 (1977), 469–86.
4. Theophanes later reports that Heraclius sailed with 'fortified ships that had on their masts reliquaries and icons of the Mother of God, as George the Pisidian relates'. See Theophanes, AM 6102, trans. Mango and Scott, pp. 427–8, referring to George of Pisidia, *Heraclias*, II.15. But George refers here not to the 'Mother of God', but to an icon of the uncorrupted Virgin (*Parthenos*), which Heraclius used

against Phocas, 'corrupter of Virgins'. See *Georgio di Pisidia, Poemi, I: Panegirici epici*, ed. Agostino Pertusi, *Studia Patristica et Byzantina* 7 (Ettal, 1959), p. 252. The only extant reference I find in George's oeuvre to Mary as Mother, not Virgin, is in a scene where she is cast as 'Mother of the Judge', presiding over a case decided by battle: Pertusi, *Georgio di Pisidia*, I, p. 193. See also B. Pentcheva, *Icons and Power: The Mother of God in Byzantium* (University Park, PA, 2006), pp. 38–40, 44–6.

5. *Easter Chronicle*, trans. Whitby, pp. 152–3; A. Cameron, *Circus Factions: Blues and Greens at Rome and Byzantium* (Oxford, 1976), pp. 116–17, reports only one instance of the factions forming part of the city's militia after the reign of Phocas. Their roles before that are explored at length throughout his still essential monograph.

6. W. Kaegi, *Heraclius: Emperor of Byzantium* (Cambridge, 2003), p. 59. See also J. D. Howard-Johnston, *Witnesses to a World Crisis: Historians and Histories of the Middle East in the Seventh Century* (Oxford, 2011).

7. S. Corcoran, 'The sins of the fathers: a neglected constitution of Diocletian on incest', *Journal of Legal History* 21 (2000), 1–34.

8. Theophanes, AM 6102–6, trans. Mango and Scott, pp. 428–31, for the births of imperial children. The *Easter Chronicle* does not record Heraclius's marriage to Martina. See Nicephorus, trans. Mango, pp. 179–80, and Kaegi, *Heraclius*, p. 106, on the date of the marriage. High office appears to have been given only to his legitimate children with Martina, born after their marriage.

9. Nicephorus, 1–3, trans. Mango, pp. 34–41; Kaegi, *Heraclius*, pp. 21–2, 68–70, 75, 83.

10. Sebeos, trans. Thomson and Howard-Johnston, pp. 68–70; F. C. Conybeare, 'Antiochus Strategos' account of the Sack of Jerusalem in AD 614', *English Historical Review* 25 (1910), 502–17.

11. *Easter Chronicle*, trans. Whitby, pp. 160–62; Sebeos, trans. Thomson and Howard-Johnston, pp. 78–80, 210–13.

12. Kaegi, *Heraclius*, pp. 86–99; Foss, 'Persians in Asia Minor', 728–30.

13. Theophanes, AM 6113, trans. Mango and Scott, pp. 435–6. Much of this is derived from George of Pisidia's lost *Bellum Avaricum*. The reference to the *themata* has drawn a great deal of attention, but surely employs the language of Theophanes's day, after 811.

14. On the seizure of church plate, dated to Easter 622, see Theophanes, AM 6113, trans. Mango and Scott, p. 435. On the hexagram, see *Easter Chronicle*, trans. Whitby, p. 158, dating the introduction to 615. Also, M. Hendy, *Studies in the Byzantine Monetary Economy, c. 300–1450* (Cambridge, 1985), pp. 494–5.

15. Theophanes, AM 6114, trans. Mango and Scott, pp. 438–9; *Georgio di Pisidia*, I, ed. Pertusi, pp. 276–7, which, however, omits the reference to 'eternal life'.

16. Theophanes, AM 6115, trans. Mango and Scott, pp. 442–3; *Georgio di Pisidia*, I, ed. Pertusi, p. 279. Kaegi, *Heraclius*, p. 129, dates this episode to 624.

17. These passages are all derived from fragments of a lost work by George of Pisidia, and therefore reflect ideas current and expressed in the 620s. Theophanes, trans. Mango and Scott, pp. xliii–lxiii, report that when reproducing material from extant works by George and others, Theophanes proves to be a reliable copyist. For the fullest commentary see J. D. Howard-Johnston, 'Herakleios' Persian campaigns and the revival of the East Roman Empire, 622–630', *War in History* 6 (1999), 1–44; Mary Whitby, 'George of Pisidia's presentation of the reign of Herakleios and his campaigns: variety and development', in G. Reinink and B. Stolte, eds, *The Reign of Herakleios (610–641): Conflict and Confrontation* (Leuven, 2002), pp. 157–73.

18. Theodore Syncellus, *De obsidione Constantinopolitana sub Heraclio imperatori*, 19; Pertusi, *Georgio di Pisidia*, I, pp. 176, 182–3; *Easter Chronicle*, trans. Whitby, pp. 168–81. See Kaegi, *Heraclius*, pp. 133–8.

19. Theophilus of Edessa, trans. Hoyland, p. 73; U. Büntgen et al., 'Cooling and societal change during the Late Antique Little Ice Age from 536 to around 660 AD', *Nature Geoscience* 9 (2016), 231–6; Sinor, 'Türk empire', pp. 308–10.

20. Theophanes, AM 6117–18, trans. Mango and Scott, pp. 446–57; Kaegi, *Heraclius*, pp. 142–68.

21. J. Howard-Johnston, 'The official history of Heraclius' Persian campaigns', in E. Dąbrowa, ed., *The Roman and Byzantine Army in the East* (Cracow, 1994), pp. 57–87.

22. *Easter Chronicle*, trans. Whitby, pp. 182–8, translation slightly altered.

23. Nicephorus, 17, trans. Mango, pp. 64–5; Greatrex and Lieu, *Roman Eastern Frontier*, II, pp. 226–8; Kaegi, *Heraclius*, pp. 188–9.

24. C. Vircillo Franklin, *The Latin Dossier of Anastasius the Persian: Hagiographic Translations and Tranformations* (Toronto, 2004).

25. Conybeare, 'Antiochus Strategus', 516–17; Kaegi, *Heraclius*, pp. 205–16.

26. *Constantine Porphyrogenitus: De Administrando Imperio*, ed. Gy. Moravcsik, trans. R. J. H. Jenkins (Washington, DC, 1985), pp. 124–5, 146–65.

27. J. Fei et al., 'Circa AD 626 volcanic eruption, climate cooling, and the collapse of the eastern Turkic empire', *Climatic Change* 81 (2007), 469–71; Sinor, 'Türk empire', p. 309. On Kuvrat, see Nicephorus, trans. Mango, pp. 70–71, and John of Nikiu, trans. Charles, p. 197. On Jabala and the Ghassanids at this time, see I. Shahid, *Byzantium and the Arabs in the Sixth Century, I.1. Political and Military History* (Washington, DC, 1995), pp. 646–51. On Jabala's flight and Ghassanid migration, see Kaegi, *Byzantium and the Early Islamic Conquests*, pp. 171–3.

28. The literature on this is vast, but a concise, critical, accessible interpretation can be found in R. Hoyland, *In God's Path: The Arab Conquest and the Creation of an Islamic Empire* (Oxford, 2015). For sources see R. Hoyland, *Seeing Islam as Others Saw It: A Survey and Evaluation of Christian, Jewish and Zoroastrian Writings on Early Islam* (Princeton, NJ, 1997).

29. See W. Kaegi, *Byzantium and the Early Islamic Conquests* (Cambridge, 1992), pp. 88–111 for the fuller contexts.

30. Kaegi, *Byzantium and the Early Islamic Conquests*, pp. 111–46.

31. John of Nikiu, trans. Charles, pp. 178–96; Nicephorus, 23, trans. Mango, pp. 70–73. See Hoyland, *In God's Path*, pp. 68–76.

32. Kaegi, *Heraclius*, pp. 229–59, 275–6; J. Haldon, *Byzantium in the Seventh Century: The Transformation of a Culture* (Cambridge, 1990), pp. 48–53.

33. *De Cerimoniis* II, 28, trans. Moffatt, pp. 628–9. See also Kaegi, *Heraclius*, pp. 267–8.

34. *De Cerimoniis* II, 29, trans. Moffatt, pp. 629–30. See also Nicephorus, trans. Mango, pp. 76–7, where a daughter, Anastasia, has become Martina. Presumably, she was named Anastasia Martina after her mother.

35. Kaegi, *Heraclius*, pp. 269–71.

36. P. Grierson, 'The tombs and obits of the Byzantine Emperors (337–1042)', *DOP* 16 (1962), 3–60, at 29–33, 48–50. Constans II died at Syracuse, Sicily, and there is no evidence that his body was returned to Constantinople.

37. Nicephorus, 28–9, trans. Mango, pp. 76–81; Theophanes, AM 6133–4, trans. Mango and Scott, pp. 474–6.

38. John of Nikiu, trans. Charles, pp. 198–9; Sebeos, trans. Thomson and Howard-Johnston, pp. 105–6, 249–55.

39. Theophanes AM 6135–7, trans. Mango and Scott, p. 476–7.

40. Theophanes AM 6143, trans. Mango and Scott, p. 480; Sebeos, trans. Thomson and Howard-Johnston, pp. 134–9, 267–8; Hoyland, *Seeing Islam*, pp. 125–6, 131–2, 642–3. See also W. Kaegi, 'The early Muslim raids into Anatolia and Byzantine reactions under Emperor Constans II', in E. Grypeou et al., eds, *The Encounter of Eastern Christianity with Early Islam* (Leiden, 2006), pp. 73–94.

41. Sebeos, trans. Thomson and Howard-Johnston, pp. 143–7, 274–6; *De Administrando Imperio*, ed. Moravcsik, trans. Jenkins, pp. 84–5. For very clear and complementary accounts of these confused years, see P. Sarris, *Empires of Faith: The Fall of Rome to the Rise of Islam, 500–700* (Oxford, 2011), pp. 275–86; Howard-Johnston, *Witnesses to a World Crisis*, pp. 216–23, 474–81. On the Colossus of Rhodes see L. Conrad, 'The Arabs and the Colossus', *Journal of the Royal Asiatic Society* 6 (1996), 165–87.

42. Hoyland, *Seeing Islam*, pp. 74–8; W. Kaegi, *Muslim Expansion and Byzantine Collapse in North Africa* (Cambridge, 2010), pp. 87–9, 261–5. On sources for the 'show trials' see Howard-Johnston, *Witnesses to a World Crisis*, pp. 157–62.

43. Howard-Johnston, *Witnesses to a World Crisis*, pp. 117–20, 481–7. As Haldon, *Byzantium in the Seventh Century*, p. 56, has noted, the Slavic population transfer was 'part of a wider change in the constitution of the population of Asia Minor ... and hints at significant changes in the recruitment and maintenance of the imperial forces in the provinces'.

44. Theophanes AM 6159, trans. Mango and Scott, pp. 488–90. For the siege of 668

and events surrounding Constans's death in Syracuse, see now M. Jankowiak, 'The first Arab siege of Constantinople', *Travaux et mémoires* 19 (2013), 237–320.

45. Howard-Johnston, *Witnesses to a World Crisis*, pp. 108–13, 224–5, 488–92, on the delegations to Mu'awiya and also the importance of a now lost 'History to 682', incorporated later into Movses's *History of the Caucasian Albanians*.

46. *De Administrando Imperio*, ed. Moravcsik, trans. Jenkins, pp. 84–93. For coins see D. Woods, 'The proclamation of peace on the coinage of Carthage under Constans II', *Greek, Roman, and Byzantine Studies* 57 (2017), 687–712. Jankowiak, 'The first Arab siege', 308–9, dates the murder of Constans to July 668, placing it in the context of Yazid's siege of Constantinople. The only source to supply a full date is the *Book of Pontiffs*, where it is given as 15 July 669.

47. Jankowiak, 'The first Arab siege', 316–17, 318–19, eliminates the siege of 674 and highlights the Roman recovery after the arrival at Constantinople of Constans's new fleet.

48. Theophanes AM 6169, trans. Mango and Scott, pp. 495–6. On the Mardaites, see *ODB*, p. 1297.

49. Nicephorus, 36, trans. Mango, pp. 88–91. See also P. Stephenson, '"About the emperor Nikephoros and how he leaves his bones in Bulgaria": a context for the controversial Chronicle of 811', *DOP* 60 (2006), 87–109.

50. Nicephorus, 37, trans. Mango, pp. 90–93; Theophanes AM 6176–7, trans. Mango and Scott, pp. 503–4; Haldon, *Byzantium in the Seventh Century*, pp. 66–70.

51. James Howard-Johnston identified the *History to 720*, viewing it as 'a rhetorical outpouring, a *psogos*', painting a portrait of Justinian that is 'blacker than it should be'. He and Warren Treadgold both identify the writer as Trajan, the patrician, who is said to have 'flourished at the time of Justinian the slit-nosed' and to have been 'very Christian and very Orthodox'. The reconstructed work of Theophilus of Edessa remains the major non-Greek source. See Howard-Johnston, *Witnesses to a World Crisis*, pp. 259, 299–307, developing an identification made by Cyril Mango, at Nicephorus, trans. Mango, pp. 15–17, and Theophanes, trans. Mango and Scott, pp. lxxxvii-xc. See also W. Treadgold, 'Trajan the Patrician, Nicephorus, and Theophanes', in D. Bumazhnov et al., eds, *Bibel, Byzanz und christlicher Orient* (Leuven, 2011), pp. 589–621.

52. Theophanes, AM 6176–80, trans. Mango and Scott, pp. 503–8; Nicephorus, 38, trans. Mango, pp. 92–5.

53. Mango, *Art*, pp. 123, 139–40; *Theophilus of Edessa*, trans. Hoyland, pp. 190–91: 'The Arabs began to strike gold, silver and copper coins with no images on them, just letters alone.' See also S. Griffith, 'Images, Islam and Christian icons: a moment in the Christian/Muslim encounter in early Islamic times', in P. Canivet and J.-P. Rey-Coquais, eds, *La Syrie de Byzance à l'Islam, VIIe-VIIIe siècles* (Damascus, 1992), pp. 121–38.

54. Theophanes, AM 6183–6, trans. Mango and Scott, pp. 509–14; Nicephorus, 38, trans. Mango, pp. 94–5.

55. Hendy, *Studies*, pp. 626–34. The whole period is now captured by J. Haldon, *The Empire that Would not Die: The Paradox of Eastern Roman Survival, 640–740* (Cambridge, MA, 2016).

56. Theophanes, AM 6187–90, trans. Mango and Scott, pp. 514–18; Nicephorus, 40–41, trans. Mango, pp. 94–101. Only Theophanes mentions the removal of Justinian's tongue. Related sources mention only his nose, on which see *Theophilus of Edessa*, trans. Hoyland, p. 190.

57. Theophanes, AM 6191–5, trans. Mango and Scott, pp. 518–20; *Theophilus of Edessa*, trans. Hoyland, pp. 192–6.

58. Theophanes, AM 6196–8, trans. Mango and Scott, pp. 520–24; Nicephorus, 42, trans. Mango, pp. 100–105. The hippodrome chant is from Psalm 90 (91):13.

59. Theophanes, AM 6203, trans. Mango and Scott, pp. 527–31; Nicephorus, 45, trans. Mango, pp. 106–13; *Theophilus of Edessa*, trans. Hoyland, pp. 202–5.

10. The End of Ancient Civilisation

1. N. Lozovsky, '*The Earth is Our Book*': *Geographical Knowledge in the Latin West, ca. 400–1000* (Ann Arbor, MI, 2000), pp. 30–31.

2. E. Albu, *The Medieval Peutinger Map*: *Imperial Roman Renewal in a German Empire* (New York and Cambridge, 2014), p. 41; S. J. Johnson, *Literary Territories: Cartographical Thinking in Late Antiquity* (Oxford, 2016); A. Merrills, 'Geography and memory in Isidore's *Etymologies*', in K. Lilley, ed., *Mapping Medieval Geographies: Geographical Encounters in the Latin West and Beyond, 300–1600* (Cambridge, 2013), pp. 45–64. The Cosmographer mentions most frequently a certain Castorius, an otherwise unknown geographer, leading some to identify him as a third- or fourth-century compiler of itineraries and even the author of a map on which the far later and famous Peutinger map would be modelled. See R. Talbert, *Rome's World: The Peutinger Map Reconsidered* (Cambridge, 2010), pp. 133–4. One of the Cosmographer's sources that we can identify was Orosius, whom he describes as 'the very wise scrutineer of the east'.

3. A. Augenti and E. Cirelli, 'From suburb to port: the rise (and fall) of Classe as a centre of trade and redistribution', in S. Keay, ed., *Rome, Portus and the Mediterranean* (London, 2012), pp. 205–21; F. H. van Doorninck Jr, 'The seventh-century Byzantine ship at Yassıadda and her final voyage: present thoughts', in D. Carlson et al., eds, *Maritime Studies in the Wake of the Byzantine Shipwreck at Yassıadda, Turkey* (College Station, TX, 2015), pp. 205–16.

4. D. Mauskopf Deliyannis, *Ravenna in Late Antiquity* (Cambridge, 2010), pp. 277–308, 391–6; E. Cirelli, 'Ravenna – rise of a late antique capital', in D. Sami and G.

Speed, eds, *Debating Urbanism Within and Beyond the Walls, AD 300–700* (Leicester, 2010), pp. 239–64.

5. Procopius, *Secret History*, 26.6–11; trans. Kaldellis, p. 114.

6. Brandes, 'Byzantine cities', pp. 54–6; Haldon and Brandes, 'Towns, tax and transformation', pp. 163–70; J. H. W. G. Liebeschuetz, *The Decline and Fall of the Roman City* (Oxford, 2001), pp. 104–36: 'Post-curial civic government'.

7. Liebeschuetz, *Decline and Fall*, pp. 137–68; Brandes, 'Byzantine cities', pp. 29, 41–4.

8. Wonderful resources and images are collected at sardisexpedition.org/

9. M. Rautman, 'Sardis in late antiquity', in O. Dalley and C. Ratté, eds, *Archaeology and the Cities of Asia Minor in Late Antiquity* (Ann Arbor, MI, 2011), pp. 1–26, at pp. 24–5.

10. Liebeschuetz, *Decline and Fall*, pp. 43–4; D. M. Metcalf, 'The Aegean coastlands under threat: some coins and coin hoards from the reign of Heraclius', *Annual of the British School at Athens* 57 (1962), 14–23; Foss, 'Persians in Asia Minor', 736–8; Foss, 'Archaeology and the "twenty cities"', 475–7; J. S. Crawford, *The Byzantine Shops at Sardis* (Cambridge, MA, 1990), pp. 126–8.

11. M. M. Fulgham and F. Heintz, 'A hoard of early Byzantine glass weights from Sardis', *American Journal of Numismatics* 10 (1998), 105–20. See now C. Entwistle and A. Meek, 'Early Byzantine glass weights: aspects of function, typology and composition', *British Museum Technical Research Bulletin* 9 (2015), pp. 1–14; N. Schibille et al., 'Comprehensive chemical characterisation of Byzantine glass weights', *PLoS ONE* 11 (2016), 1–16.

12. Liebeschuetz, *Decline and Fall*, p. 44; Foss, 'Persians in Asia Minor', 738–9; Foss, 'Archaeology and the "twenty cities"', 472–5.

13. F. Stock et al., 'In search of the harbours: new evidence of late Roman and Byzantine harbours of Ephesus', *Quaternary International* 312 (2013), 57–69.

14. Foss, 'Persians in Asia Minor', 735–6; C. Foss, 'Late antique and Byzantine Ankara', *DOP* 31 (1977), 29–87, at 71–7.

15. C. Foss, 'The Lycian coast in the Byzantine age', *DOP* 48 (1994), 1–52.

16. C. Lightfoot, 'The survival of cities in Byzantine Anatolia: the case of Amorion', *Byzantion* 68 (1998), 56–71.

17. R. M. Harrison, 'Amorium 1987: a preliminary survey', *Anatolian Studies* 38 (1988), 175–84; R. M. Harrison, 'Amorium 1988: the first preliminary excavation', *Anatolian Studies* 39 (1989), 167–74; R. M. Harrison and N. Christie, 'Excavations at Amorium: 1992 interim report', *Anatolian Studies* 43 (1993), 147–62; C. Lightfoot et al., 'The Amorium project: excavation and research in 2003', *DOP* 61 (2007), 353–85; C. Lightfoot, 'Trade and industry in Byzantine Anatolia: the evidence from Amorium', *DOP* 61 (2007), 269–86; C. Lightfoot and E. Ivison, eds, *Amorium Reports 3: the Lower City Enclosure* (Istanbul, 2012).

18. J. Haldon et al., 'Euchaïta', in G. McMahon and S. Steadman, eds, *The Archaeology*

of Anatolia: Recent Discoveries (2011–2014) (Cambridge, 2015), pp. 328–51, at pp. 330, 333.

19. F. Curta, 'Amphorae and seals: the "sub-Byzantine" Avars and the quaestura exercitus', in A. Bollók et al., eds, *Between Byzantium and the Steppe* (Budapest, 2016), pp. 307–34.

20. A. Gândilă, 'A hoard of sixth-century coppers and the end of Roman Capidava', in I. C. Opriş and A. Raţiu, eds, *Capidava II* (Cluj-Napoca, 2017), pp. 161–74; A. Gândilă, 'Reconciling the "step sisters": early Byzantine numismatics, history and archaeology', *Byzantinische Zeitschrift* 111 (2018), 103–34; G. Custurea and I. Nastasi, 'Un nou depozit de monede bizantine descoperit la Histria', in *Histria: histoire et archéologie en Mer Noire, Pontica* 47, supplementum III (Constanţa, 2014), pp. 343–51. See also I. C. Opriş and A. Raţiu, 'An early Byzantine building next to the main gate at Capidava', in A. Panaite et al., eds, *Moesia et Christiana: Studies in Honour of Professor Alexandru Barnea* (Brăila, 2016), pp. 193–217.

21. J. Wiseman et al., eds, *Studies in the Antiquities at Stobi*, 3 vols (Boston and Princeton, 1975–81); R. Kolarik, 'Seasonal animals in the narthex mosaic of the large basilica, Heraclea Lyncestis', *Niš and Byzantium* 10 (2012), 105–18.

22. W. Bowden and R. Hodges, 'An "Ice Age settling on the Roman Empire": post-Roman Butrint between strategy and serendipity', in N. Christie and A. Augenti, eds, *Vrbes Extinctae: Archaeologies of Abandoned Classical Towns* (Aldershot, 2012), pp. 207–41, at p. 209.

23. S. Kamani, 'Butrint in the mid-Byzantine period: a new interpretation', *Byzantine and Modern Greek Studies* 35 (2011), 115–33.

24. D. Pettegrew, 'The end of ancient Corinth? Views from the landscape', in W. Caraher et al., eds, *Archaeology and History in Roman, Medieval and Post-Medieval Greece* (Aldershot, 2008), pp. 249–66; E. Weiberg et al., 'The socio-environmental history of the Peloponnese during the Holocene: towards an integrated understanding of the past', *Quaternary Science Reviews* 136 (2016), 40–65.

25. C. Morrisson, 'La fin de l'antiquité dans les Balkans à la lumière des trésors monétaires des VIe et VIIe siècles', *Comptes rendus des séances de l'Académie des Inscriptions et Belles-Lettres* 151 (2007), 661–84, summarising C. Morrisson et al, *Les trésors monétaires byzantins des Balkans et d'Asie Mineure, 491–713* (Paris, 2006), which brings together the seminal studies of V. Popović; F. Curta, 'The beginning of the middle ages in the Balkans', *Millennium* 10 (2013), 145–214, at 150; N. Bakirtzis, 'The practice, perception and experience of Byzantine fortification', in P. Stephenson, ed., *The Byzantine World* (London and New York, 2010), pp. 352–71.

26. D. M. Metcalf, 'The mint of Thessalonica in the early Byzantine period', in *Villes et peuplement dans l'Illyricum protobyzantin: actes du colloque de Rome* (12–14 mai 1982) (Rome, 1984), pp. 111–29, observes that 'The later seventh, eighth, and ninth centuries witnessed, it is true, a decline in the Byzantine currency almost everywhere except in the region of Constantinople itself. But if one attempted to

compare the situation in Illyricum with that in north Africa, Egypt, Syria, or Meso-potamia, one would have to conclude that the collapse came sooner and was more complete in Illyricum.'

27. F. Curta, *The Making of the Slavs: History and Archaeology of the Lower Danube Region* (Cambridge, 2001), pp. 90–119.

28. C. Walter, *The Warrior Saints in Byzantine Art and Tradition* (Aldershot, 2003), pp. 67–93.

29. E. Fentress et al., 'Accounting for ARS: fineware and sites in Sicily and Africa', in S. Fontana et al., eds, *Side by Side Survey: Comparative Regional Studies in the Mediterranean World* (Oxford, 2004), pp. 147–62; P. Bes and J. Poblome, 'African red slip ware on the move: the effects of Bonifay's *Études* for the Roman east', in J. H. Humphrey, ed., *Studies on Roman Pottery of the Provinces of Africa Proconsularis and Byzacena (Tunisia)* (Portsmouth, RI, 2009), pp. 73–91; L. Saguì et al., 'Nuovi dati ceramologici per la storia economica di Roma tra VII e VIII secolo', in G. Démians d'Archimbaud, ed., *La céramique medieval en Méditerranée* (Aix-en-Provence, 1997), pp. 35–48; P. Mirti and A. Lepora, 'Scientific analysis of seventh-century glass fragments from the Crypta Balbi in Rome', *Archaeometry* 42 (2000), 359–74; P. Mirti et al., 'Glass fragments from the Crypta Balbi in Rome: the composition of eighth-century fragments', *Archaeometry* 43 (2001), 491–502.

30. C. Fenwick, 'From Africa to Ifrīqiya: settlement and society in early medieval north Africa', *Al-Masāq* 25 (2013), 9–33; C. Fenwick, 'Early medieval urbanism in Ifrīqiya and the emergence of the Islamic city', in S. Panzram and L. Callegarin, eds, *Entre civitas y madīna: el mundo de las ciudades en las Península Ibérica y el norte de África (siglos IV-IX)* (Madrid, 2018), pp. 203–20; E. Vaccaro, 'Sicily in the eighth and ninth centuries AD: a case of persisting economic complexity', *Al-Masāq* 25 (2013), 34–69.

31. P. Pensabene, 'Il contributo degli scavi 2004–2014 alla storia della Villa del Casale di Piazza Armerina tra IV e VIII secolo', in C. Giuffrida and M. Cassia, eds, *Silenziose rivoluzioni: la Sicilia dalla tarda antichità al primo medioevo* (Catania, 2016), pp. 223–71; A. Castrorao Barba, 'Sicily before the Muslims: the transformation of the Roman villas between late antiquity and the early middle ages, fourth to eighth centuries CE', *Journal of Transcultural Medieval Studies* 3 (2016), 145–89.

32. I. Baldini, 'Early Byzantine churches in Crete and Cyprus: between local identity and homologation', *Cahiers du Centre d'Études Chypriotes* 43 (2013), 31–49; A. Kouremenos, 'In the heart of the wine-dark sea: Cretan insularity and identity in the Roman period', in A. Kouremenos, ed., *Insularity and Identity in the Roman Mediterranean* (Oxford, 2018), pp. 41–64.

33. P. Pirazzoli et al., 'Earthquake clustering in the eastern Mediterranean during historical times', *Journal of Geophysical Research: Solid Earth* 101 (1996), 6083–97.

34. J. F. Harrison and G. Harrison, 'Gortyn: first city of Roman Crete', *American Journal of Archaeology* 107 (2003), 487–92; E. Giorgi, 'Water technology at Gortyn

in the 4th–7th c. AD: transport, storage and distribution', in L. Lavan, ed., *Technology in Transition, AD 300–650* (Leiden, 2007), pp. 287–320.

35. According to al-Maqdisi, quoted in translation by Mango, *Art,* p. 132. On the establishment of the condominium in 685–6, which affected Cyprus, Armenia and Caucasian Iberia, see Theophanes, AM 6178, trans. Mango and Scott, p. 506. On coins, see D. M. Metcalf, *Byzantine Cyprus, 491–1191* (Nicosia, 2009), pp. 141–213.

36. L. Zavagno and B. Kızılduman, 'A countryside in transition: the Galinoporni-Kaleburnu Plain (Cyprus) in the passage from Late Antiquity to the early Middle Ages (ca. 600–ca. 850)', *Památky Archaeologické* 109 (2018), 233–51; W. Caraher et al., 'Surveying late antique Cyprus', *Near Eastern Archaeology* 71 (2008), 82–9; W. Caraher et al., *Pyla-Koutsopetria I: Archaeological Survey of an Ancient Coastal Town* (Boston, 2014); I. Randall, 'Continuity and change in the ceramic data: "the Byzantine problem" and Cyprus during the treaty centuries', *Cahiers du Centre d'Études Chypriotes* 43 (2013), 273–84.

37. B. Shilling, *Apse Mosaics of the Virgin Mary in Early Byzantine Cyprus*, unpublished PhD dissertation (Johns Hopkins University, Baltimore, MD, 2013), pp. 85–159.

38. I. Freestone et al., 'The origins of Byzantine glass from Maroni Petrera, Cyprus', *Archaeometry* 44 (2002), 257–72; T. Papacostas, 'Neapolis/Nemesos/Limassol: the rise of a Byzantine settlement from late antiquity to the time of the Crusades', in *Lemesos: A History of Limassol in Cyprus from Antiquity to the Ottoman Conquest* (Cambridge, 2015), pp. 96–188, at pp. 98–106; D. M. Metcalf, *Byzantine Lead Seals from Cyprus* (Nicosia, 2004), pp. 54–5, 114, 360.

39. D. Baarkan et al., 'The "Dor 2006" shipwreck: the ceramic material', *Tel Aviv* 40 (2013), 117–43.

40. O. Barkai et al., 'The Tantura F shipwreck: the ceramic material', *Levant* 42 (2010), 88–101; O. Barkai, 'The Tantura F ceramic assemblage and the fish remains: evidence of trade in the eastern Mediterranean in the early Islamic period', *Skyllis* 11 (2011), 18–20; S. Grainger, 'Roman fish sauce: fish bones residues and the practicalities of supply', *Archaeofauna* 22 (2013), 13–28.

41. The literature is now too extensive to cite adequately. For themes and references as the discussion has developed, see G. Ostrogorsky, 'Byzantine cities in the early middle ages', *DOP* 13 (1959), 45–66; M. Whittow, 'Ruling the late Roman and early Byzantine city: a continuous history', *Past & Present* 129 (1990), 3–29.

42. N. Roberts et al, 'Not the end of the world? Post-classical decline and recovery in rural Anatolia', *Human Ecology* 46 (2018), 305–22, which also provides data for the medieval recovery, and correlative data from archaeological surveys. See also M. Whittow, 'Early medieval Byzantium and the end of the ancient world', *Journal of Agrarian Change* 9 (2009), 134–53, at 146–7; A. Izdebski et al., 'The environmental, archaeological and historical evidence for regional climatic change and their societal impacts in the eastern Mediterranean in late antiquity', *Quaternary Science Reviews* 30 (2015), 189–208.

43. J. Haldon, *The Empire that Would not Die: The Paradox of Eastern Roman Survival, 640–740* (Cambridge, MA, 2016), pp. 215–48, at pp. 232–6.

44. J. Grattan et al., 'The local and global dimensions of metalliferous pollution derived from a reconstruction of an eight thousand year record of copper smelting and mining at a desert-mountain frontier in southern Jordan', *Journal of Archaeological Science* 34 (2007), 83–110.

11. Apocalypse and the End of Antiquity

1. Marcellinus Comes, trans. Croke, p. 25; Cassiodorus, *Variae*, 4.50; *The Letters of Cassiodorus*, trans. T. Hodgkin (London, 1886), pp. 261–3. See also C. Kostick and F. Ludlow, 'The dating of volcanic events and their impact upon European society, 400–800 CE', *European Journal of Postclassical Archaeologies* 5 (2015), 7–30, at 8–13.

2. Suggestions that it was a volcano in the equatorial zone or in Iceland have been superseded by the observation that the 'geochemistry of tephra filtered from the NEEM-2011-S1 ice core at a depth corresponding to 536 CE indicated multiple North American volcanoes as likely candidates for a combined volcanic signal'. See M. Sigl et al., 'Timing and climate forcing of volcanic eruptions for the past 2,500 years', *Nature* 523 (2015), 543–9, at 547.

3. Cassiodorus, *Variae*, 12.25; trans. Hodgkin, p. 519.

4. Procopius, *Wars*, 4.14; trans. Dewing and Kaldellis, pp. 220–21. An account by John of Ephesus, preserved by Michael the Syrian, records that: 'The sun was dark and its darkness lasted for eighteen months; each day it shone for about four hours; and still this light was only a feeble shadow; the fruits did not ripen and the wine tasted like sour grapes.'

5. A. Martínez Cortizas et al., 'Atmospheric Pb deposition in Spain during the last 4600 years recorded by two ombrotrophic peat bogs and implications for the use of peat as archive', *Science of the Total Environment* 292 (2002), 33–44; D. Fuks et al., 'Dust clouds, climate change and coins: consiliences of paleoclimate and economic history in the late antique southern Levant', *Levant* 49 (2017), 205–23.

6. Sigl et al., 'Timing and climate forcing', 53–9; U. Büntgen et al., 'Cooling and societal change during the Late Antique Little Ice Age from 536 to around 660 AD', *Nature Geoscience* 9 (2016), 231–6. The same data suggest further that there may even have been a third volcanic eruption in 547. An alternative theory, with far less support, identifies the source of the dust veil as a comet strike.

7. Theophilus of Edessa, trans. Hoyland, p. 73; J. Fei et al., 'Circa AD 626 volcanic eruption, climate cooling, and the collapse of the eastern Turkic empire', *Climatic Change* 81 (2007), 469–71; Sigl et al., 'Timing and climate forcing', 547, fig. 5. See also R. Stothers, 'Cloudy and clear stratospheres before AD 1000 inferred from written sources', *Journal of Geophysical Research* 107 (2002), 4718 (online, 10

pages); R. Stothers, 'Volcanic dry fogs, climate cooling, and plague pandemics in Europe and the Middle East', *Climatic Change* 42 (1999), 713–23, including the observation that there was 'severe frost damage to the growth rings in California bristlecone pines (*P. aristata*) in AD 628'; M. Toohey et al., 'Disproportionately strong climate forcing from extratropical explosive volcanic eruptions', *Nature Geoscience* 12 (2019), 100–107, suggests that since both the 536 and 626 eruptions were northern hemispherical, they had a more profound impact on climate, and therefore civilisations, in that hemisphere than would tropical (or southern hemispherical) eruptions.

8. Procopius, *Wars*, 2.22; trans. Dewing and Kaldellis, pp. 120–23; A. Kaldellis, *Procopius of Caesarea: Tyranny, History, and Philosophy at the End of Antiquity* (Philadelphia, 2004), pp. 210–13; P. Sarris, 'Bubonic plague in Byzantium: the evidence of non-literary sources', in L. Little, ed., *Plague and the End of Antiquity: The Pandemic of 541–750* (Cambridge, 2007), pp. 119–32. See now K. Sessa, 'The new environmental fall of Rome: a methodological consideration', *Journal of Late Antiquity* 12 (2019), 211–55, which is critical of 'scientistic' studies by historians but also highlights the absence of scientific evidence for widespread plague mortality. B. Schmid et al., 'Climate-driven introduction of the Black Death and successive plague reintroductions into Europe', *Proceedings of the National Academy of Sciences of the USA* 112 (2015), 3020–25, argues for repeated climate-driven introductions of disease into Europe as rodents fled 'climate-sensitive plague reservoirs'. That is to say, abnormally warm and wet conditions fostered growth in rodent populations, and sudden changes led to the collapse of rodent populations, forcing fleas to seek new hosts. However, the scholars admit 'We have found no evidence supporting the existence of a climate-sensitive wildlife plague reservoir in medieval Europe'. D. Wagner et al., '*Yersinia pestis* and the Plague of Justinian 541–543 AD: a genomic analysis', *The Lancet Infectious Diseases* 14 (2014), 319–26, constructed a 'maximum likelihood phylogenic tree' based on analysis of DNA extracted from the teeth of two plague victims in an early medieval Bavarian cemetery. Radiocarbon dating of the bodies suggested they had died in the first wave of bubonic plague, and certainly before the end of the sixth century. M. Feldman et al., 'A high-coverage *Yersinia pestis* genome from a sixth-century Justinianic Plague victim', *Molecular Biology and Evolution* 33 (2016), 2911–23, confirmed and enhanced the findings by analysis of a body of the same date and region.

9. R. Sallares, 'Ecology, evolution, and epidemiology of plague', in Little, ed., *Plague*, pp. 232–89. Pneumonic plague can be transmitted from human to human, and for that reason is far more effectively spread and far more deadly unless identified and treated early.

10. John of Ephesus, 17.1. See L. Little, 'Life and afterlife of the first plague pandemic', in Little, ed., *Plague*, pp. 3–32, at p. 7.

11. Sessa, 'New environmental fall of Rome', 234–5, highlights the absence of bodies as

a useful mismatch between scientific data and literary sources. However, for more bodies, still too few, see the paper to which she refers, now peer reviewed, M. Keller et al., 'Ancient *Yersinia pestis* genomes from across western Europe reveal early diversification during the first pandemic (541–750)', *Proceedings of the National Academy of Sciences of the USA* 116 (2019), 12363–75. A maximalist position is taken by M. Meier, 'The "Justinianic Plague": the economic consequences of the pandemic in the eastern Roman empire and its cultural and religious effects', *Early Medieval Europe* 24 (2016), 267–92. The minimalist position is stated effectively in M. Eisenberg and L. Mordechai, 'Rejecting catastrophe: the case of the Justinianic Plague', *Past & Present* 244 (2019), 3–50; L. Mordechai et al., 'The Justinianic Plague: an inconsequential pandemic?', *PNAS* 116 (December, 2019), 25546–54.

12. D. Stathakopoulos, 'Crime and punishment: the plague in the Byzantine empire, 541–749', in Little, ed., *Plague*, pp. 99–118, at pp. 100–105.

13. Theophanes, AM 6238; trans. Mango and Scott, pp. 585–6.

14. Until recently, very few mass graves from this period had been identified. However, see M. McCormick, 'Tracking mass death during the fall of Rome's empire (I)', *Journal of Roman Archaeology* 28 (2015), 325–57; M. McCormick, 'Tracking mass death during the fall of Rome's empire (II): a first inventory of mass graves', *Journal of Roman Archaeology* 29 (2016), 1004–46, where mass graves are identified as the resting place of five or more individuals.

15. P. Pirazzoli et al., 'Earthquake clustering in the eastern Mediterranean during historical times', *Journal of Geophysical Research: Solid Earth* 101 (1996), 6083–97; N. Evelpidou and P. Pirazzoli, 'Did the early Byzantine tectonic paroxysm (EBTP) also affect the Adriatic area?', *Geomorphology* 295 (2017), 827–30; G. Kelly, 'Ammianus and the Great Tsunami', *Journal of Roman Studies* 94 (2004), 141–67.

16. G. Downey, 'Earthquakes at Constantinople and vicinity', *Speculum* 30 (1955), 596–600; Averil Cameron, *Agathias* (Oxford, 1970), Appendix A, 'The date of the earthquake which destroyed Berytus'. Theophanes then reports earthquakes in 740 and 742, 790 and 796.

17. Malalas 17.16, trans. Jeffreys et al., p. 238.

18. See L. Becker and C. Kondoleon, eds, *The Arts of Antioch: Art Historical and Scientific Approaches to Roman Mosaics and a Catalogue of the Worcester Art Museum Antioch Collection* (Princeton, 2005), p. 203, where for the exhibition catalogue the six surviving fragments of the bird rinceau (decorative border) are digitally reconnected. Also, in the same book, A. Gonosová, 'Mosaic with peacocks', pp. 238–43.

19. A. A. Eger, '(Re)Mapping medieval Antioch: urban transformations from the early Islamic to the middle Byzantine periods', *DOP* 67 (2013), 95–134.

20. D. Krueger, *Symeon the Holy Fool: Leontius's* Life *and the Late Antique City* (Berkeley, 1996), suggests that the life reveals more about urban life in seventh-century Cyprus than about Emesa. See also M. Guidetti, *In the Shadow of the Church: The Building of Mosques in Early Medieval Syria* (Leiden, 2017), pp. 36–9.

21. See the papers collected in J. Haldon, ed., *Money, Power and Politics in Early Islamic Syria: A Review of Current Debates* (Farnham, 2010); R. Schick, *The Christian Communities of Palestine from Byzantine to Islamic Rule: A Historical and Archaeological Study* (Princeton, NJ, 1995); G. Avni, *The Byzantine–Islamic Transition in Palestine: An Archaeological Approach* (Oxford, 2014), pp. 41–55.

22. M. Piccirillo, *The Mosaics of Jordan* (Amman, 1992), pp. 81–95 (Madaba map), 232–3, 238–9 (Umm ar-Rasas), 273, 288–9 (Gerasa), and many additional figures. See also G. W. Bowersock, *Mosaics as History: The Near East from Late Antiquity to Islam* (Cambridge, MA, 2006), pp. 1–29, 64–88.

23. H. Kennedy, 'From *polis* to *madina*: urban change in late antique and early Islamic Syria', *Past & Present* 106 (1985), 3–27, at 9–10, 12–13; Avni, *Byzantine–Islamic Transition*, pp. 93–8; A. Ostrasz, 'The hippodrome of Gerasa: a report on excavations and research, 1982–1987', *Syria* 66 (1989), 51–77.

24. A. Ucatescu and M. Martín-Bueno, 'The *macellum* of Gerasa (Jerash, Jordan): from a market place to an industrial area', *Bulletin of the American Schools of Oriental Research* 307 (1997), 67–88; H. Kennedy, 'Gerasa and Scythopolis: power and patronage in the Byzantine cities of Bilad al-Sham', *Bulletin d'Études Orientales* 52 (2000), 199–206; A. Walmsley and K. Damgaard, 'The Umayyad congregational mosque of Jarash in Jordan and its relationship to early mosques', *Antiquity* 79 (2005), 362–78.

25. A. Lichtenberger and R. Raja, 'A hoard of Byzantine and Arab-Byzantine coins from the excavations at Jerash', *Numismatic Chronicle* 175 (2015), 299–308; M. Phillips and T. Goodwin, 'A seventh-century Syrian hoard of Byzantine and imitative copper coins', *Numismatic Chronicle* 157 (1997), 61–87.

26. Doris Behrens-Abouseif, 'The Islamic history of the lighthouse of Alexandria', *Muqarnas* 23 (2006), 1–14; El Sayed Abdel Aziz Salem, 'The influence of the lighthouse of Alexandria on the minarets of north Africa and Spain', *Islamic Studies* 30 (1991), 149–56; C. Foss, 'Egypt under Muʿāwiya part II: Middle Egypt, Fusṭāṭ and Alexandria', *Bulletin of SOAS* 72 (2009), 259–78; P. Sijpesteijn, *Shaping a Muslim State: The World of a Mid-Eighth-Century Egyptian Official* (Oxford, 2013), pp. 49–113.

27. U. Avner and J. Magness, 'Early Islamic settlement in the southern Negev', *Bulletin of the American Schools of Oriental Research* 310 (1998), 39–57. See also Foss, 'Egypt under Muʿāwiya part II', 262, picked up in later literature, including Sijpesteijn, *Shaping a Muslim State*, p. 91.

28. J. Patrich, 'Caesarea in transition: the archaeological evidence from the southwest zone (areas CC, KK, NN)', in H. Lapin and K. Holum, eds, *Shaping the Middle East* (Bethesda, MD, 2011), pp. 33–4. See also Avni, *Byzantine–Islamic Transition*, pp. 41–55.

29. I. Taxel, 'Rural settlement processes in central Palestine, ca. 640–800 CE: the Ramla-Yavneh region as a case study', *Bulletin of the American Schools of Oriental Research*

369 (2013), 157–99; Avni, *Byzantine–Islamic Transition*, pp. 159–88; R. Grafman and M. Rosen-Ayalon, 'Two great Syrian Umayyad mosques: Jerusalem and Damascus', *Muqarnas* 16 (1999), 1–15, argues that the Damascus mosque was modelled on the original Aqsa mosque in Jerusalem. See J. Johns, 'Archaeology and the history of early Islam: the first seventy years', *Journal of the Economic and Social History of the Orient* 46 (2003), 411–36, at 416–18, on the absence of mosques before the start of the eighth century before the sudden proliferation.

30. T. Jonson et al., 'The Byzantine mint in Carthage and the Islamic mint in North Africa: new metallurgical findings', *Revue Numismatique* 171 (2014), 655–99; C. Morrisson, '*Regio dives in omnibus bonis ornata*: the African economy from the Vandals to the Arab conquest in light of coin evidence', in S. Stevens and J. Conant, eds, *North Africa under Byzantium and Early Islam* (Cambridge, MA, 2016), pp. 173–98.

31. Grafman and Rosen-Ayalon, 'Two great Syrian Umayyad mosques', 2–5, summarising archaeological work published by R. W. Hamilton in 1949. Johns, 'Archaeology and the history of early Islam', 416, n. 7, somewhat grudgingly accepts this exception to the rule that mosques were not built before the eighth century.

32. O. Grabar, *The Dome of the Rock* (Cambridge, MA, 2006), pp. 59–119, quotation from pp. 118–19. See also Johns, 'Archaeology and the history of early Islam', 424–6, on the debate over the origins and purpose of the Dome of the Rock; and 427–9, on the long inscription and its relationship to the Quran.

33. *The Testament of Our Lord*, trans. J. Cooper and A. J. Maclean (Edinburgh, 1902), pp. 51–7.

34. R. Hoyland, *Seeing Islam as Others Saw It: A Survey and Evaluation of Christian, Jewish and Zoroastrian Writings on Early Islam* (Princeton, NJ, 1997), pp. 257–335, 'Apocalypses and visions'.

35. S. Brock, 'North Mesopotamia in the late seventh century: Book XV of John Bar Penkāyē's Rīš Mellē', *Jerusalem Studies in Arabic and Islam* 9 (1987), 51–75. I have followed also an online translation by Alphonse Mingana and by Roger Pearse at www.tertullian.org/

36. S. Brock 'Apocalypse of Pseudo-Methodius', in *The Seventh Century in the West-Syrian Chronicles*, trans. A. Palmer (Liverpool, 1993), pp. 222–50. There is a full English translation of the Syriac Pseudo-Methodius in P. Alexander, *The Byzantine Apocalyptic Tradition* (Berkeley, 1985), pp. 36–51.

37. G. Reinink, 'Ps.-Methodius: a concept of history in response to the rise of Islam', in Averil Cameron and L. Conrad, eds, *The Byzantine and Early Islamic Near East, I: Problems in the Literary Sources* (Princeton, 1992), pp. 149–87. See also G. Reinink, 'The Romance of Julian the Apostate as a source for seventh century Syriac Apocalypses', in P. Canivet and J.-P. Rey-Coquais, eds, *La Syrie de Byzance à l'Islam, VIIe-VIIIe siècles: actes du colloque international Lyon – Maison de l'Orient méditerranéen* (Damascus, 1992), pp. 75–86.

38. Visual depictions of the apocalypse have not survived from this period and representations may have been restricted to the separation of those to the left and right of the throne, the 'sheep' and 'goats' of Matthew 25. See D. Kinney, 'The Apocalypse in early Christian monumental decoration', in R. K. Emmerson and B. McGinn, eds, *The Apocalypse in the Middle Ages* (Ithaca, 1992), pp. 200–16, at p. 200: 'no early Christian apse or wall painting represents the Apocalypse per se', although many motifs 'make a sudden and profuse appearance' after AD 350. Kinney is concerned principally with the West, and Italy in particular, but lists fifth- to seventh-century examples from Greece (Thessalonike), Egypt (Mt Sinai, Bawit, Saqqara) and Bulgaria (Sofia). See also H. Kessler, *Spiritual Seeing* (Philadelphia, 2000), pp. 88–103, on images of the Second Coming, with reference to the Vatican *Christian Topography* of Cosmas Indicopleustes (Bibliotheca Apostolica, Cod. gr. 699, fol. 89r.).

39. P. Crone, *Slaves on Horses: The Evolution of the Islamic Polity* (Cambridge, 1980); D. Pipes, 'Mawlas: freed slaves and converts in early Islam', in R. Hoyland, ed., *Muslims and Others in Early Islamic Society* (Aldershot, 2004), pp. 277–322.

40. Wadād al-Qāḍī, 'Population census and land surveys under the Umayyads (41/132/661–750)', *Der Islam* 83 (2008), 341–416, at 366.

41. Johns, 'Archaeology and the history of early Islam', 421–3; al-Qāḍī, 'Population census', 381–2.

42. Sahner, '"The monasticism of my community is jihad": a debate on asceticism, sex, and warfare in early Islam', *Arabica* 64:2 (2017), 165–74; *The Book of Strangers*, trans. P. Crone and S. Moreh (Princeton, 2000), pp. 58–62; T. Sizgorich, *Violence and Belief in Late Antiquity: Militant Devotion in Christianity and Islam* (Philadelphia, 2009), pp. 231–71.

43. See for example the *surah* on ascending stairways (Quran 70), that on the resurrection (75), on the striking calamity (101), and, most notably, those on the overthrowing and the cleaving (81, 82), 'When the sun is wrapped up, and when the stars fall ... when the sky breaks apart ... and when the seas are erupted, and when the graves are scattered'. Some have suggested that for Muhammad and his fellow believers, what became the Islamic faith was in origin very simple beliefs about 'one God, the Last Judgment, God's messengers, the book, and the angels', on which see F. Donner, *Muhammad and the Believers at the Origins of Islam* (Cambridge, MA, 2010), p. 61. See also Crone, *Slaves on Horses*, p. 18; D. Cook, *Studies in Muslim Apocalyptic* (Princeton, 2002). See also D. Peterson, 'Eschatology', *Oxford Encyclopedia of the Islamic World*, online, for a summary and for recent literature.

44. Averil Cameron, 'Byzantium in the seventh century: the search for redefinition', in J. Fontaine and J. Hillgarth, eds, *The Seventh Century* (London, 1992), pp. 250–76.

45. John of Damascus, *Fountain of Knowledge*, 2:101.

46. J. Elsner, 'The viewer and the vision: the case of the Sinai apse', *Art History* 17 (1994), 81–102; *John Climacus, The Ladder of Divine Ascent*, trans. C. Luibhead

and N. Russell (New York, 1982); J. R. Martin, *The Illustration of the Heavenly Ladder of John Climacus* (Princeton, 1954).

47. J. Haldon, 'The works of Anastasius of Sinai: a key source for the history of seventh-century east Mediterranean society and belief', in Cameron and Conrad, eds, *Byzantine and Early Islamic Near East*, pp. 107–47, at p. 112.

48. Haldon, 'The works of Anastasius of Sinai', pp. 129–39.

49. James Howard-Johnston has made a remarkable effort to evaluate the range of historical materials that are available for this period in his *Witness to a World Crisis: Historians and Histories of the Middle East in the Seventh Century* (Oxford, 2010).

12. Emperors of New Rome

1. A. Kaldellis, *The Byzantine Republic: People and Power at New Rome* (Cambridge, MA, 2015).

2. The translation is adapted from E. Barker, *Social and Political Thought in Byzantium* (Oxford, 1961), pp. 54–5, 57. See also *Three Political Voices from the Age of Justinian*, trans. P. Bell (Liverpool, 2009), pp. 99–122. Agapetus later urges Justinian to 'impose on yourself the compulsion of observing the laws'.

3. *CJ* 1.14.4. See M. Anastos, 'CI I.14.4 and the emperors' exemption from the laws', in his *Aspects of the Mind of Byzantium* (Aldershot, 2001), III, pp. 1233–43, emphasises that 'the emperors ... clung to their exemption from the laws as set forth in the familiar excerpt from the thirteenth book of Ulpian's commentary *Ad legem Juliam et Papiam* (D. I. 3.31: *Princeps legibus solutum est*): "The emperor is exempt from the laws"'. See however Jones, *Later Roman Empire*, I, 321: 'Though an absolute, [the Roman emperor] was not an arbitrary monarch. There was no sanction for this principle, but it was generally respected by the emperors ...'

4. Agapetus, *Ekthesis*, ch. 2, enforcement of law; ch. 9, the soul as a mirror; ch. 15, piety; ch. 17, philosophy and the love of wisdom; ch. 18, temperance and justice; ch. 20, courage. For the allusion to Proverbs, see P. Henry, 'A mirror for Justinian: the *Ekthesis* of Agapetus Diaconus', *Greek, Roman, and Byzantine Studies* 8 (1967), 281–308, at 296.

5. My translation (from PG 86: 1169) does not much resemble that of Barker, *Social and Political Thought*, p. 57. There is here a distinction drawn between the Greek terms *philanthropia* and *agape*, although the former had by now largely replaced the latter in embracing the meaning of the Latin *caritas*, which we call 'charity'; hence we translate *agape* as 'love'. On this see Henry, 'Mirror', 300–302, citing G. Downey, '*Philanthropia* in religion and statecraft in the fourth century after Christ', *Historia* 4 (1955), 199–208, where a competing pagan notion of *philanthropia* is also identified.

6. At the start of our period, a law indicated that praetorians 'alone may truly be said

to judge in place of the emperor' (*CTh* 11.30.16). However, over time the number of their subordinates and responsibilities had declined steadily until the office ceased to exist.

7. *De Cerimoniis* II, 27, trans. Moffatt, pp. 627–8; *ODB*, pp. 705 (eparch of the city), pp. 1710–11 (praetorian prefect), pp. 2144 (urban prefect); J. Haldon, *The Empire that Would not Die: The Paradox of Eastern Roman Survival, 640–740* (Cambridge, MA, 2016), pp. 164–77.

8. 'Great Palace', *ODB*, pp. 869–70, for summary and references.

9. The quotation from John of Ephesus is adapted from Mango, *Art*, p. 128.

10. Averil Cameron, 'Notes on the Sophiae, the Sophianae and the Harbour of Sophia', *Byzantion* 37 (1967), 11–20. Cf. Theophanes, AM 6071, trans. Mango and Scott, pp. 369–70, wrongly transferring construction of the Sophianae to Tiberius. On the aqueduct, see Theophanes, AM 6068, trans. Mango and Scott, p. 367.

11. M. Whitby, *The Emperor Maurice and his Historian* (Oxford, 1998), pp. 17–24, offers a full description of Maurice's building. On Leontius and Apsimar, see Theophanes AM 6190, trans. Mango and Scott, pp. 517–18; J. Haldon, *Byzantium in the Seventh Century: The Transformation of a Culture* (Cambridge 1990), pp. 74–6.

12. Theophanes, AM 6186, trans. Mango and Scott, pp. 512–14; *De Cerimoniis* I, 64, trans. Moffatt, pp. 284–93, at pp. 287–8; *ODB*, p. 2116. See also P. Magdalino, 'The culture of water in the "Macedonian Renaissance"', in B. Shilling and P. Stephenson, eds, *Fountains and Water Culture in Byzantium* (Cambridge, 2016), pp. 130–44.

13. A. Cameron, *Circus Factions: Blues and Greens at Rome and Byzantium* (Oxford, 1976), pp. 230–70; *Flavius Cresconius Corippus, In laudem iustini Augusti minoris, libri IV,* ed. and trans. Averil Cameron (London, 1976), pp. 12–14, 173–4.

14. A. Cameron, *Porphyrius the Charioteer* (Oxford, 1973), p. 239. This is one of a series of important monographs by the late Alan Cameron that studies the complex of institutions, sports teams, theatre claques, political structures, events and personalities that defined this period. Cameron observed that 'From 491 on faction riots were serious, expensive, and above all regular threats to life and property in Constantinople, Antioch, and other eastern cities. Many thousands died, scores of buildings were burned and burned again.'

15. *Easter Chronicle*, trans. Whitby, p. 99, translation modified.

16. *Anthologia Latina*, 100, quoted by R. Webb, *Demons and Dancers: Performance in Late Antiquity* (Cambridge, MA, 2008), p. 61; Marcellinus Comes, trans. Croke, p. 33; Cameron, *Porphyrius*, p. 231. The law of 414 is *CTh*. 15.11.1.

17. Marcellinus Comes, trans. Croke, p. 41. See also Cameron, *Porphyrius*, p. 229; Justinian, Novel 105, on which see A. Cameron and D. Schauer, 'The last consul: Basilius and his diptych', *JRS* 72 (1982), 126–45. The diptych of Flavius Anastasius is explored elegantly by Webb, *Demons and Dancers*, pp. 24, 43, the first and last pages of a chapter devoted to 'Theater and society'.

18. Theophanes, AM 6101, trans. Mango and Scott, p. 426. The translations and analyses, however, are here taken from Cameron, *Circus Factions*, pp. 252–4, 288–9.

19. *Easter Chronicle*, trans. Whitby, pp. 149–54; Theophanes, AM 6103–5, trans. Mango and Scott, pp. 429–31.

20. U. Gehn and B. Ward-Perkins, 'Constantinople', in R. R. R. Smith and B. Ward-Perkins, eds, *The Last Statues of Antiquity* (Cambridge, 2016), pp. 136–44.

21. N. Fıratlı, 'A late antique imperial portrait recently discovered at Istanbul', *American Journal of Archaeology* 55 (1951), 67–71. *LSA database*: LSA-337, now Istanbul Archaeological Museum, inv. 5028; K. Schade, 'Women', in Smith and Ward-Perkins, eds, *The Last Statues of Antiquity*, pp. 249–58 at p. 255, where the photograph is reversed.

22. B. Ward-Perkins, 'The end of the statue habit, AD 284–620' in Smith and Ward-Perkins, eds, *The Last Statues of Antiquity*, pp. 295–308.

23. Socrates Scholasticus, 7.22.

24. G. Dennis, 'Religious services in the Byzantine army', in *Eulogema: Studies in Honor of Robert Taft* (Rome, 1993), pp. 107–17 at p. 108; S. Mergiali-Sahas, 'Byzantine emperors and holy relics', *Jahrbuch der Österreichischen Byzantinistik* 51 (2001), 41–60.

25. *Greek Anthology* XVI. 62.

26. A. Cutler, 'The making of the Justinian diptychs', *Byzantion* 54 (1984), 75–115; A. Cutler, 'Barberiniana: notes on the making, content, and provenance of Louvre, OA. 9063', in E. Dassmann and K. Thraede, eds, *Tesserae: Festschrift für Josef Engemann* (Jena, 1991), pp. 329–39.

27. See R. Viladesau, *The Beauty of the Cross* (Oxford, 2006), pp. 37–41. As Vidaescu, p. 41, notes: 'The cross is God's means of glorious victory in battle (*Pange Lingua* 1; *Vexilla Regis* 7)'. See also H. Kessler, 'Evil Eye(ing): Romanesque art as a shield of faith', in C. Hourihane, ed., *Romanesque Art and Thought in the Twelfth Century* (Princeton, 2008), pp. 107–35. The Douce ivory is the upper cover to Oxford, Bodleian MS Douce 176.

28. J. W. Drijvers, 'Heraclius and the *restitutio crucis*: notes on symbolism and ideology', in G. J. Reinink and B. H. Stolte, eds, *The Reign of Heraclius (610–641): Crisis and Confrontation* (Leuven, 2002), pp. 175–90; C. Mango, 'Deux études sur Byzance et la Perse sassanide, II: Héraclius, Šahrvaraz et la Vraie Croix', *Travaux et mémoires* 9 (1985), 105–18; D. Woods, 'Mu'awiyah, Constans II and coins without crosses', *Israel Numismatic Research* 19 (2015), 169–235, citing John of Nikiu at 176.

29. M. De Groote, 'Andrew of Crete's *Homilia de exaltatione s. crucis* (*CPG* 8199; *BHG* 434f.): editio princeps', *Harvard Theological Review* 100 (2007), 443–87, at 473–5.

30. A. Grabar, *Les peintures de l'évangéliaire de Sinope (Bibliothèque nationale, suppl. gr. 1286* (Paris, 1948); K. Weitzmann, 'The mosaic in St. Catherine's Monastery on Mount Sinai', *Proceedings of the American Philosophical Society* 110/6 (1966),

392–405. The information on Paul the Silentiary (lines 429–37) is drawn from the exegesis by R. Macrides and P. Magdalino, 'The architecture of *ekphrasis*: construction and context of Paul the Silentiary's poem on Hagia Sophia', *Byzantine and Modern Greek Studies* 12 (1988), 47–82, at 65–7, where it is noted that by the ninth century David was celebrated on the Sunday after Easter, which in 562 fell on 31 December (their proposed date for the delivery of the oration).

31. K. Weitzmann, *The Age of Spirituality: Late Antique and Early Christian Art, Third to Seventh Century* (New York, 1977), pp. 475–83 (David Plates), 491–2 (Sinope Gospels); S. Spain Alexander, 'Heraclius, Byzantine imperial ideology, and the David Plates', *Speculum* 52 (1977), 217–37. R. E. Leader, 'The David Plates revisited: transforming the secular in early Byzantium', *Art Bulletin* 82 (2000), 407–27, offers a non-imperial reading of the David Plates, which has not proven compelling to viewers or scholars of the plates, now kept at the Met in New York (six, including the largest) and the Cyprus Museum, Nicosia (three). The quotations about Heraclius are from George of Pisidia, preserved by Theophanes, AM 6116, 6118, trans. Mango and Scott, pp. 445, 449.

32. See *Georgio di Pisidia, Poemi, I: Panegirici epici*, ed. Agostino Pertusi, *Studia Patristica et Byzantina* 7 (Ettal, 1959), p. 249. See also Mary Whitby, 'A new image for a new age: George of Pisidia on the emperor Heraclius', in E. Dąbrowa, ed., *The Roman and Byzantine Army in the East* (Cracow, 1994), pp. 197–225, especially at pp. 215 and 219.

33. Averil Cameron and J. Herrin, eds, *Constantinople in the Early Eighth Century: the Parastaseis Syntomoi Chronikai* (Leiden, 1984); C. Mango, 'Antique statuary and the Byzantine beholder', *DOP* 17 (1963), 55–75; S. Bassett, *The Urban Image of Late Antique Constantinople* (Cambridge, 2004); B. Anderson, 'Classified knowledge: the epistemology of statuary in the *Parastaseis Syntomoi Chronikai*', *Byzantine and Modern Greek Studies* 35 (2011), 1–19; P. Chatterjee, 'Viewing the unknown in eighth-century Constantinople', *Gesta* 56 (2017), 137–49. The *Greek Magical Papyri*, dating from the second to fifth centuries AD, preserve many magical uses of statues. See also P. Stephenson, *The Serpent Column: A Cultural Biography* (Oxford, 2016).

INDEX

Italic page numbers refer to text illustrations; colour plates are indicated by 'Pl.'

E

earthquakes and seismic activity, 3, 309–12, 375*n*33; Antioch, 161, 194, 200, 234, 310–11; Constantinople, 90, 104, 105, 106, 107, 113, 154, 168, 214, 310; Crete, 296; Cyprus, 297, 298; Gerasa, 316; Stobi, 290

Easter Chronicle (Chronicon Paschale), 133, 219, 308; on Chosroes II, 248; on Constantinople, 113, 246–7, 308, 337; on court of Theodosius II, 147–8; on Heraclius, 243; on Maurice, 237, 238

ecumenical councils: first *see* Nicaea, Council of (325); second *see* Constantinople, Council of (381); third *see* Ephesus, Council of (431); fourth *see* Chalcedon, Council of (451); fifth *see* Constantinople, Council of (553); sixth *see* Constantinople, Council of (680);

Edessa (Mesopotamia), 188–9, 210, 222, 226, 237–8, 253, 322; School of, 77

Edfu *see* Apollonopolis Magna

'Edict of Reunion' *see* Henoticon

Edirne *see* Adrianople

Edix Hill (Cambridgeshire), 306

education *see* Christianity, and education; literary culture and education; schooling

Egnatian Way, 82, 290

Elaia (Turkey), 120

Elaisussa Sebaste (Turkey), 54, 361*n*19

Elias, governor of Cherson, 271

Elijah (biblical figure), 7–8

Elpidius, coup leader under Phocas, 238

Emesa (Homs; Syria), 253, 254, 312

emperors: inauguration ceremonies of, 181–2; role of, 330–37

end time *see* apocalypticism

Ephesus (Turkey), 23, 54–6, *55*, 64, 138, 160, 191, 243, 283–4, Pl. 3; agora, 283; harbour (Panormus), 284; Temple of Artemis, 105; walls, 283–4

Ephesus, Council of (431), 54, 63, 151–2, 166, 178

Ephesus, Council of (449), 54, 167

epidemics *see* plagues and pandemic disease

Epiphania Eudocia *see* Eudocia, Epiphania

Erythrae (Ionia), 160

Erzurum *see* Theodosiopolis

eschatology *see* apocalypticism

'eternal peace' (532), 196, 201, 209, 231, 311

Euchaita (Anatolia), 187, 287

Eudaemon, prefect, 197

Eudocia, daughter of Valentinian III, 153, 166, 201–2

Eudocia, wife of Justinian II, 257

Eudocia, Aelia, wife of Theodosius II, 21, 120, 141, 148–50, 345, 376*n*15

Eudocia, Epiphania, daughter of Heraclius, 241, 247, 251, 256, 340

Eudocia, Fabia, wife of Heraclius, 239, 241, 256, 340

Eudoxia, Aelia, wife of Arcadius, 21, 121, 136–7, 138–9, 375*n*8

Eudoxia, Licinia, daughter of Theodosius II, 21, 67, 148; marriage to Valentinian III, 146, 150, 161, 165–6

Eugene I, Pope, 261

Eugenius, usurper, 122, 123–5, 131, 135

Eunapius of Sardis, 78, 133

Eunomians, 155

eunuchs, 47–8, 97, 136, 212, 256, 336

Euphemia, St, 374*n*30

Euphemius, patriarch of Constantinople, 181–2, 185, 187

Euphrasius, patriarch of Antioch, 194

Eusebius of Caesarea, 60, 61, 95, 134

Eusebius of Nicomedia, 61

Eutropius, city councillor, 66–7

Eutropius, consul, 135–6, 137, 375*n*6

Eutyches, abbot, 166–7

Eutychides of Sicyon, 56; statue of Tyche of Antioch, 56, *57*

Eutychius, exarch of Ravenna, 279

Eutychius, patriarch of Constantinople, 214, 215, 218, 219, 234

Evagrius Ponticus, 62

Evagrius Scholasticus, 166, 221, 225, 226, 234, 305–6, 379*n*2

Excubitors (imperial bodyguard), 103, 169, 189, 219, 238, 257